APPLIED RELIABILITY

Second Edition

APPLIED RELIABILITY

Second Edition

Paul A. Tobias

David C. Trindade

VAN NOSTRAND REINHOLD

I(T)P   A Division of International Thomson Publishing Inc.

New York • Albany • Bonn • Boston • Detroit • London • Madrid • Melbourne
Mexico City • Paris • San Francisco • Singapore • Tokyo • Toronto

A division of International Thomson Publishing Inc.
The ITP logo is a trademark under license

Printed in the United States of America
For more information, contact:

Van Nostrand Reinhold
115 Fifth Avenue
New York, NY 10003

International Thomson Publishing GmbH
Königswinterer Strasse 418
53227 Bonn
Germany

International Thomson Publishing Europe
Berkshire House 168–173
High Holborn
London WCIV 7AA
England

International Thomson Publishing Asia
221 Henderson Road #05–10
Henderson Building
Singapore 0315

Thomas Nelson Australia
102 Dodds Street
South Melbourne, 3205
Victoria, Australia

International Thomson Publishing Japan
Hirakawacho Kyowa Building, 3F
2-2-1 Hirakawacho
Chiyoda-ku, 102 Tokyo
Japan

Nelson Canada
1120 Birchmount Road
Scarborough, Ontario
Canada M1K 5G4

International Thomson Editores
Campos Eliseos 385, Piso 7
Col. Polanco
11560 Mexico D.F. Mexico

1 2 3 4 5 6 7 8 9 10   QEBFF 01 00 99 98 97 96 95 94

Library of Congress Cataloging-in-Publication Data

Tobias, Paul A.
  Applied reliability / Paul A. Tobias, David C. Trindade—2nd ed.
    p.   cm.
  Includes bibliographical references and index.
  ISBN 0-442-00469-9
  1. Reliability (Engineering).   2. Quality control—Statistical methods.   I. Trindade, David C.   II. Title.
  TA169.T63   1994   94-42855
  620'.00452—dc20   CIP

Contents

8.   SYSTEM MODELS AND RELIABILITY ALGORITHMS/219

9.   QUALITY CONTROL IN RELIABILITY: APPLICATIONS OF THE BINOMIAL DISTRIBUTION/251

Preface

Over the years following the publication of *Applied Reliability,* we received many communications assuring us that our intended audience—individuals designing or evaluating the reliability of components or hardware systems—found the book readable and useful. We also received feedback from a number of professors teaching reliability at engineering schools telling us they frequently referred to the text and recommended it to their students.

Along with the compliments, however, we got numerous comments best described as "Why didn't you . . .?" statements. "Why didn't you include exercises with answers and additional problems?"; "Why didn't you give more details about the important topic of maximum likelihood estimation and likelihood ratio tests?"; "Why didn't you cover the most important practical topic of all: repairable system reliability?"; "Why didn't you include topics such as Bayesian reliability methods, or reliability growth models, or more information on simulation methods?"

Those "why didn't you" criticisms led to the second edition. Of course, there are still numerous topics we could have included and expanded further. In the final analysis, we chose material based on our personal experiences working with industrial engineers and statisticians and making judgments about what was *needed* and what *worked.*

The second edition has three new chapters and more than 40 new examples with detailed solutions. There are almost 200 new exercises located throughout the text (many with answers given at the end of the book) and additional problems following each chapter.

While the size of the book (and, at times, the level of mathematical complexity) has increased significantly to include new material and exercises, we have tried very hard to keep the basic "flavor" consistent with the first edition. Our goal remains that the text be application oriented, with numerous practical examples and graphical illustrations. At the same time, we inform the reader about the more complex "state-of-the-art" techniques offering analysis improvements justifying their additional complexity or software expense. Recent classes taught by the authors using the new material give us confidence that we are still on target toward our goal.

Finally, we want to thank the numerous readers of the first edition who offered corrections or made suggestions to improve the clarity or scope of various sections. The second edition has also benefited heavily from comments made by reviewers of pre-publication drafts. In particular, we owe much to the comprehensive suggestions made by Wayne Nelson, and the review critiques offered by Doug Montgomery, Ed Russell, and Bill Heavlin.

PAUL A. TOBIAS, PH.D.
Manager of Statistical Methods Group
SEMATECH Corporation
Austin, Texas

DAVID C. TRINDADE, PH.D.
Senior Fellow and
Corporate Director of Applied Statistics
Advanced Micro Devices, Inc.
Sunnyvale, California

November 9, 1994

Chapter 1

Basic Descriptive Statistics

One of the most useful skills a reliability specialist can develop is the ability to convert a mass (mess?) of data into a form suitable for meaningful analysis. Raw numbers by themselves are not useful; what is needed is a distillation of the data into information.

In this chapter we discuss several important concepts and techniques from the field of descriptive statistics. These methods will be used to extract a relevant summary from collected data. The goal is to describe and understand the random variability that exists in all measurements of real world phenomena and experimental data.

The topics we shall cover include: populations and samples; frequency functions, histograms, and cumulative frequency functions; the population cumulative distribution function (CDF) and probability density function (PDF); elementary probability concepts; random variables, population parameters, and samples estimates; theoretical population shape models; and data simulation.

POPULATIONS AND SAMPLES

Statistics is concerned with variability, and it is a fact of nature that variation exists. No matter how carefully a process is run, an experiment is executed, or a measurement is taken, there will be differences in repeatability due to the inability of any individual or system to completely control all possible influences. If the variability is excessive, the study or process is described as lacking control. If, on the other hand, the variability appears reasonable, we accept it and continue to operate. How do we visualize variability in order to understand if we have a controlled situation?

Consider the following example:

1

Example 1.1   Automobile Fuse Data

A manufacturer of automobile fuses produces lots containing 100,000 fuses rated at 5 A. Thus, the fuses are supposed to open in a circuit if the current through the fuse exceeds 5 A. Since a fuse protects other elements from potentially damaging electrical overload, it is very important that fuses function properly. How can the manufacturer assure himself that the fuses do indeed operate correctly and that there is no excessive variability? Obviously, he cannot test all fuses to the rated limit, since that act would destroy the product he wishes to sell. However, he can sample a small quantity of fuses (say, 100 or 200) and test them to destruction to measure the opening point of each fuse. From the sample data, he could then infer what the behavior of the entire group would be if all fuses were tested.

In statistical terms, the entire set or collection of measurements of interest (e.g., the blowing values of all fuses) define a *population*.

A population is the entire set or collection of measurements of interest.

Note that a population may be finite, as in the case of the fuses, or it may be infinite, as occurs in a manufacturing process where the population could be all product that has been or could ever be produced in a fabricating area.

The *sample* (e.g., the 100 or 200 fuses tested to destruction) is a subset of data taken from the population.

A sample is a subset of data from the population.

The objective in taking a sample is to make inference about the population.

Note that data may exist in one of two forms. In *variables data,* the actual measurement of interest is taken. In *attribute data,* the results exist in one of two categories: either pass-fail, go-no go, in spec-out of spec, etc. Both types of data will be treated in this text.

In the fuse data example, we record variables data, but we could also transform the same results into attribute data by stating whether a fuse opened before or after the 5 A rating. Similarly, in reliability work one can measure the actual failure time of an item (variables data) or record the number of items failing before a fixed time (attribute data). Both types of data occur frequently in reliability studies. Later, we will discuss such topics as choosing a sample size, drawing a sample randomly, and the "confidence" in the data from a sample. For now, however, let's assume that the sample has been properly drawn and consider what to do with the data in order to present an informative picture.

HISTOGRAMS AND FREQUENCY FUNCTIONS

In stating that a sample has been randomly drawn, we imply that each measurement or data point in the population has an equal chance or probability of being selected for the sample. If this requirement is not fulfilled, the sample may be "biased" and correct inference about the population might not be possible.

What information does the manufacturer expect to obtain from the sample measurements of 100 fuses? First, the data should cluster about the rated value of 5 A. Second, the spread in the data (variability) should not be large, because the manufacturer realizes that serious problems could result for users of the fuses if some blow at too high a value. Similarly, fuses opening at too low a level could cause needless repairs or generate unnecessary concerns.

The reliability specialist randomly samples 100 fuses and records the data shown in Table 1.1. It is easy to determine the high and low values from the sample data and see that the measurements cluster roughly about the number 5. Yet, there is still difficulty in grasping the full significance of this set of data.

Let's try the following procedure:

1. Find the *range* of the data by subtracting the lowest from the highest value. For this set, the range is $5.46 - 4.43 = 1.03$.

2. Divide the range into ten or so equally spaced intervals such that readings are uniquely classified into each cell. Here, the cell width is $1.03/10 \approx 0.10$, and we choose the starting point to be 4.395, a convenient value below the minimum of the data and carried out one digit more precise than the data to avoid any confusion in assigning readings to individual cells.

3. Increment the starting point by multiples of the cell width until the maximum value is exceeded. Thus, since the maximum value is 5.46, we generate the numbers 4.395, 4.495, 4.595, 4.695, 4.795, 4.895, 4.995, 5.095, 5.195, 5.295, 5.395, and 5.495. These values will represent the end points or boundaries of each cell, effectively dividing the range of the data into equally spaced class intervals covering all the data points.

TABLE 1.1 Sample Data on 100 Fuses

4.64	4.95	5.25	5.21	4.90	4.67	4.97	4.92	4.87	5.11
4.98	4.93	4.72	5.07	4.80	4.98	4.66	4.43	4.78	4.53
4.73	5.37	4.81	5.19	4.77	4.79	5.08	5.07	4.65	5.39
5.21	5.11	5.15	5.28	5.20	4.73	5.32	4.79	5.10	4.94
5.06	4.69	5.14	4.83	4.78	4.72	5.21	5.02	4.89	5.19
5.04	5.04	4.78	4.96	4.94	5.24	5.22	5.00	4.60	4.88
5.03	5.05	4.94	5.02	4.43	4.91	4.84	4.75	4.88	4.79
5.46	5.12	5.12	4.85	5.05	5.26	5.01	4.64	4.86	4.73
5.01	4.94	5.02	5.16	4.88	5.10	4.80	5.10	5.20	5.11
4.77	4.58	5.18	5.03	5.10	4.67	5.21	4.73	4.88	4.80

4. Construct a *frequency table* as shown in Table 1.2, which gives the number of times a measurement falls inside a class interval.

5. Make a graphical representation of the data by sketching vertical bars centered at the midpoints of the class cells with bar heights proportionate to the number of values falling in that class. This graphical representation shown in Figure 1.1 is called a *histogram*.

A histogram is a graphical representation in bar chart form of a frequency table or frequency distribution.

Note that the vertical axis may represent the actual count in a cell, or it may state the percentage of observations in the total sample occurring in that cell. Also, the range here was divided by the number 10 to generate a cell width, but any convenient number (usually between 8 and 20) could be used. Too small a number would not reveal the shape of the data, and too large a number would result in many empty cells and a flat appearing distribution. Sometimes a few tries are required to arrive at a suitable choice.

In summary, the histogram provides us with a picture of the data from which we can intuitively see the center of the distribution, the spread, and the shape. The shape is important because we usually have an underlying idea or model as to how the entire population should look. The sample shape either confirms this expectation or gives us reason to question our assumptions. In particular, a shape that is symmetric about a center, with most of the observations in the central region, might reflect data from certain symmetric distributions, like the normal or Gaussian distribution. Alternatively, an asymmetric appearance would imply the

TABLE 1.2   Frequency Table of Fuse Data

Cell Boundaries	Number in Cell
4.395 to 4.495	2
4.495 to 4.595	2
4.595 to 4.695	8
4.695 to 4.795	15
4.795 to 4.895	14
4.895 to 4.995	13
4.995 to 5.095	16
5.095 to 5.195	15
5.195 to 5.295	11
5.295 to 5.395	3
5.395 to 5.495	1
Total Count	***100***

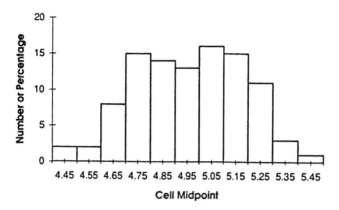

Figure 1.1   Histogram of Measurements

existence of data points spaced farther from the center in one direction than in the other. For the data presented in the fuse example, we note that the distribution appears reasonably symmetric. Hence, based on the histogram and the way the ends of the distribution taper off, the manufacturer believes that values much greater or much less than about 10 percent of the central target are not likely to occur. This variability he accepts as reasonable.

CUMULATIVE FREQUENCY FUNCTION

There is another way of representing the data which can be very useful. By reference to Table 1.2, let us accumulate the number of observations less than or equal to a given value as shown in Table 1.3. Such a means of representing data is called a *cumulative frequency function*.

TABLE 1.3   Frequency Table of Fuse Data

Upper Cell Boundary (UCB)	Number of Observations ≤ UCB
4.495	2
4.595	4
4.695	12
4.795	27
4.895	41
4.995	54
5.095	70
5.195	85
5.295	96
5.395	99
5.495	100

The graphical rendering of the cumulative frequency function is shown as Figure 1.2. Note that the cumulative frequency distribution is never decreasing, and it starts at zero and goes to the total sample size. It is often convenient to represent the cumulative count in terms of a fraction or percentage of the total sample size used. In that case, the cumulative frequency function will range from 0 to 1.00 in fractional representation or to 100 percent in percentage notation. In this text, we will often employ the percentage form. Table 1.3 and Figure 1.2 make it clear that the cumulative frequency curve is obtained by summing the frequency function count values. This summation process will later be generalized by integration when we discuss the population concepts underlying the frequency function and the cumulative frequency function in the next section.

THE CUMULATIVE DISTRIBUTION FUNCTION AND THE PROBABILITY DENSITY FUNCTION

The frequency distribution and the cumulative frequency distribution are calculated from sample measurements. Since the samples are drawn from a population, what can we state about this population? The typical procedure is to assume a mathematical formula which provides a theoretical model for describing the way the population values are distributed. The sample histograms and the cumulative frequency functions are then estimates of these population models.

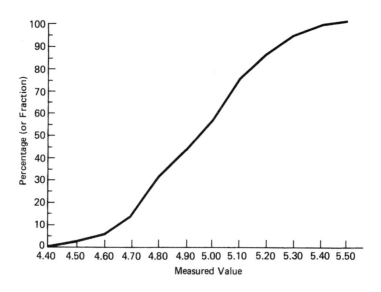

Figure 1.2   Plot of Cumulative Frequency Function

The model corresponding to the frequency distribution is the *probability density function* (PDF), denoted by $f(x)$, where x is any value of interest. The PDF may be interpreted in the following way: $f(x)dx$ is the fraction of the population values occurring in the interval dx. In reliability work, we often have time, t, as the variable of interest. Therefore, $f(t)dt$ is the fraction of failure times of the population occurring in the interval dt. A very simple example for $f(t)$ is called the *exponential distribution,* and it is given by the equation

$$f(t) = \lambda e^{-\lambda t}, \quad 0 \le t < \infty$$

where λ is a constant. The plot of $f(t)$ is shown in Figure 1.3. The exponential distribution is a widely applied model in reliability studies and forms the basis of Chapter 3.

The cumulative frequency distribution similarly corresponds to a population model called the *cumulative distribution function* (CDF) and denoted by $F(x)$. The CDF is related to the PDF via the following relationship

$$F(x) = \int_{-\infty}^{x} f(y)\, dy$$

where y is the dummy variable of integration. $F(x)$ may be interpreted as the fraction of values in the population less than or equal to x. Alternatively, $F(x)$ gives the probability of a value less than or equal to x occurring in a single random

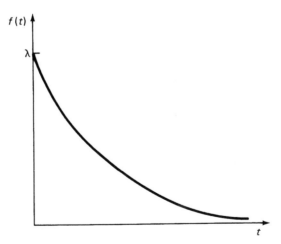

Figure 1.3   Plot of Probability Density Function for Exponential Distribution

draw from the population described by $F(x)$. Since, in reliability work, we usually deal with failure times, t, which are non-negative, the CDF for population failure times is related to the PDF by

$$F(t) = \int_0^t f(y)\,dy, \quad 0 \le t < \infty$$

For the exponential distribution,

$$F(t) - \int_0^t \lambda e^{-\lambda y}\,dy = -e^{-\lambda y}\Big]_0^t = 1 - e^{-\lambda t}$$

The CDF for the exponential distribution is plotted in Figure 1.4.

When we calculated the cumulative frequency function in the fuse example, we worked with grouped data (that is, data classified by cells). However, another estimate of the population CDF could have been generated by ordering the individual measurements from smallest to largest and then plotting the successive fractions

$$\frac{1}{n}, \frac{2}{n}, \frac{3}{n}, \ldots, \frac{n}{n}$$

versus the ordered points. Such a representation is called the *empirical distribution function* and is shown in Figure 1.5 for the data from the fuse example. The

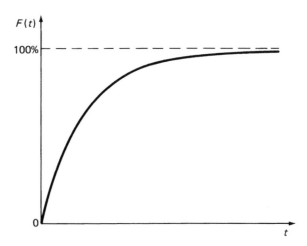

Figure 1.4   The CDF for the Exponential Distribution

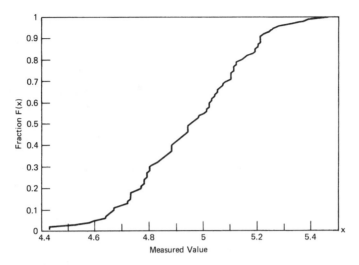

Figure 1.5   EDF for Fuse Data

advantage of using the empirical distribution function instead of grouping the data is obviously that all data points are pictured; the disadvantage is that more computational effort is involved. However, computers are often available to perform the calculations.

Since $F(x)$ is a probability, all the rules and formulas for manipulating probabilities can be used when working with CDFs. Some of these basic rules will be described in the next section.

Exercise 1.1

For the data in Table 1.1, construct a frequency table using the starting point 4.395 and 0.2 as the interval width. Sketch a histogram of this frequency table. Compare it to Figure 1.1.

Exercise 1.2

Using the results from Exercise 1.1, construct a cumulative frequency table and sketch a plot of the cumulative frequency function. How does it compare to Figure 1.2?

Exercise 1.3

Take columns 2, 5, and 8 (left to right) from Table 1.1, for a total of 30 points. Assume a random sample and arrange the points in order from smallest to largest and plot the empirical distribution function. Compare to Figure 1.5.

PROBABILITY CONCEPTS

In the classical sense, the term *probability* can be thought of as the expected relative frequency of occurrence of a specific event in a very large collection of possible outcomes. For example, if we toss a balanced coin a large number of times, we expect the number of occurrences of the event "heads" to comprise approximately half of the number of outcomes. Thus, we say the probability of heads on a single toss is 0.5 or 50 percent, or 50-50. It is typical to express probabilities as either a fraction between 0 to 1 or as a percentage between 0 to 100 percent.

There are two very useful relations often invoked in probability theory. These rules relate to the occurrence of two or more events. In electrical engineering terms, we are defining "and" and "or" relations. The first rule states that if $P(A)$ is the probability of event A occurring and $P(B)$ is the probability of event B occurring, then the probability of events A and B occurring simultaneously, denoted $P(AB)$, is

$$P(AB) = P(A)P(B|A)$$

or

$$P(AB) = P(B)P(A|B)$$

where $P(A|B)$ is the designator for the "conditional" probability of A, given that event B has occurred.

Let's explain conditional probability further. We imply by the terminology that one event may be affected by the occurrence of another event. For example, suppose we ask what is the probability of getting two black cards in a row in successive draws from a well shuffled deck of cards, without replacing the first card drawn. Obviously, the probability of the first card being a black card (call this event A) is

$$P(A) = \frac{\text{favorable outcomes}}{\text{total outcomes}} = \frac{26}{52} = \frac{1}{2}$$

The probability of the second card being a black card (event B) changes depending on whether the first card drawn was a black card. If yes, then the probability of the second card being a black card is

$$P(B|A) = \frac{25}{51}$$

So the probability of two successive black cards is

$$P(AB) = P(A)P(B|A) = \frac{1}{2} \times \frac{25}{51} = \frac{25}{102}$$

Two events, A and B, are said to be *independent* if the occurrence of one does not affect the probability of the other happening. The formal definition states that events A and B are independent *if and only if*

$$P(AB) = P(A)P(B)$$

This expression is sometimes referred to as the *multiplication rule* for the probability of independent events occurring simultaneously. In general, the probability of independent events occurring is just the product of the individual probabilities for each event. For example, in the card situation replacing the first card drawn and reshuffling the deck will make event B independent of event A. Thus, the probability of two successive black cards, with replacement and reshuffling between draws, is

$$P(AB) = P(A)P(B) = \frac{26}{52} \times \frac{26}{52} = \frac{1}{4}$$

Similarly, the probability of simultaneously getting 6 on one roll of a die and an ace in one draw from a deck of cards, apparently independent events, is

$$P(AB) = \frac{1}{6} \times \frac{4}{52} = \frac{1}{78}$$

The extension of these conditional probability principles to three or more events is possible. For example, the rule for the joint probability of three events, A,B, and C, is

$$P(ABC) = P(A)P(B|A)P(C|AB)$$

For independent events, the formula becomes

$$P(ABC) = P(A)P(B)P(C)$$

The second important probability formula relates to the situation in which either of two events, A or B, may occur. The expression for this "union" is

$$P(A \cup B) = P(A) + P(B) - P(AB)$$

If the events are independent, then the relation becomes

$$P(A \cup B) = P(A) + P(B) - P(A)P(B)$$

The last term in the above expressions corrects for double counting of the same outcomes. For example, what is the probability of getting either an ace (event A) or a black card (event B) in one draw from a deck of cards? The events are independent (See Exercise 1.4.), and so

$$P(A \cup B) = P(A) + P(B) - P(A)P(B)$$

$$= \frac{4}{52} + \frac{26}{52} - \left(\frac{4}{52} \times \frac{26}{52}\right) = \frac{28}{52} = \frac{7}{13}$$

Note that the term $P(A)P(B)$ subtracts out the probability for black aces. This probability has already been counted twice—once in the $P(A)$ term and once in the $P(B)$ term.

When events A and B are mutually exclusive or disjoint (that is, both events cannot occur simultaneously), then $P(AB) = 0$, and

$$P(A \cup B) = P(A) + P(B)$$

Furthermore, if both events are also exhaustive in that at least one of them must occur when an experiment is run, then

$$P(A \cup B) = P(A) + P(B) = 1$$

Thus, event A is the complement of event B. Event B can be viewed as the nonoccurrence of A and designated as event $\bar{A}$. Hence, the probability of occurrence of any event is equal to one minus the probability of occurrence of its complementary event. This *complement rule* has important applications in reliability work because a component may either fail (event A) or survive (event $\bar{A}$), resulting in

$$P(\text{Failure}) = 1 - P(\text{Survival})$$

As another example, we note that the event "at least one occurrence" and the event "zero occurrences" are mutually exclusive and exhaustive events. Therefore, the probability of at least one occurrence is equal to (1 − probability of no occurrences).

An extension to three or more events is also possible. For three events, A, B, and C, the formula is

$$P(A \cup B \cup C) = P(A) + P(B) + P(C) - P(AB) - P(BC) - P(AC) + P(ABC)$$

For independent events, the relation becomes

$$P(A \cup B \cup C)$$

$$= P(A) + P(B) + P(C) - P(A)P(B) - P(B)P(C) - P(A)P(C) + P(A)P(B)P(C)$$

For mutually exclusive, exhaustive events, we have

$$P(A \cup B \cup C) = P(A) + P(B) + P(C) = 1$$

Example 1.2   Conditional Probabilities

A tricky word problem that appears often and in many forms can be stated as follows: A computer hack visits the surplus store and sees two 400 MB hard drives in the case. The sign says: "Specially Reduced: 50-50 Chance of Working." He asks the dealer if the hard drives operate properly. The dealer replies: "At least one of them is working." What is the probability that both drives are functional? Does the probability change if the dealer says, "The one on the left works."?

Solution

The first question asks for the probability that both drives work, given that at least one is working; that is, P(both work | at least one works). Let A be the event "both drives work," and let B be the event "at least one drive works." We want the $P(A|B)$. From our conditional probability formula, we can rewrite the expression as

$$P(A|B) = \frac{P(AB)}{P(B)}$$

Now $P(AB)$ is the probability that both drives work and at least one drive works. That joint event is the same as the probability of event A alone, since event B is included in event A; that is, if both drives work, then at least one works. So $P(AB) = P(A) = [(0.5)(0.5)] = 0.25$, assuming the drives are independent. Since the event B ("at least one drive works") and the event "both drives not working" are mutually exclusive and exhaustive events, then the denominator $P(B) = P$(at least one works) $= 1 - P$(both not working) $= 1 - (0.5)(0.5) = 0.75$. Hence, the desired probability is $P(A|B) = (0.25)/(0.75) = 1/3$.

This result surprises many individuals who incorrectly assume that the conditional probability of two working drives given at least one works should be 1/2, since they figure that the other disk drive is equally likely to work or not work. However, let us list the sample space of possible outcomes. With no dealer information, there are the four equally likely outcomes (work, work), (work, not work), (not work, work), (not work, not work) for the left and right drive, respectively. Thus, the probability is only 1/4 that both drives work. When we are told that at least one drive works, we eliminate the outcome (not work, not work). So we have only three equally likely outcomes remaining: (work, work), (work, not work), and (not work, work). Consequently, the probability that both drives work has increased from 1/4 to 1/3 with the added data. Alternatively, the probability that at least one of the drives does not work has decreased from 3/4 to 2/3. On the other hand, if the dealer points out the working drive (maybe he didn't have the time to test both drives), the probability that both drives work does change. Let event A be "both drives work" and C, "the left drive works." Now, $P(A|C)$ = $(0.5)(1)/(1) = 0.5$. In this case, there are only two possible outcomes (work, work) and (work, not work), where the first position indicates the left drive, and only one outcome of the two has both drives working.

For a set of events, $E_1, E_2, \ldots, E_k$, that are mutually exclusive and exhaustive, we can apply another useful relationship, sometimes called the *law of total probabilities*. Any event A can be written as

$$P(A) = \sum_{j=1}^{k} P(A|E_j) P(E_j)$$

In words, $P(A)$ is the weighted average of conditional probabilities, each weighted by the probability of the event on which it is conditioned. This expression is often easier to calculate than $P(A)$ directly.

Example 1.3   Total Probabilities

A computer manufacturer purchases equivalent microprocessor components from three different distributors. Each computer assembly utilizes one microprocessor randomly chosen from in-house inventory. Typically, the inventory of this component type consists of 30 percent from distributor A, 50 percent from distributor B, and 20 percent from distributor C. Historical records show that components from distributors A and C are twice as likely to cause a system failure as those from B. The probability of system failure with component B is 0.5 percent. What is the probability that a computer system will experience failure?

Solution

Since there are three distributors and we randomly chose a component from one of the distributors, we have three mutually exclusive and exhaustive events. The theorem of total probability is the basis for the solution:

$$P(\text{failure}) = P(\text{failure}|D_A) P(D_A) + P(\text{failure}|D_B) P(D_B) + P(\text{failure}|D_C) P(D_C)$$

$$= 2(0.005)(0.3) + (0.005)(0.5) + 2(0.005)(0.2) = 0.0075 \text{ or } 0.75\%$$

where

$$D_A, D_B, D_C = \text{distributors } A, B, \text{ and } C, \text{ respectively}$$

A final key probability formula, know as Bayes' Rule, allows us to "invert" conditional probabilities; that is, determine which one of the conditioning events E_j also occurred, given that event A has occurred. Again, for a set of mutually exclusive and exhaustive events, $E_1, E_2, \ldots, E_k$, Bayes' Rule states that

$$P(E_j|A) = \frac{P(E_j A)}{P(A)} = \frac{P(A|E_j) P(E_j)}{\sum_{j=1}^{k} P(A|E_j) P(E_j)}$$

Note that the denominator of this final expression made use of the law of total probabilities.

Example 1.4   Bayes' Rule

The probability that a batch of incoming material from any supplier is rejected is 0.1. Typically, material from supplier S_1 is rejected 8 percent of the time; from supplier S_2, 15 percent; and S_3, 10 percent. We know that 50 percent of the incoming material comes from S_1, 20 percent from supplier S_2, and 30 percent from S_3. Given that the latest lot of incoming material is rejected, what is the probability the supplier is S_1?

Solution

Let A denote the event that the batch is rejected. Then, by Bayes' Rule,

$$P(S_1|A) = \frac{P(A|S_1) P(S_1)}{P(A)} = \frac{(0.08)(0.5)}{(0.1)} = 0.4$$

In this example, the starting (that is, before we know the batch is rejected) probability of the event S_1 is 0.5. This knowledge is sometimes referred to as the "a priori" probability of S_1. After the batch rejection, Bayes' Rule allows us to calculate 0.4 for the new (conditional) probability of S_1. The result is sometimes called the "a posteriori" probability of S_1.

Exercise 1.4

Show that the probability of getting either an ace (event A) or a black card (event B) in one draw from a deck or cards are independent events.

Exercise 1.5

Three assembly plants produce the same type of parts. Plant A produces 25 percent of the volume and has a shipment defect rate of 1 percent; Plant B produces 30 percent of the volume and ships 1.2 percent defectives. Plant C makes the remainder and ships 0.6 percent defectives. Given that the component picked at random from the warehouse stocked by these plants is defective, what are the probabilities it was manufactured by plant A or B or C?

Exercise 1.6

An electronic card has three components, A, B, and C on it. Component A has a probability of 0.02 of failing in three years. Component B has a probability of 0.01 of failing in three years, and component C has a probability of 0.10 of failing in three years. What is the probability that the card survives three years without failing? What assumptions were made for this calculation?

RANDOM VARIABLES

In reliability studies, the outcome of an experiment may be numerical (e.g., time to failure of a component), or the result may be other than numerical (e.g., type of failure mode associated with a nonfunctional device). In either case, analysis is made possible by assigning a number to every point in the space of all possible outcomes—called the *sample space*. Examples of assigning numbers are: the time to failure is assigned the elapsed hours of operation, or the failure mode may be assigned a category number 1, 2, etc. Any rule for assigning a number creates a random variable.

A random variable is a function for assigning real numbers to points in the sample space.

The practice is to denote the random variable by a capital letter (X, Y, Z, etc.) and the realization of the random variable (i.e., the real number or piece of sample data) by the lower-case letter (x, y, z, etc.). Since this definition appears a bit abstract, let's consider a simple example using a single die, with six faces, each face having one to six dots. The experiment consists of rolling the die and observing the upside face. The random variable is denoted X and assigns numbers matching the number of dots on the upside face. Thus, $(X = x)$ is an event in the sample space, and $X = 6$ refers to the realization where the face with six dots is upside. It is also common to refer to the probability of an event occurring using the notation $P(X = x)$. In this example, we usually assume all six possible outcomes are equally likely (fair die), and therefore, $P(X = x) = 1/6$ for $x = 1, 2, 3, 4, 5$, or 6.

Example 1.5   Random Variable Notation for CDF

The CDF $F(x)$ can be defined as $F(x) = P(X \le x)$; that is, $F(x)$ is the probability that the random variable X has a value less than or equal to x.

SAMPLE ESTIMATES OF POPULATION PARAMETERS

We have discussed descriptive techniques such as histograms to represent observations. However, in order to complement the visual impression given by the frequency histogram, we often employ numerical descriptive measures called *parameters* for a population and *statistics* for a sample. These measures summarize the data in a population or sample and also permit quantitative statistical analysis. In this way, the concepts of central tendency, spread, shape, symmetry, etc. take on quantifiable meanings.

For example, we stated that the frequency distribution was centered about a given value. This central tendency could be expressed in several ways. One simple method is just to cite the most frequently occurring value, called the *mode*. For grouped data, the mode is the midpoint of the interval with the highest frequency. For the fuse data in Table 1.1, the mode is 5.05.

Another procedure involves selecting the *median;* that is, the value that effectively divides the data in half. For individual readings, the n data points are first ranked in order from smallest to largest, and the median is chosen according to the following algorithm: the middle value if n is odd, and the average of the two middle values if n is even. Alternatively, the location of the median is found by counting $(n + 1)/2$ observations up from the bottom of the list of ordered data. For grouped data, the median occurs in the interval for which the cumulative frequency distribution registers 50 percent; that is, a vertical line through the median divides the histogram into two equal areas. For grouped data with n points, to get the median, one first determines the number of observations in the class containing the middle measurement $(n+1)/2$ and the number of observations in the class to get to that measurement. For example, for the fuse data, $n = 100$, and the middle value is the 50.5 point, that is, between the 50th and 51st observations. The 50th point occurs in the class marked 4.895 to 4.995 (width 0.1). There are 41 data points before the interval and 13 points in this class. We must count the 9.5 values to get the median. Hence, the median is

$$4.895 + \left(\frac{9.5}{13}\right) \times 0.1 = 4.968$$

(In reliability work, it is common terminology to refer to the median as the "T_{50}" value, meaning the time to 50 percent failures.)

The most common measure of central tendency, however, is called the *arithmetic mean* or *average*. The sample mean is simply the sum of the observations divided by the number of observations. Thus, the mean, denoted by $\overline{X}$, of n readings is given by the statistic

$$\overline{X} = \frac{X_1 + X_2 + X_3 + \dots + X_n}{n} = \frac{\sum\limits_{i=1}^{n} X_i}{n}$$

This expression is called a statistic because its value depends on the sample measurements. Thus, the sample mean will change with each sample drawn, which is another instance of the variability of the real world. The sample mean estimates the population mean. In contrast, the population mean depends on the entire set of measurements and thus is a fixed quantity which we call a *parameter*.

For a discrete (i.e., countable) population, the mean is just the summation over all discrete values where each value is weighted by the probability of its occurrence. For a continuous (i.e., measurable) population, the mean parameter is expressed in terms of the PDF model as

$$\mu = \int\limits_{-\infty}^{\infty} xf(x)\, dx$$

For reliability work involving time, the population mean is

$$\mu = \int\limits_{0}^{\infty} tf(t)\, dt$$

An alternate expression for the mean of a lifetime distribution is sometimes easier to evaluate. The form of the equation, when a finite mean exists, is

$$\mu = \int_0^\infty [1 - F(t)] \, dt$$

(See Feller, 1968, p. 148, for a proof.)

A common practice in statistics is to refer to the mean for both discrete and continuous random variables as the expected value of the random variable and use the notation $E(X) = \mu$ or $E(T) = \mu$. We shall occasionally use this terminology in this text.

Knowing the center of the distribution is not enough; we are also concerned about the spread of the data. The simplest concept for variability is the range, the difference between the highest and lowest readings. However, the range does not have very convenient statistical properties, and so another measure of dispersion is more frequently used. This numerical measure of variation is called the *variance*. The variance has certain statistical properties which make it very useful for analysis and theoretical work. The variance of a random variable X is defined as the expected value of $(X - \mu)^2$; that is, $V(x) = E[(X - \mu)^2]$. For continuous data, the population variance for common reliability analysis involving time is

$$V(t) = \sigma^2 = \int_0^\infty (t - \mu)^2 f(t) \, dt$$

In engineering terms, we see that the variance is the expected value of the second moment about the mean.

The square root of the variance is called the *standard deviation*. The standard deviation is expressed in the same units as the observations. The sample variance is denoted by S^2, and the calculating formula is

$$S^2 = \frac{\displaystyle\sum_{i=1}^{n} \left(X_i - \overline{X} \right)^2}{n-1}$$

The $n-1$ term occurs because statistical theory shows that dividing by $n-1$ gives a better (i.e., unbiased) estimate of the population variance (denoted by $\hat{\sigma}^2$) than just dividing by n. Alternatively, we may state that one degree of freedom has been taken to estimate the population mean μ using $\overline{X}$. However, if the mean is known, replace $\overline{X}$ by known mean and divide by n instead of $n-1$.

We have defined numerical measures of central tendency ($\overline{X}$, μ) and dispersion (S^2, σ^2). It is also valuable to have measures of symmetry about the center and a measure of how peaked the data is over the central region. These measures are called skewness and kurtosis, and are respectively defined as expected values of the third and fourth moments about the mean; that is,

$$\text{skewness: } \mu_3 = E\left[(X-\mu)^3\right], \text{ and kurtosis: } \mu_4 = E\left[(X-\mu)^4\right]$$

Symmetric distributions have skewness equal to zero. A unimodal (that is, single peak) distribution with an extended right "tail" will have positive skewness and will be referred to as skewed right; skewed left implies a negative skewness and a corresponding extended left tail. For example, the exponential distribution in Figure 1.3 is skewed right. Kurtosis, on the other hand, indicates the relative flatness of the distribution or how "heavy" the tails are. Both measures are usually expressed in relative (that is, independent of the scale of measurement) terms by dividing μ_3 by σ^3 and μ_4 by σ^4. Sample estimates are calculated using the formulas

$$\text{Skewness estimate} = \frac{\left[\sum_{i=1}^{n}(X_i - \overline{X})^3/n\right]}{\left[\sum_{i=1}^{n}(X_i - \overline{X})^2/n\right]^{3/2}}$$

$$\text{Kurtosis estimate} = \frac{\left[\sum_{i=1}^{n}(X_i - \overline{X})^4/n\right]}{\left[\sum_{i=1}^{n}(X_i - \overline{X})^2/n\right]^{2}}$$

For further information, consult the text by Hahn and Shapiro (1967).

These various measures allow us to check the validity of assumed models. Ott (1993) shows applications to the normal distribution. Table 1.4 contains a listing of properties of distributions frequently used in reliability studies.

The important statistical concept involved in sample estimates of population parameters (e.g., mean, variance, etc.) is that the population parameters are fixed quantities, and we infer what they are from the sample data. For example, the fixed constant θ in the exponential model $F(t) = 1 - e^{-t/\theta}$, where $\theta = 1/\lambda$, can be shown to be the mean of the distribution of failure times for an exponential population. The sample quantities, on the other hand, are random statistics which may change with each sample drawn from the population.

TABLE 1.4   Properties of Distributions Frequently Used in Reliability Studies

Property	Uniform	Normal	Weibull	Exp	Lognormal
Symmetric	yes	yes	no	no	no
Bell shaped	no	yes	no	no	no
Skewed	no skew = 0	no skew = 0	yes (right)	yes (right) skew = 2	yes (right)
Kurtosis	1.8	3		9	
Log data is symmetric and bell shaped	no	no	no	no	yes
Cumulative distribution	straight line	"S" shape		exponential curve	

Two other useful distributions are:
Rayleigh: skew = 0.63, kurtosis = 3.26, a Weibull with shape = 2 and a linear failure rate
Extreme value: skew = −1.14 (skewed left), kurtosis = 5.4

We also mention here a notation common in statistics and reliability work. The sample estimate of a population parameter is often designated by the symbol "∧" over the population quantity. Thus, $\hat{\mu}$ is an estimate of the population mean μ, and $\hat{\sigma}^2$ estimates σ^2, the population variance.

Example 1.6   The Uniform Distribution

The uniform distribution is a continuous distribution with probability density function for the random variable T given by

$$f(t) = \frac{1}{\theta_2 - \theta_1}, \quad \theta_1 \le t \le \theta_2$$

and zero elsewhere, where θ_1 and θ_2 are the parameters specifying the range of T. The rectangular shape of this distribution is shown in Figure 1.6.

We note that $f(t)$ is constant between θ_1 and θ_2. The cumulative distribution function (CDF) of T, denoted by $F(t)$, for the uniform case is given by

$$F(t) = \frac{t - \theta_1}{\theta_2 - \theta_1}$$

Thus, $F(t)$ is linear in t in the range $\theta_1 \le t \le \theta_2$, as shown in Figure 1.7.

Exercise 1.7

Show that the uniform distribution has expected value $E(t) = (\theta_1 + \theta_2)/2$ and variance $V(t) = (\theta_2 - \theta_1)^2/12$.

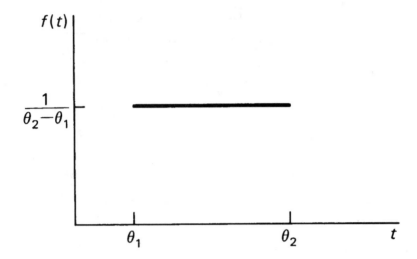

Figure 1.6   The Uniform PDF

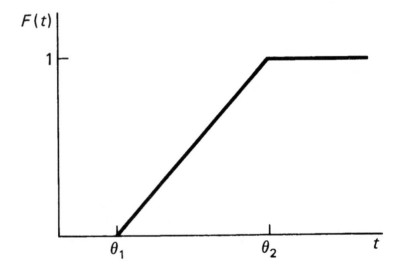

Figure 1.7   The CDF for the Uniform Distribution

Exercise 1.8

The uniform distribution defined on the unit interval [0, 1] is a popular and useful model—so much so that the name *uniform distribution* is often taken to refer to this special case. Find $f(u)$, $F(u)$, $E(u)$, and $V(u)$ for this distribution.

Exercise 1.9

Let $F(t) = 1 - (1 + t)^{-1}, 0 \leq t \leq \infty$. This is a legitimate CDF that goes from 0 to 1 continuously as t goes from 0 to ∞. Find the PDF, and the T_{50} for this distribution. Try to calculate the mean. (Hint: Use either integration by parts or the alternate formula given in the text for calculating means.)

HOW TO USE DESCRIPTIVE STATISTICS

At this point it is important to emphasize some considerations for the analyst. No matter what summary tools or computer programs are available, the researcher should always "look" at the data, preferably in several ways. For example, many data sets can have the same mean and standard deviation and still be very different—and that difference may be of critical significance. See Figure 1.8 for an illustration of this effect.

Generally, the analyst will start out with an underlying model in his mind based on the type of data, where the observations came from, previous experience, familiarity with probability models, etc. However, after obtaining the data, it is necessary that the analyst go through a verification stage before he blindly plunges ahead with his model. This requirement is where the tools of descriptive statistics

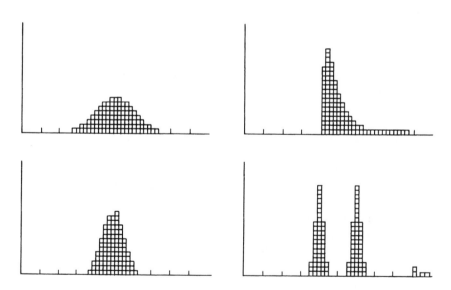

Distributions with the same mean and sigma

Figure 1.8   Mean and Sigma Do Not Tell Us Enough

are very useful. Indeed, in many cases we utilize descriptive statistics to help us choose an appropriate model right at the start of our studies. Other useful graphical techniques include boxplots, dot plots, stem and leaf plots, 3-D plots, and so on. (See Chambers et al, 1983, for further information on graphical analysis.)

In this text, we will focus on several key distributions that are most applicable to reliability analysis. These are: the exponential, the Weibull, the normal, and the lognormal distributions. By learning what these distributions should look like, we can develop a yardstick by which to measure our data for appropriateness. Graphics (frequency histograms, cumulative frequency curves) and numbers (mean, median, variance, skewness, etc.) are the means by which the characteristics of distributions are understood. In later chapters, we shall introduce other valuable descriptive procedures such as probability plotting.

DATA SIMULATION

Many different PDFs (and CDFs) exist, and reliability studies are often concerned with determining what model is most appropriate for the analysis. In reliability work one may wish to simulate data from various distributions in order to

1. calculate percentiles of complicated distributions that are functions of common distributions
2. evaluate the effectiveness of different techniques and procedures for analyzing sample data
3. test the potential effectiveness of various experimental designs and sample size selections
4. illustrate statistical concepts, especially to understand the effects of variability in data

Computer programs are available that will generate random variables from almost any desired distribution. However, there is a simple and general technique that allows us to produce what are called *pseudo-random numbers* from many common distributions. (The term "pseudo-random" is used because a specific computer algorithm generates the numbers to be as nearly random as possible.) To begin, we need a good table of random numbers or the kind of random number generator found in many hand calculators.

For simplicity, we consider only distribution functions $F(x)$ that are continuous and map one-to-one onto the unit interval $(0, 1)$; that is, $0 \le F(x) \le 1$. This class includes all the life distributions discussed in this text. Let $F(x) = u$. Then we can define an inverse function, $F^{-1}(u) = x$, that provides the specific percentile corresponding to the CDF value in the unit interval. For example, given $F(x) = 0.5$, then $F^{-1}(0.5) =$ the median, which is the 50th percentile. F and its inverse have the following properties: $F(F^{-1}(u)) = u$ and $F^{-1}(F(x)) = x$.

To generate a random sample $x_1, x_2, \ldots, x_n$ from $F(x)$, first generate a random sample $u_1, u_2, \ldots, u_n$ from the uniform distribution defined on $[0, 1]$. This procedure is done with random numbers. For example, if a five-digit random number is obtained from a table or a calculator, divide the number by 100,000 to obtain a pseudo-random number from the uniform distribution. (Many hand calculators provide random numbers directly in the unit interval.) Next set $x_1 = F^{-1}(u_1), x_2 = F^{-1}(u_2), \ldots, x_n = F^{-1}(u_n)$. It is easy to show that the x samples are distributed according to the $F(x)$ distribution. (See the hint to problem 1.4 at the end of this chapter.)

Example 1.7   Data Simulation

Let F be the distribution given in Exercise 1.9 Generate a sample of 5 random times from this distribution.

Solution

We obtain F^{-1} by solving for t in $F(t) = u = 1 - (1 + t)^{-1}$. This procedure provides $t = u/(1 - u) = F^{-1}(u)$. Next, we use a random number generator on a calculator to obtain the uniform distribution sample $(0.880, 0.114, 0.137, 0.545, 0.749)$. Transforming each of these by F^{-1} gives the values $t_1 = 0.880/(1 - 0.880) = 7.333, t_2 = 0.129, t_3 = 0.159, t_4 = 1.198, t_5 = 2.984$. The sample $(t_1, t_2, t_3, t_4, t_5)$ is the desired random sample from F.

More examples of data simulation will be given in later chapters. Many of the examples worked out in this text will contain sets of data simulated from our key distributions. We shall use such simulated data to illustrate our procedures and methods.

SUMMARY

In this chapter we have introduced descriptive statistical techniques including histograms and cumulative frequency curves. We have discussed the concepts of populations and samples. Probability rules have been illustrated. Simple concepts of probability have been treated. Numerical measures of central tendency and dispersion have been presented. The importance of visualizing the observations has been emphasized. We have mentioned several important reliability distribution models and discussed how to simulate data from a distribution. The next chapter will begin an in depth presentation of the applications and uses of these concepts in the study of reliability.

PROBLEMS

1.1 The following American experience mortality table gives the proportion living as a function of age, starting from age 10 in increments of 10 years.

Age:	10	20	30	40	50	60	70	80	90	100
Living:	1.00	0.926	0.854	0.781	0.698	0.579	0.386	0.145	0.008	0.000

a. Calculate the percentage dying in each ten year interval and plot the histogram.

b. Calculate the average life-span of ten-year-olds. Use the midpoints of the interval as the age at death.

1.2 For the electronic card assembly in Exercise 1.6, component C (with a probability of failing of 0.10) clearly was the major source of failures. If a second component C was added to the card in parallel with the first component C, so that both had to fail in order for the card to fail, what would be the probability of the card lasting three years now be? Note the card survives if both A and B survive and at least one of the components C survives.

1.3 Show that Bayes' Rule follows almost directly from the definition of conditional probability and the Law of Total Probability.

1.4 Show that a sample generated using the general simulation method given in this chapter does, in fact, have the distribution $F(x)$. Hint: Let $X = F^{-1}(U)$ be the random variable generated by applying F^{-1} to a uniform random variable. Now use $P(X \leq x) = P(F(X) \leq F(x))$ and the properties of inverse distributions, combined with the uniform distribution property that $P(U \leq u) = u$ for $0 \leq u \leq 1$, to show that $P(X \leq x) = F(x)$.

1.5 Each column in the following table contains 25 data points simulated from a distribution with shape and skewness and kurtosis described in Table 1.4. For each of the five sets of data, construct a histogram and calculate the mean, standard deviation, skewness, and kurtosis. Then try to identify the type of distribution from which each sample came. See if looking at the natural log of the data helps when it is difficult to decide between a lognormal model or a Weibull model.

DATA1	DATA2	DATA3	DATA4	DATA5
3565	3752	4485	2673	628
2713	2294	6248	2386	9116
1903	2804	1752	1900	27475
2694	3572	5992	2385	2793
4138	3573	1202	2112	9384
4395	3220	1159	2511	1936
4853	3404	4102	2679	2971
3531	3640	6024	6721	4870
2196	4494	7808	2841	12547
6195	3556	21335	4827	6782
2626	3008	7861	4238	5991
4632	3643	81	2368	4457
2334	2825	6715	5801	3787
636	1984	946	2064	6163
1366	3295	9483	3177	11175
3423	2597	4385	3047	44581
5610	2809	2424	2889	6800
1925	3950	8247	4176	12906
1080	3691	897	2353	1087
1178	3632	8793	5528	1051
1646	3220	58	5351	3044
5000	2054	6466	3370	689
2273	3407	6923	1538	736
1047	3323	10885	1802	22
3174	2550	7222	3033	4126

Chapter 2

Reliability Concepts

This chapter introduces the terms and concepts needed to describe and evaluate product reliability. These are the *reliability* function, the *hazard* and *cumulative hazard* functions, the *failure rate* and *average failure rate*, the *renewal rate*, the *mean time to failure*, and the well known "bathtub curve." In addition, we will look at the kinds of data a reliability analyst typically obtains from laboratory testing or a customer environment (*uncensored, censored, and multicensored data*).

THE RELIABILITY FUNCTION

The theoretical population models used to describe device lifetimes are known as "life distributions." For example, if we are interested in a particular type of transistor, then the population might be all the lifetimes obtainable from transistors of this type. Alternatively, we might want to restrict our population to just transistors from one particular manufacturer made during a set time period. In either case, the CDF for the population is called a *life distribution*. If we denote this CDF by $F(t)$, then $F(t)$ has two useful interpretations.

1. $F(t)$ is the probability that a random unit drawn from the population fails by t hr.
2. $F(t)$ is the fraction of all units in the population that fail by t hr.

Pictorially, $F(t)$ is the area under the probability density function $f(t)$ to the left of t. This area is shown in Figure 2.1. The total area under $f(t)$ is unity (i.e., the probability of failure approaches 1 as t approaches infinity).

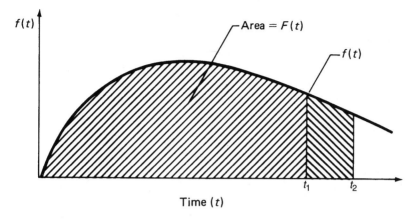

Figure 2.1   CDF $F(t)$

Since $F(t)$ is a probability, the shaded region has an area equal to the probability of a new unit failing by t hr of operation. This equivalence of area to probability generalizes so that the area under $f(t)$ between two vertical lines drawn at time t_1 and a later time t_2 corresponds to the probability of a new unit surviving to time t_1 and then failing in the interval between t_1 and t_2. This area can be obtained by taking all the area to the left of t_2 and subtracting the area to the left of t_1, which is just $F(t_2) - F(t_1)$.

$F(t_2) - F(t_1)$ *is the probability that a new unit survives to time t_1 but fails before time t_2. It is also the fraction of the entire population that fails in that interval.*

Since it is often useful to focus attention on the unfailed units, or survivors, we define the reliability function (survival function) by

$$R(t) = 1 - F(t)$$

(We could also call $F(t)$ the "unreliability" function.) The reliability function may be thought of in either of two ways:

1. as the probability a random unit drawn from the population will still be operating after t hours
2. as the fraction of all units in the population that will survive at least t hours

If n identical units are operating and $F(t)$ describes the population they come from, then $nF(t)$ is the expected (or average) number of failures up to time t, and $nR(t)$ is the number expected to be still operating.

Example 2.1   Life Distribution Calculations

Suppose that a population of components is described by the life distribution $F(t)$ $= 1 - (1 + 0.001t)^{-1}$. What is the probability that a new unit will fail by 1000 hr? By 4000 hr? Between 1000 and 4000 hr? What proportion of these components will last more than 9000 hr? If we use 150 of them, how many do we expect to fail in the first 1000 hr? In the next 3000 hr?

Solution

By substitution, $F(1000) = 1 - (1 + 1)^{-1} = 0.5$, and $F(4000) = 1 - (1 + 4)^{-1} = 0.8$. These are the probabilities of failing in the first 1000 and first 4000 hr, respectively. The probability of failing between 1000 and 4000 hr is $F(4000) - F(1000)$, or 0.3. The proportion surviving past 9000 hr, $R(9000)$, is $(1 + 9)^{-1}$, or 0.1. Finally, the expected failures in the first 1000 hr are 150×0.5, or 75. In the next 3000 hr, an additional 150×0.3, or 45, are expected to fail.

Exercise 2.1

Suppose that a population of components follows the following life distribution model:

$$F(t) = 1 - e^{-(t/2000)}$$

a. What is the probability a new unit will fail by 500 hours? By 3000 hours?
b. Between 500 and 3000 hours?
c. What proportion of new units will last more than 5000 hours?
d. If we use 200 of them, how many do we expect to fail in the first 500 hours? In the next 2500 hours?

SOME IMPORTANT PROBABILITIES

Because $F(t)$ and $R(t)$ are probabilities, a few simple but powerful formulas can be derived easily using the basic rules for calculating probabilities of events presented in Chapter 1. The two rules needed in this section are as follows:

1. **Multiplication Rule**: The probability that several independent events will all occur is the product of the individual event probabilities.
2. **Complement Rule**: The probability that an event does not occur is 1 minus the probability of the event.

In our application, the independent events are the failure or survival of each of n randomly chosen, independently operating units. If we want the probability of all of them still operating after t hours, we apply the Multiplication Rule and multiply $R(t)$ by $R(t)$ n times. In other words,

The probability that n independent identical units, each with reliability R(t), all survive past t hours is $[R(t)]^n$.

If we want the probability that at least one of the n units fails, we apply the Complement Rule and obtain:

The probability that at least one of n independent identical units fails by time t is given by

$$1 - [R(t)]^n = 1 - [1 - F(t)]^n$$

The power of these formulas is readily apparent if we consider a simple system composed of n identical components all operating independently (in terms of working or failing). If the life distribution for each of these components is $F(t)$, then the probability that the system does not have a failure by time t is $[R(t)]^n$. If the system fails when the first of its components fails, and we denote the life distribution function for a population of these systems by $F_s(t)$, then the complement rule gives us

$$F_s(t) = 1 - [R(t)]^n$$

This equation shows how, in this simple case, system reliability is built up using a bottoms-up approach starting with the individual component reliabilities. This is a key concept, which will be discussed in detail in Chapter 8.

Example 2.2   System Reliability

A tragic illustration of how a system's reliability can be much worse than any of its components is given by the 1986 Challenger space shuttle disaster. The reliability of a single O ring (the failing component) had been estimated at 0.99. However, there were eight of them in the shuttle, and all had to work properly. The system reliability is given by $0.99^8 = 0.89$; therefore the potential for failure was 0.11, or approximately 1 in 9. The Challenger mishap occurred on the twelfth shuttle launch.

Exercise 2.2

Assume a computer contains 3,000 identical components, all operating independently and each critical to the operation of the computer. If each component has a reliability estimated at 0.9995, what is the reliability of the computer?

We turn now from the probabilities of failure to the various ways of defining rates at which failures occur.

THE HAZARD FUNCTION OR FAILURE RATE

Consider a population of 1000 units that start operating at time zero. Over time, units fail one by one. Say that at 5000 hours the fourth unit has already failed, and another unit fails in the next hour. How would we define a "rate of failure" for the units operating in the hour between 5000 and 5001? Since 996 units were operating at the start of that hour, and one failed, a natural estimate of the failure rate (for units at 5000 hr of age) would be 1/996 per hour.

If we look closely at that calculation, we see that we have calculated a conditional rate of failure, or rate of failure for the survivors at time t.

We can make this definition more precise by using the concept of conditional probability discussed in Chapter 1. There, we used the notation $P(A)$ to denote the probability of event A occurring and $P(B|A)$ to denote the conditional probability that event B will occur, given that A is known to have occurred. $P(B|A)$ was defined as follows:

$$P(B|A) = \frac{P(B \text{ and } A \text{ both occur})}{P(A)}$$

Using this formula, we can calculate the probability of failing after surviving up to time t in a small interval of time, Δt, as follows:

$$P(\text{fail in next } \Delta t | \text{survive to } t) = \frac{F(t + \Delta t) - F(t)}{R(t)}$$

We divide this by Δt to convert it to a rate and obtain

$$\frac{F(t + \Delta t) - F(t)}{R(t) \Delta t}$$

If we now let Δt approach zero, we obtain the derivative of $F(t)$, denoted by $F'(t)$, divided by $R(t)$. Since $F'(t) = f(t)$ (see Chapter 1), we have derived the expression for the instantaneous failure rate or hazard rate $h(t)$:

$$h(t) = \frac{f(t)}{R(t)}$$

For the remainder of the text, the terms *failure rate, instantaneous failure rate,* and *hazard rate* will all be equivalent and have the above definition.

The units for the rate we have just defined are failures per unit time. It is the failure rate of the survivors to time t in the very next instant following t. It is not a probability, and it can have values greater than 1 (although it is always non-negative). In general, it is a function of t and not a single number or constant.

The reader should be cautioned that not all sources use the same definition when talking about failure rates. Some authors define the failure rate to be $f(t)$, which is the rate of failure of the original time zero population at time t.

THE CUMULATIVE HAZARD FUNCTION

Just as the probability density function $f(t)$ can be integrated to obtain the cumulative distribution function $F(t)$, we can integrate the hazard function $h(t)$ to obtain the cumulative hazard function $H(t)$.

$$H(t) = \int_0^t h(t) \, dt$$

This integral can be expressed in closed form as

$$H(t) = -\ln R(t)$$

where the notation ln denotes natural logarithms or logarithms to the base e.

Exercise 2.3

Verify the closed form expression for $H(t)$ is correct by taking derivatives of both sides and obtaining the definition of the hazard function $h(t)$.

By taking antilogarithms in the above equation for $H(t)$, a well known and useful identity relating failure rates and CDFs is obtained:

$$F(t) = 1 - e^{-H(t)} = 1 - e^{-\int_0^t h(y) \, dy}$$

This expression shows that, given $H(t)$, we can calculate $F(t)$, and vice-versa. So, in a sense, all the quantities we have defined give the same amount of information: with any one of $F(t), f(t), h(t)$, or $H(t)$, we can calculate all of the others.

$H(t)$ will be particularly useful later when we discuss graphical plotting methods for estimating life distribution parameters from failure data.

Exercise 2.4

Derive the equation for $F(t)$ given that $h(t)$ is the constant λ, i.e., $h(t) = \lambda$ for all t.

THE AVERAGE FAILURE RATE

Since the failure rate $h(t)$ varies over time, it is useful to define a single average number that typifies failure rate behavior over an interval. This number might be used in an engineering specification for a component, or it might be an input to service cost and stock replacement calculations.

A natural way to define an average failure rate (AFR) between time t_1 and t_2 is to integrate the (instantaneous) failure rate over the interval and divide by $t_2 - t_1$.

$$AFR(t_1, t_2) = \frac{\int_{t_1}^{t_2} h(t)\, dt}{t_2 - t_1} = \frac{H(t_2) - H(t_1)}{t_2 - t_1} = \frac{\ln R(t_1) - \ln R(t_2)}{t_2 - t_1}$$

If the time interval is from 0 to T, the AFR simplifies to

$$AFR(T) = \frac{H(T)}{T} = \frac{-\ln R(T)}{T}$$

and this quantity is approximately equal to $F(T)/T$ for small $F(T)$, that is, $F(T)$ less than about 0.10. Conversely, in the time interval 0 to T, if we know the average failure rate $AFR(T)$, the CDF $F(T)$ for any distribution is given exactly by the relation

$$F(T) = 1 - e^{-AFR(T)}$$

which is approximately equal to $T \times AFR(T)$, for small $F(T)$.

The AFR finds frequent use as a single-number specification for the overall failure rate of a component that will operate for T hr of useful life. For example, if the desired lifetime is 40,000 hr, then AFR(40,000) is the single average lifetime failure rate.

It should be noted that the AFR is not generally defined or used in most of the literature on reliability, despite its usefulness.

UNITS

Failure rates for components are often so small that units of failures per hour would not be appropriate. Instead, the scale most often used for failure rates is percent per thousand hours (%/K). One percent per thousand hours would mean an expected rate of 1 fail for each 100 units operating 1000 hours. Another scale rapidly becoming popular for highly reliable components is parts per million per thousand hours (PPM/K). One part per million per thousand hours means 1 fail is expected out of 1 million components operating for 1000 hr. Another name for PPM/K is FIT for *fails in time* (other authors have stated that the name FIT comes from *failure unit*). This name will be used for the rest of this text.

The factors to convert $h(t)$ and the AFR to %/K or FIT are given below:

$$\text{Failure rate in \%/K} = 10^5 \times h(t)$$

$$\text{AFR in \%/K} = 10^5 \times \text{AFR}(T_1, T_2)$$

$$\text{Failure rate in FITs} = 10^9 \times h(t)$$

$$= 10^4 \times \text{failure rate in \%/K}$$

$$\text{AFR in FITs} = 10^9 \times \text{AFR}(T_1, T_2)$$

Example 2.3   Failure Rate Calculations

For the life distribution $F(t) = 1 - (1 + 0.001t)^{-1}$, derive $h(t)$ and calculate the failure rate at 10, 100, 1000 and 10,000 hr. Give the last failure rate in both %/K and FIT. What is AFR(1000)? What is the average failure rate between 1000 and 10,000 hr? If five components, each having this life distribution, are starting operation, what is the probability that they will experience no failures in the first 1000 hr?

Solution

$$F(t) = 1 - (1 + 0.001t)^{-1}$$

and, by taking the derivative,

$$f(t) = 0.001(1 + 0.001t)^{-2}$$

By definition,

$$h(t) = f(t)/(1 - F(t)) = 0.001(1 + 0.001t)^{-1}$$

By substitution,

$$h(10) = 0.001/1.01 = 0.00099$$

Similarly,

$$h(100) = 0.001/1.1 = 0.00091$$

$$h(1000) = 0.0005$$

$$h(10,000) = 0.000091$$

This last failure rate is 9.1%/K and 91,000 FITs.

$$AFR(1000) = -\ln R(1000)/1000 = -\ln(1+1)^{-1}/1000 = 0.0007$$

$$AFR(1000,10,000) = [\ln 2^{-1} - \ln 11^{-1}]/9000 = 0.00019$$

The probability that five components operate for 1000 hr without any failures is $[R(1000)]^5$ by applying the Multiplication Rule. This expression yields

$$[(1+1)^{-1}]^5 = 2^{-5} = 1/32$$

Exercise 2.5

Using the life distribution from Example 2.3, calculate the failure rate at 10, 100, 1000, 10,000 and 100,000 hours. Give the answers in both %/K and FITs.

Exercise 2.6

What is the average failure rate for the life distribution in Example 2.3, between 500 and 3000 hours, expressed in %/K.

BATHTUB CURVE FOR FAILURE RATES

Any non-negative function $h(t)$ whose integral $H(t)$ approaches infinity as time approaches infinity can be a failure rate function. However, in practice most components, or groups of components operating together as subassemblies or systems, tend to have failure rate curves with a similar kind of appearance.

This typical shape for failure rate curves, known as the *bathtub curve,* is shown in Figure 2.2. The first part of the curve, known as the *early failure period*, has a decreasing failure rate. During this period, the weak parts that were marginally functional are weeded out. In Chapter 8, we will see that it also makes sense to include the discovery of defects that escape to the field in the front end of this curve.

The long, fairly flat portion of the failure rate curve is called the *stable failure period* (also known as the *intrinsic failure period*). Here, failures seem to occur in a random fashion at a uniform or constant rate that doesn't seem to depend on how long the part has been operating. For example, failures of many electronic components may be associated with electrical overstress due to lightning storms or other external sources of high-voltage transients. Most of the useful life of a component should take place in this flat region of the curve.

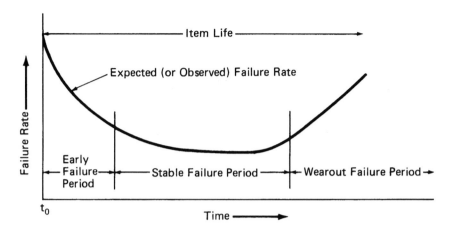

Figure 2.2   Bathtub Curve for Failure Rates

The final part of the curve, where the failure rate is increasing, is known as the *wearout failure period*. Here, degradation failures occur at an ever increasing pace. One of the main purposes of reliability testing is to assure that the onset of the wearout period occurs far enough out in time as not to be a concern during the useful life of the product. (Another key purpose of reliability testing is to establish the value of the failure rate during the long, stable period.)

It is interesting to note that, while every experienced reliability analyst has come across many examples of real failure data exhibiting the shape shown in Figure 2.2, none of the prominent life distributions discussed in the literature or in this book has that shape for $h(t)$. These distributions, as we shall see in later chapters, only fit one (or at most, two) regions of the curve reasonably well. In Chapter 8, algorithms for obtaining the entire curve will be developed.

RENEWAL RATES

A kind of failure rate that takes into account the fact that failed parts are replaced with parts that may, in turn, later fail can be defined. If the replacements are always new parts from the same population as the original parts, we call the process a renewal process. The mean or expected number of failures per unit interval at time t is known as the renewal rate $r(t)$. Chapter 10 contains a detailed discussion of renewal processes and renewal rates.

Which failure rate is better to use—the hazard rate we are calling the failure rate or the renewal rate? The hazard rate (or the AFR) corresponds to a process where, instead of replacing parts with brand new parts when they fail, a part similar to the failed part in terms of remaining life is used.

Typically, however, the replacement part is new but comes from a different time or vintage of manufacture from that of the original part. As a result, it may have a different failure rate curve, and neither the renewal rate concept nor the failure rate we have defined applies exactly.

Fortunately, when failure rates are nearly constant, as on the long, flat portion of the bathtub curve, it does not matter which rate is used; the renewal rate, the failure rate, and the AFR have the same value when the failure rate is constant. For this reason, and because the renewal rate is very difficult to work with mathematically, we will use failure rates and AFRs predominantly throughout this text. (A method for estimating $F(t)$ from renewal data will be given in Chapter 10, however.)

Example 2.4   Examples of Various Rates

In Figure 2.3, 10 similar components operating in a system are represented by horizontal lines. An X on the line shows that a failure occurred at the corresponding time. After each failure, the failing component is replaced by a new component, and the system continues to operate. The first failure for each component is shown by an X with a circle around it.

How would we estimate the CDF and reliability function at 500 and at 550 hr? What would an estimate of the failure rate or average failure rate be for the inter-

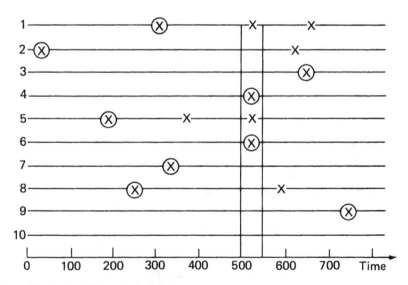

Figure 2.3   Bathtub Curve for Failure Rates

val 500 to 550 hr? How does this compare to an estimate of the renewal rate based on the same interval? What would an estimate of $f(t)$ over this interval be?

Solution

The various estimates derived from Figure 2.3 are

$$\hat{F}(500) = 5/10$$

$$\hat{R}(500) = 5/10$$

$$\hat{F}(550) = 7/10$$

$$\hat{R}(550) = 3/10$$

$$\hat{h}(500 \text{ to } 550) = (2/5)/(50) \text{ hr} = \text{AFR}(500,550)$$

$$\hat{r}(500 \text{ to } 550) = (4/10)/(50) \text{ hr}$$

$$\hat{f}(500 \text{ to } 550) = (2/10)/(50) \text{ hr}$$

This example clearly illustrates the differences between F, f, and the failure rate and renewal rate. To estimate F at a given time, the denominator is the starting number of components, and the numerator is all the first-time-only failures. To estimate f over an interval, the same denominator is used, but only the first-time failures in that interval are in the numerator. This estimate approaches the true f in value if the population is large and the interval is small.

The failure rate or $h(t)$ estimate uses a different denominator. Instead of the time zero original population, only the survivors at the start of the interval are used in the calculation. Since five of the original units had failed by 500 hr, the denominator is $10 - 5 = 5$. The numerator is the number of first time failures in the interval, or 2. This process actually estimates the average failure rate over the interval; as with the estimate of $f(t)$, the larger the original population and the smaller the interval, the better an estimate of $h(t)$ this method yields.

The renewal rate estimate uses the total number of operating units in the denominator. Under the renewal concept, this number will also equal the starting population. The numerator now includes all the failures in the interval, whether from the starting population (circled), or failures of replacement components (uncircled). This number is 4 in the example.

Note that all the rate estimates also have a division by 50 hr to convert to failures per unit hours. This result could then be multiplied by the appropriate constant, as explained in the section on units, if %/K or FIT were desired.

Exercise 2.7

For the mortality table given in Problem 1.1 at the end of Chapter 1, calculate the hazard (or failure) rate for each ten-year period (as a % per year) and plot it.

TYPES OF DATA

The statistical analysis of reliability data is often more complicated and difficult than the analysis of other experimental data because of the diversity of forms this data may take. Most other experiments yield straightforward sample data: a random sample of size n gives n numbers that can be used in a simple way to make inferences about the underlying population. This kind of data is seldom available in reliability evaluation.

The kinds of data generally encountered fall into the categories of exact failure times with censoring (Type I or Type II), readout (or interval or grouped) data, and the most general form of all: multicensored data.

Exact Times—Censored Type I

Suppose n units are put on test for a fixed planned duration of T hr. Say that r of them fail, and the exact failure times $t_1 \leq t_2 \leq t_3 \leq \ldots \leq t_r \leq T$ are recorded. At the end of the test, there are $n - r$ survivors (unfailed units). All that is known about them is that the times of failure they will eventually record are beyond T.

This kind of testing is called censored Type I. It has the advantage of ensuring that schedules are met, since the duration T is set in advance. A fixed test time is valuable in situations where there may be serious impacts if schedules are not rigidly met.

There is one problem with this kind of testing: the number of failures is not known in advance. This number, r, is a random quantity. Since, as we shall see, the precision of failure rate estimates depend on the number of failures, r, and not on the number of units, n, on test, a bad choice of sample size or test conditions may result in insufficient information obtained from the test. The test time may fit within the allotted schedule, but the test results will be inadequate.

Exact Times—Censored Type II

Again, we place n units on test and record the exact times when failures occur. However, instead of ending the test at a predetermined time, we wait until exactly r failures occur and then stop. Since we specify r in advance, we know exactly how much data the test will yield.

This procedure has obvious advantages in terms of guaranteeing adequate data. However, the length of test time is random and open ended. Based on practical scheduling considerations, Type II testing is usually ruled out in favor of Type I.

Readout Time Data

Both of the above testing schemes require instruments that can record exact times of failure. Thus when testing electronic components, continuous in situ monitoring is recommended. However, his kind of test setup may be impractical in many cases.

The following is a practical, commonly used testing procedure: n components are put on test at time zero. T_1 hr later, a "readout" takes place whereby all the components are examined and failures are removed. Let's say r_1 failures are found. Then $n - r_1$ components go back on test. At time T_2, after $T_2 - T_1$ more hours of test, another readout takes place. This time r_2 failures are found and $n - r_1 - r_2$ units go back on test. The process is continued with readouts at time T_3, T_4, and so on. The last readout, at time $T_k = T$, takes place at the end of the test. Figure 2.4 illustrates this kind of data.

This type of data is called *readout* or *interval* or *grouped* data. The readout times are predetermined, as is the end of test time. The number of failures that will occur in an interval is not known until the readout takes place. The exact times of failure are never known.

Readout data experiments have the same problem that censored type data has: the experiment may end before a sufficient number of failures take place. In addition, precision is lost by not having exact times of failure recorded. Even if there are many failures, the data may be inadequate if these failures are spread out over too few intervals.

Despite the above drawbacks, and the difficulties analyzing readout data, it is probably the most common type of reliability data. In situ monitoring usually requires expensive test equipment that is often neither available nor cost justified. The reliability analyst must learn to make the best possible use of readout data experiments; this goal involves careful planning of the experiment (i.e., sample sizes and times of readout) as well as use of the good analysis methods.

Multicensored Data

In the most general case, every unit on test may have a specified interval during which it is known to have failed, or a censoring time past which it is known to have survived (often called a *run-time*.). These intervals and censoring times might be different for each unit. An exact time of failure would be a "degenerate"

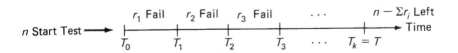

Figure 2.4   Readout Data

or very small interval. This kind of data is called *multicensored*. In the laboratory, it is rare that every unit has a different readout time or censoring time, but less complicated examples of multicensored data are common. In the field, all types of multicensored data frequently occur.

Example 2.5   Multicensored Data

Capacitors are to be tested on fixtures mounted in ovens. While the test is on, the parts will be subjected to a fixed high voltage and high temperature. At the end of each day, they will be removed from the ovens and tested to determine which ones have failed. Then the unfailed units will be put back on test. The plan is to continue like this for 1000 hr.

This test plan would normally yield standard readout data, but several unexpected things might happen to change this. Two possibilities will illustrate this point.

Assume that the test starts with 200 capacitors in four ovens, each containing 50 units. Halfway through the test, one of the ovens malfunctions, causing all further data on its 50 parts to be invalid. This event results in multicensored data with the capacitors in the bad oven "taken off test" at the time of malfunction. The other ovens and parts continue for the full test time.

Instead of a malfunction, the same situation might occur if priorities change, and the test engineer must give up one of his ovens before his capacitor experiment is completed.

These examples show that a mild form of multicensoring may occur even when straightforward readout data was expected. The next example shows a case where heavily multicensored data arises as a natural consequence of the data collection scheme.

Example 2.6   Multicensored Data

Data obtained from components operating in customer environments often has a characteristic form of multicensoring. The data, in this case, consists of failure information on several groups of components. Each group is on test for a possibly different interval of time, and data is read out once at the end of that interval. A concrete example will make this clear. Assume that a machine uses 100 components of a type that is of interest. We are able to examine the field history of 10 machines. Each of these machines has operated a different length of time, based on the date of customer installation. For each machine, only the total number of component failures, up to the date of the investigation, is available. All failures are assumed to come from the original 100 components, and either no repairs were necessary, despite failures, or repair time is considered negligible. (Error correction circuitry in computer memories is an example where a machine continues to operate despite component failures.)

Each machine provides data on a group of 100 components that were "on test" for just one readout interval. That interval is the amount of time the machine has operated and varies from machine to machine. There are also 10 different censoring times: one for each machine's group of unfailed units.

The kind of multicensored data described in this example does not lend itself to graphical methods and can be very difficult to analyze unless the exponential distribution (described in Chapter 3) applies, or sophisticated computer analysis programs are available.

FAILURE MODE SEPARATION

Multicensored data can also come about when test components fail for more than one reason. For example, a corrosion failure mode and a metal migration failure mode might both take place when testing a semiconductor chip. Each mode of failure might follow a completely different life distribution, and it is desirable to analyze the data for each mode separately.

If the failure modes act independently of one another, we can analyze the corrosion data alone by treating the times when units failed due to migration as censoring times. After all, if migration failure had not occurred, these units would have eventually failed due to corrosion. Instead they were "taken off test" at the time of their migration failure.

The analysis of migration failures is done in a similar fashion, by treating the times of corrosion failures as censoring times. This approach can be extended to more than two independent modes of failure. This topic is treated further in the section on data analysis in Chapter 8.

SUMMARY

This chapter defined the reliability function $R(t) = 1 - F(t)$ and the (instantaneous) failure rate $h(t) = f(t)/R(t)$. This failure rate applies at the instant of time, t, for the survivors of the starting population still in operation at that time. An average failure rate over the interval (t_1, t_2) can also be defined by integrating the failure rate over this interval and dividing by $(t_2 - t_1)$. This expression turns out to equal $[\ln R(t_1) - \ln R(t_2)]/(t_2 - t_1)$.

A useful identity that shows how the CDF can be reconstructed from the failure rate is

$$F(t) = 1 - e^{-\int_0^t h(t)\,dt}$$

A plot of failure rate versus time for most components or systems yields a curve with the so-called "bathtub" shape. The front decreasing portion shows early life fallout. Then a long, fairly flat region occurs. Finally, at some point in time, wearout failures due to degradation mechanisms start to dominate, and the failure rate rises.

One common aspect of reliability data causes analysis difficulties. This feature is the censoring that takes place because, typically, not all units on test fail before the test ends.

Reliability data may consist of exact times of failure up to the end of a fixed-length test. This type of data is called Type I censored (also referred to as *time censored*). Type II censoring (also called *failure censoring*) refers to a test that lasts until a prespecified number of failures occur. While Type II censoring may lead to better data, it is less popular because of the open-ended nature of the test duration.

Perhaps the most typical kind of data consists of numbers of failures known only to have occurred between test readouts. This data is called *readout* or *interval* or *grouped* data. If all the units do not have the same readout intervals or the same end of test time, the data is called *multicensored* and is the most difficult of all to analyze.

Multicensored data may also come about when some units on test are damaged, removed prematurely, or fail due to more than one failure mode. Field (or customer operation) data can also come in a multicensored form.

PROBLEMS

2.1 Show that an alternate, equivalent definition of the hazard rate is the negative derivative of the natural logarithm of the reliability function.

2.2 Use the data in Figure 2.3 to estimate the CDF and reliability function at 600 and 700 hours. Estimate the failure rate for the interval 600 to 700 hours. Also estimate the renewal rate and $f(t)$ for this period.

2.3 Fifty units are tested for 1000 hours. Thirty exact times of failure are recorded. After failure analysis has been completed, it is known that there were three different failure modes. Mode A had 13 failures, mode B had 9 failures and mode C had 8 failures. Describe how you would summarize the data first to analyze failure mode A, then B, then C. What assumption are you making?

2.4 From the mortality table given in Problem 1.1 at the end of Chapter 1, generate a new mortality table for 50-year-olds. (Note that the fraction living at age 50 should start with 1.000.) Calculate and plot the hazard function for 50-year-olds and compare it with that for 10-year-olds. What general conclusions do you draw?

2.5 One thousand units are stressed on test for 1,000 hr. Readouts to check for failures take place at 24, 48, 96, 168, 500, and 1,000 hr. Cumulative failures noted at these readouts are 19, 28, 32, 34, 35, and 35, respectively. The remaining 965 test units survive the 1,000 hr test. Estimate the CDF and the reliability function at the end of each readout interval. Estimate the hazard rate (or the average failure rate) during each interval.

Chapter 3

The Exponential Distribution

The exponential distribution is one of the most common and useful life distributions. In this chapter we will discuss the properties and areas of application of the exponential, and then we will look at how to estimate exponential failure rates from data. Tables of factors to calculate upper and lower confidence bounds are included. Detailed examples illustrate the use of these tables to solve many important problems of experimental planning.

EXPONENTIAL DISTRIBUTION BASICS

The PDF for the exponential

$$f(t) = \lambda e^{-\lambda t}$$

and the CDF

$$F(t) = 1 - e^{-\lambda t}$$

were introduced in Chapter 1. Figures 1.3 and 1.4 gave plots of these functions.

In both of these equations, λ is the single unknown parameter that defines the exponential distribution. If λ is known, values of $F(t)$ can be calculated easily for any t. Only an inexpensive calculator or a table of natural logarithms is needed.

Since $R(t) = 1 - F(t) = e^{-\lambda t}$, the failure rate function [using the definition of $h(t)$ given in Chapter 2] for the exponential distribution is

$$h(t) = \frac{f(t)}{R(t)} = \frac{\lambda e^{-\lambda t}}{e^{-\lambda t}} = \lambda$$

This result shows that the exponential failure rate function reduces to the value λ for all times. This is a characteristic property of the exponential distribution; the only distribution with a constant failure rate function is the exponential (see Example 3.2). Figure 3.1 shows how this failure rate looks when plotted against time.

The units for λ are failures per unit time—consistent with whatever units time is measured in. Thus, if time is in hours, λ is in failures per hour. If time is in thousand hour units (abbreviated as K), then λ is in failures per K. Often, however, failure rates are expressed in percent per thousand hour units; this rate has to be converted to failures per unit time before making any calculations using the exponential formula. The same is true if failure rates are measured in FITs, which are failures per 10^9 device hours, as explained in the section on units in Chapter 2. Table 3.1 shows some examples of how to convert between these systems of units.

Exercise 3.1

The average failure rate, or AFR, between time t_1 and time t_2 was defined in Chapter 2 to be $[\ln R(t_1) - \ln R(t_2)]/(t_2 - t_1)$. Show that for the exponential, this expression reduces to the constant λ.

Example 3.1   Exponential Probabilities

A certain type of transistor is known to have a constant failure rate with $\lambda = 0.04\%/K$. What is the probability one of these transistors fails before 15,000 hr of use? How long do we have to wait to expect 1% failures?

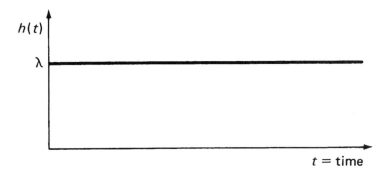

Figure 3.1   The Exponential Distribution Failure Rate $h(t)$

TABLE 3.1   Equivalent Failure Rates in Different Units

Failures per Hour	%/K	FIT
.00001	1.0	10,000
.000001	.1	1,000
.0000001	.01	100
.00000001	.001	10
.000000001	.0001	1

Failures per hour $\times\ 10^5$ = %/k
Failures per hour $\times\ 10^9$ = FIT
%K $\times\ 10^4$ = FIT

Solution

First we convert λ to failures per hour units by multiplying by 10^{-5} to get $\lambda = 0.0000004$. The probability of failure by 15,000 hr is $F(15,000)$ or

$$1 - e^{-0.0000004\ \times\ 15000} = 0.006 = 0.6\%.$$

We find the time corresponding to any % by inverting the formula for the CDF [solving for t in terms of the proportion $F(t)$].
This inversion gives

$$t = \frac{-\ln\left(1 - F\left(t\right)\right)}{\lambda}$$

Substituting 0.01 for $F(t)$ and 0.0000004 for λ gives $t = 25{,}126$ hr.

Example 3.2   Constant Failure Rate

Show that a constant failure rate implies an exponential distribution model.

Solution

The basic identity relating failure rates to CDFs was derived in Chapter 2. The general formula is

$$F\left(t\right) = 1 - e^{-\int_0^t h\left(y\right)dy}$$

If $h(y) = \lambda$, the integral becomes

$$\int_0^t \lambda\,dt = \lambda t$$

Then we have $F(t) = 1 - e^{-\lambda t}$, or $F(t)$ is the exponential CDF.

The theoretical model for the shape we expect exponential data to resemble, when plotted in histogram form, is that of the PDF $f(t) = \lambda e^{-\lambda t}$. This was shown originally as Figure 1.3. We repeat the plot in Figure 3.2, for easy reference.

Sample data that are suspected of coming from an exponential distribution can be plotted in histogram form, as described in Chapter 1, and compared in shape to this ideal form. An alternative way of "looking at data" to see if it appears exponential will be discussed in Chapter 6 (graphical plotting).

Example 3.3   Exponential Data

The Lifetime Light Bulb company makes an incandescent filament that they believe does not wear out during an extended period of normal use. They want to guarantee it for 10 years of operation. The Quality Department is given three months to determine what such a guarantee is likely to cost.

Fortunately, the engineer who has to come up with a test plan has a verified way of stressing light bulbs (using higher than normal voltages) that can simulate a month of typical use by a buyer in 1 hr of laboratory testing. He is able to take a random sample of 100 bulbs and test them all until failure in less than three months. He does this experiment and records the equivalent typical user month of failure for each bulb. The sample data is given in Table 3.2.

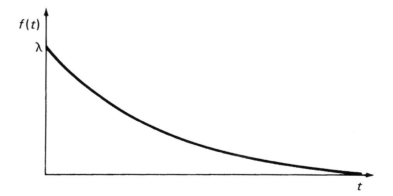

Figure 3.2   The Exponential PDF

TABLE 3.2   Sample Data of Equivalent Month of Bulb Failure

1	2	2	3	4	5	7	8	9	10
11	13	15	16	17	17	18	18	18	20
20	21	21	24	27	29	30	37	40	40
40	41	46	47	48	52	54	54	55	55
64	65	65	65	67	76	76	79	80	80
82	86	87	89	94	96	100	101	102	104
105	109	109	120	123	141	150	156	156	161
164	167	170	178	181	191	193	206	211	212
214	236	238	240	265	304	317	328	355	363
365	369	389	404	427	435	500	522	547	889

The test engineer wants to use an exponential distribution for bulb failures, based on his past experience. His first step in analyzing this data is, therefore, to decide whether an exponential is a reasonable model.

Using the techniques of Chapter 1, we will construct a histogram for this data. First, however, we note how much larger our intervals will have to be in order to include the last point, because of its distance from the other data points. So we ignore the last point and use $547 - 1 = 546$ instead of the actual range of the data. Dividing this by 10 gives 54.6 or 55 for a cell width. For ease of calculation, we will make the first interval from 0 to and including 55. The second interval is greater than 55 to and including 110, and so on. Table 3.3 contains the frequency table for these intervals.

By comparing the shape of the histogram in Figure 3.3 to the exponential PDF shape in Figure 3.2, we see that an exponential model is a reasonable choice for

TABLE 3.3   Frequency Table of Bulb Data

Cell Boundaries	Number in Cell
0 to 55	40
Greater than 55 to 110	23
Greater than 110 to 165	8
Greater than 165 to 220	10
Greater than 220 to 275	4
Greater than 275 to 330	3
Greater than 330 to 385	4
Greater than 385 to 440	4
Greater than 440 to 495	0
Greater than 495 to 550	3
Greater than 550	1
Total Count	100

this data. Later in this Chapter, we will show how to choose an estimate of λ from the data, and test statistically whether an exponential model with that λ is an acceptable fit.

This example also introduced a very useful concept: that of accelerated testing. How does one verify that testing one hour at a given condition is equivalent to ten or one hundred or some other number of typical use condition? What is the mathematical basis behind "acceleration," and what data analysis tools are needed? Chapter 7 will deal with this important topic.

THE MEAN TIME TO FAIL

The mean for a life distribution, as defined in Chapter 1, may be thought of as the population average or mean time to fail. In other words, a brand new unit has this expected lifetime until it fails. We abbreviate mean time to fail with *MTTF*. For the exponential, the definition is

$$MTTF = \int_0^\infty t\lambda e^{-\lambda t}\,dt$$

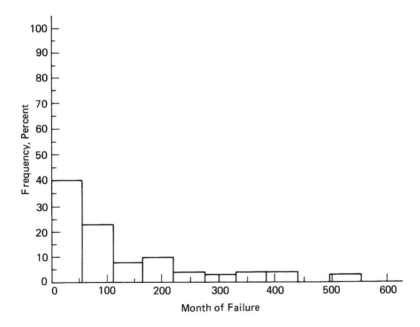

Figure 3.3   Histogram of Bulb Failure Data

This expression can be integrated by parts to yield

$$MTTF = 1/\lambda$$

We interpret this result as follows: The $MTTF$ for a population with a constant failure rate λ is the reciprocal of that failure rate or $1/\lambda$.

Even though $1/\lambda$ is the average time of failure, it is not also the time when half the population will have failed. This median time to failure, or T_{50}, was described in general in Chapter 1. For the entire population, the median is defined to be the point where the CDF function first reaches the value 0.5. For the exponential we have

$$F(T_{50}) = 0.5 = 1 - e^{-\lambda T_{50}}$$

Taking natural logarithms and solving for T_{50} yields

$$T_{50} = \frac{\ln 2}{\lambda} \approx \frac{0.693}{\lambda}$$

This is less than the $MTTF$, since the numerator is only 0.693 instead of 1. In fact, when time has reached the $MTTF$, we have

$$F(MTTF) = 1 - e^{-\lambda/\lambda} = 1 - e \approx 0.632$$

This shows that approximately 63.2% of an exponential population with failure rate λ has failed by the time the $MTTF$ $1/\lambda$ is reached.

Example 3.4   Mean Time to Fail

A company manufactures resistors which are known to have an exponential failure rate with $\lambda = 0.15\%/K$. What is the probability any one resistor will survive 20,000 hr of use? What is the probability it fails in the next 35,000 hr? What is the $MTTF$? At what point in time will 10% of these resistors be expected to fail? 50%? When will 63.2% have failed?

Solution

First we convert λ to failures per unit of time from %/K using the 10^{-5} conversion factor (if λ had been given in FITs, the factor would be 10^{-9}). Then $\lambda = 0.15 \times 10^{-5}$ or 0.0000015.

The probability of surviving 20,000 hr is $R(20,000) = e^{-0.03}$ or 0.97. The probability of failing in the next 35,000 hr after surviving to 20,000 hr is a conditional probability with value $[F(55,000)-F(20,000)]/R(20,000)$ or $(0.079 - 0.030)/0.97 = 0.051$.

The $MTTF$ is $1/\lambda$ or $1/0.0000015 = 666,667$ hr.

We find out when 10% will have failed (known as the tenth percentile) by solving for t in $F(t) = 0.1$.

$$1 - e^{-0.0000015t} = 0.1$$

$$-0.0000015t = \ln 0.9$$

$$t = 0.10536/0.0000015 = 70,240 \text{ hr}$$

Finally, the time when 50% of the population has failed is the median or T_{50} point, given by $0.693/\lambda = 462,000$ hr, and the 63.2% point is the $MTTF$ or 666,667 hr.

In the last example, we calculated the probability of failing in 35,000 hr, after surviving 20,000 hr. This turned out to be 0.051. But the probability of a new resistor failing in its first 35,000 hr, or $F(35,000)$, also equals 0.051. Previous stress time doesn't seem to make any difference. This property of the exponential distribution is discussed in the next section.

Exercise 3.2

Consider the exponential distribution with $MTTF = 50,000$ hr. What is the failure rate in %/k hr? What is the time to 10% failure? To 50% failure?

LACK OF MEMORY PROPERTY

The constant failure rate was one of the characteristic properties of the exponential: closely related is another key property: the exponential "Lack of Memory." A component following an exponential life distribution does not "remember" how long it has been operating. The probability that it fails in the next hour of operation is the same if it is new, one month old, or several years old. It does not age or wear out or degrade with time or use (at least in a probabilistic sense). Failure is a chance happening—always at the same constant rate, unrelated to accumulated power–on.

The equation that describes this property says that the conditional probability of failure in some interval of time of length h, given survival up to the start of that interval, is the same as the probability of a new unit failing in its first h.

$$P(\text{fail in next } h \mid \text{survive } t) = P(\text{new unit fails in } h)$$

In terms of the CDF, this becomes

$$\frac{F(t+h) - F(t)}{1 - F(t)} = F(h)$$

This equation holds if $F(t)$ is the exponential CDF, as the reader can easily verify. It can also be shown that only the exponential has this property for all t and h (see Feller, 1968, page 459).

The implications of this concept from a testing point of view are highly significant. We gain equivalent information from testing 10 units for 20,000 hr or from testing 1,000 units for 200 hr (or even 1 unit for 200,000 hr). If a unit fails on test, we have the option of repairing or replacing it and continuing the test without worrying about the fact that some of the test units have a different "age" from that of other units.

If engineering judgement says that the above testing equivalences seem wrong in a particular case, then we are really saying that we do not believe the exponential distribution applies. However, when the exponential can be used as a reasonable model for the data and the type of item on test, these advantages in test design apply.

Another consequence of the "Lack of Memory" property is that the renewal rate (defined in Chapter 2) and the failure rate (as well as the average failure rate) are all equal and have value λ. This fact takes away concern over the issue discussed in Chapter 2: namely, which rate is the best to use in a particular situation. We can even define an expected time between failures for a repairable, exponentially distributed unit. This mean time between failures, or MTBF, is again $1/\lambda$.

Exercise 3.3

Given that a battery lifetime follows the exponential distribution with $MTTF = 36$ months, find the probability of failure in 10 months. Next find the probability of failure in the following 10 months, given survival to 10 months.

AREAS OF APPLICATION

If we feel that a unit under test has no significant wearout mechanisms, at least for its intended application life, and either we do not expect many early defect failures or we intend to separate these out and treat them separately, then the exponential is a good initial choice for a life distribution model.

Note the words "no significant wearout mechanisms, at least for its intended application life." Even though we know that, in every imaginable real-world case, some kind of wearout eventually takes place, we can ignore this consideration if we feel it is not a practical concern. Having a nearly constant failure rate over the region of time we are interested in (and confining our testing time to this region or its equivalent under test acceleration) is all we need to use the exponential.

Another application of the exponential is in modeling the long, flat portion of the bathtub curve. Since actual failure data on many systems, subassemblies, or even individual components has a nearly constant failure rate for most of the product life, exponential methods of data analysis can be used successfully.

In fact, if we specify product performance over an interval of time using an average failure rate, and if we are not particularly concerned with how the failures spread out over that interval, then we can use an exponential assumption and the exponential confidence bound factors to be described later in this chapter. This feature extends the usefulness of the exponential to many cases where it would not be the correct theoretical model. Just as any smooth curve can be approximated piece-wise by straight lines to any degree of accuracy required, we can consider a changing failure rate curve to be constructed of many piece-wise constant exponential portions. Then we can analyze data from within any one piece, or interval, as if it were exponential.

Example 3.5   Piece-Wise Exponential Approximation

Figure 3.4 illustrates how a typical "bathtub" curve showing the actual failure rate for a given product might be piece-wise approximated by constant failure rate (exponential distribution) segments. Each constant failure rate value equals the average failure rate over that section of the "bathtub" curve. The time points might be months or quarters in the field. The average failure rate estimate would be the number of failures occurring from units beginning that age range divided by the total hours of operation within that age range of all shipped units (as described in the next section).

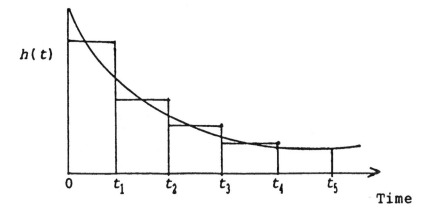

Figure 3.4   Piece-Wise Approximation of Actual Failure Rate

Later in this chapter, we will see that, when the exponential assumption applies, preplanning experiments is a very straightforward process. For example, sample sizes, confidence levels, and precision can be decided in advance. It will be much more difficult—or even impossible—to do similar exact planning for other life distribution models. For this reason, the exponential is also useful as a trial model in the experimental planning stage, even if we do not expect it to actually apply later on.

One must be careful, however, when setting up an experiment designed to accelerate a known wearout mechanism. If the purpose is to determine whether wearout will, indeed, start well beyond useful product life, then we are clearly interested in the non–flat rising portion of the bathtub curve. Exponential analysis methods would not apply, and preplanning should be based on the actual life distribution model, if at all possible.

ESTIMATION OF λ

When data comes from an exponential distribution, there is only one parameter, λ, to estimate. The best estimate for complete or censored samples is

$$\hat{\lambda} = \frac{\text{number of failures}}{\text{total unit test hours}}$$

The denominator is obtained by adding up all the operating hours on test of all the units tested, including both those that failed and those that completed the test without failing.

For a complete sample (everything fails and exact times of failure are recorded), this expression reduces to the reciprocal of the sample mean. Thus, we have = 1/(sample mean time to failure), just as we had $\lambda = 1/MTTF$.

For censored Type I data (fixed test time), with r failures out of n on test

$$\hat{\lambda} = \frac{r}{\sum_{i=1}^{r} t_i + (n-r)\,T}$$

T is the pre–fixed end of test time and $t_1, t_2, t_3, \ldots, t_r$ are the exact failure times of the r units that fail before the test ends.

If the test is censored Type II (ends at rth failure time), the same rule yields

$$\hat{\lambda} = \frac{r}{\sum_{i=1}^{r} t_i + (n-r)\,t_r}$$

Lognormal (data) $\sqrt{s}$ Probit
ln (data)

If new units are put on test to replace failed units, or to increase the sample size part way through the testing, then we have multicensored data. Applying the general form for λ, the denominator is the sum of each test unit's time on test.

When we have readout data, we can no longer exactly calculate the numerator in order to estimate λ. In this case, the graphical methods described in Chapter 6 can be used. (They also apply, as an alternate approach, when exact times are available.) More precise techniques, based on the method of maximum likelihood estimation, will be described in Chapter 4.

We can also apply the simple procedure described in this section to readout data, often with little loss of accuracy. For example, if many units are on test with only a few failures, the error in assuming all failures occur in the middle of the readout interval will be negligible. With this assumption, the simple confidence bound procedures described later in this chapter can be used.

It should be noted that when exact times are available, the estimation rule described here yields estimates that are also maximum likelihood estimates (see definition in Chapter 4 and Example 4.4).

Example 3.6   Failure Rate and *MTTF*

Returning to the light bulb data of Example 3.3 and Table 3.2, the estimate of λ is $\hat{\lambda} = 100/13{,}563 = 0.0074$ bulb fails per month. The *MTTF* estimate $1/\hat{\lambda}$ is 135.6 months.

On the other hand, if we had used the summarized readout data version of bulb failure months given in Table 3.3, we would have calculated $\hat{\lambda} = 100/13{,}505 = 0.0074$ bulb fails per month and *MTTF* = 135.1 months. For this calculation, the numerator would be obtained by multiplying the 40 failures in the first interval by the middle of the interval, or 27.5, and adding 23 × 82.5 for the second interval, and so on. For the one fail time after 550 months, we pretend our 55-month intervals go on to 935 months, where the last fail is recorded. Placing this fail in the center of that last interval (880 to 935) at 907.5 months gives the total test time of 13,505 months.

Exercise 3.4

Perform the calculation described above, using the interval data in Table 3.3 to estimate λ.

This example shows that using the failures divided by total test time formula on readout data is not likely to cause much error.

One immediate application we can make of this λ estimate is to use it to calculate values of the PDF to compare to our histogram in Figure 3.3. In other words, the histogram should have the same shape as $f(t)$, which we are estimating by $f(t) = 0.0074\,e^{-0.0074t}$.

Before graphing $f(t)$, however, and putting it on the same chart with the histogram, we have to adjust scales. The total area under the $f(t)$ curve is always 1. The histogram we plotted, because the intervals have width 55 and the height units are in percent, has an area of 100 × 55 or 5500 (if we put one more box of height 1 between 880 and 935). In Figure 3.5, we plot 5500 × $f(t)$ along with the light bulb data histogram. This graph gives us a direct shape comparison to show that an exponential model applies. In the section after the next, we will test this fit with the "chi–square goodness of fit" test.

Exercise 3.5

Using the data in Table 3.2, construct a histogram based on cell width 35. Estimate λ from the interval data. Fit $f(t)$ to the histogram and compare to Figure 3.5

EXPONENTIAL DISTRIBUTION CLOSURE PROPERTY

The exponential distribution possesses a convenient closure property that applies to assemblies or systems made up of exponentially distributed components. If the system fails when the first component fails, and all the components operate independently, then the system life distribution is also exponential. The failure rate parameter for the system is equal to the sum of the failure rate parameters of the components.

A system model where n components operate independently and the system fails with the first component failure is called a *series model*. This model is discussed in Chapter 8, where it is shown that the system failure rate function $h_s(t)$ is the sum of the n component failure rate functions $h_1(t)$, $h_2(t)$, ..., $h_n(t)$. When the

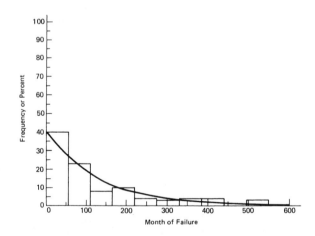

Figure 3.5   Light Bulb Data Histogram Compared to $f(t)$ Shape

components have exponential lifetimes with parameters $\lambda_1, \lambda_2, \ldots, \lambda_n$, then the system has a constant failure rate equal to

$$\lambda_s = \sum_{i=1}^{n} \lambda_i$$

This establishes the exponential closure property, since a constant failure rate implies an exponential distribution.

If the components are all the same, each having failure rate λ, then the system has failure rate $n\lambda$ and a mean time to failure ($MTTF$) of $1/n\lambda$.

This exponential closure property is almost unique. In general, the minimum lifetime of n components will not have the same type life distribution as the components themselves. (There is another case where a similar closure property applies: when all the components have Weibull distributions with the same shape parameter. [See Chapter 4.])

Exercise 3.6

A system consists of 20 serially connected independent components, each of which has a lifetime described by the exponential distribution with a $MTTF = 1500$ hr. What is the probability of the system failing in the first 100 hr of operation?

TESTING GOODNESS OF FIT

There is a standard statistical test, known as the chi-square (χ^2) goodness of fit test, for deciding whether sample data is consistent with a preconceived model. Our basic, or null, hypothesis is that the model is adequate. We then pick a *confidence level* such as 90% or 95%. The higher the level means we want very strong negative evidence from the sample data before we will be willing to reject our hypothesis. For example, if we set our confidence level at 90%, we are saying that we need sample data that is so unusual it would occur less than 10% of the time if our model really is correct. If we obtain such unlikely sample data, we will reject the model. If we set our confidence level to 95%, we require data so unlikely that it occurs less than 5% of the time if the model is correct, before we feel comfortable rejecting the model.

In other words, what we really control by picking a confidence level is the probability of making a mistake by rejecting a good model. This probability of error is known as the Type I error, and it has value less than $0.01 \times (100 - \text{confidence level})$. (A more detailed treatment of the probabilities of making decision errors based on sample data is given in Chapter 9).

In the goodness of fit test, we go through a set procedure (enumerated below) and calculate a number. If our chosen model is correct, this random quantity we have calculated ("random" because its value depends on the random sample data) will have an approximate chi–square distribution. This is a well known distribution which has percentile values that are in a table in the Appendix. If the calculated number turns out so large that the table indicates it to be a highly unlikely value, then we reject the model. Any value higher than the confidence level percentile is defined to be "highly unlikely" enough.

The above discussion may sound somewhat abstract and academic. However, it does have a very serious consequence that is not always understood or appreciated by those who use statistical hypothesis testing methods. When we perform a statistical test and end up not rejecting the assumed hypothesis, we have not proved the hypothesis to be correct. The terminology "accept" the hypothesis and "90 percent confidence level" may mislead us into thinking we are confident that our hypothesis is true. The statistical test gives us no such confidence. We started with what we thought was a reasonable model, and all we have shown is that the sample data has not changed our minds.

On the other hand, when we reject, we are making a strong statement of belief. We are saying that the sample evidence was so overwhelmingly against the model we had chosen that we have to reconsider our choice.

Readers who desire more background on the theory of hypothesis testing should consult a basic statistics textbook such as Mendenhall, Scheaffer, and Wackerly (1990). We now turn to how to test whether sample data is consistent with a distributional assumption such as exponential.

The steps for testing goodness of fit are as follows:

1. Group the data, if necessary, into intervals as if preparing to plot a histogram. The intervals need not be equal, however. Form a table with the interval and the observed frequency, as in Table 3.3, for the light bulb data.
2. If the model is a completely specified distribution (i.e., no unknown parameters), go on to step 3. Otherwise, estimate all the unknown parameters using the method of maximum likelihood. In the case of the exponential, we estimate λ using the method previously described. At this point, the model is completely specified.
3. Use the CDF implied by the model to calculate the probability of failure in each of the intervals listed in the table. If the interval is $(I_1\ I_2)$, this probability is $F(I_2)-F(I_1)$. Add a column with these probabilities to the table.
4. Multiply the probabilities just calculated (which should total to 1) by the sample size on test. This calculation results in an expected number of failures for each interval. Add a column of these numbers to the table.
5. If any intervals have less than 5 expected failures (not actual) as just calculated, it is generally better to combine them with other intervals until every number in the expected column is 5 or greater.
6. Now calculate the following:

$$\frac{(\text{actual number of failures} - \text{expected number of failures})^2}{\text{expected number of failures}}$$

for each of the remaining intervals on the table. Add these numbers together. This total is the χ^2 test statistic.

7. Take one less than the number of separate intervals in the table (or the number of terms added together in step 6 less one) and subtract the number of parameters estimated in step 2. The number obtained is the degrees of freedom of the χ^2 test statistic.

8. Compare the χ^2 statistic to the table values in the Appendix for the calculated degrees of freedom. Look to see if the test statistic is higher than the value in the table for the chosen confidence level. If it is higher, reject the model. If not, continue to use the model, but note just how close you were to rejecting it.

The grouping in step five is necessary because the test statistic calculated in step 6 has only an approximate chi–square distribution. As long as each interval has at least five expected failures, the approximation is good enough to be useful; for under five expected failures the chi–square approximation may be inappropriate.

Example 3.7   Chi-Square Goodness of Fit

Table 3.4 shows the worksheet obtained from applying the first four steps in calculating a chi–square test statistic to the light bulb failure month data in Table 3.3. The estimate of = 0.0074 obtained in Example 3.5 has been used. The probability value for the 55 to 110 interval will be calculated in detail to illustrate the procedure.

To have 5 or more failures in every interval, we can combine 275 to 330 and 330 to 385 to have an expected of 7.3 versus an actual of 7. We also have to combine the last three intervals into a single "greater than 385" interval with $1.9 + 1.3 + 0.9 + 1.7 = 5.8$ expected and $4 + 0 + 3 + 1 = 8$ observed.

The calculation described in step 6 becomes

$$\frac{(40 - 33.4)^2}{33.4} + \frac{(23 - 22.3)^2}{22.3} + \frac{(8 - 14.8)^2}{14.8} + \frac{(10 - 9.9)^2}{9.9}$$

$$+ \frac{(4 - 6.6)^2}{6.6} + \frac{(7 - 7.3)^2}{7.3} + \frac{(8 - 5.8)^2}{5.8}$$

$$= 1.30 + 0.02 + 3.12 + 0 + 1.02 + 0.01 + 0.83 = 6.3$$

TABLE 3.4   Chi-Square Goodness of Fit Worksheet for the Light Bulb Data

Interval (I_1, I_2)	Actual Failures	P(fail in interval) $F(I_2) - F(I_1)$	Expected Failures
0 to 55	40	0.334	33.4
55 to 110	23	0.223	22.3
110 to 165	8	0.148	14.8
165 to 220	10	0.099	9.9
220 to 275	4	0.066	6.6
275 to 330	3	0.044	4 4
330 to 385	4	0.029	2.9
385 to 440	4	0.019	1.9
440 to 495	0	0.013	1.3
495 to 550	3	0.009	0.9
>550	1	0.017	1.7

This test statistic was a sum of 7 terms, and the rule for degrees of freedom given in step 7 says to subtract 1 and subtract another 1 because λ was estimated from the data. This gives 5 for the degrees of freedom. The chi–square table in the Appendix has 11.1 for a 95% point and 9.24 for a 90% point. The test statistic is only 6.3, which is a little higher than the 70% point. That value means there is no reason to reject an exponential model for the population from which this data is derived.

Exercise 3.7

Fifty components are placed on a life test. The recorded times to failure, in hours, are:

16.2	18.9	68.4	84.7	112
143	154	201	212	229
231	292	366	371	412
439	441	490	548	561
609	610	662	662	673
703	836	957	1020	1045
1047	1130	1131	1137	1179
1215	1240	1275	1410	1512
1635	1651	1694	1698	1973
2301	2496	2827	3142	4966

The sample data are suspected of coming from a population of lifetimes that can be adequately modeled by an exponential distribution. Analyze the sample data as follows:

a. Construct a histogram using cell intervals of size 500 (from 0 to 5000).
b. Estimate λ and the *MTTF* using the actual data points.
c. Estimate the median time to fail using the population formula given in the text and the estimate just calculated for λ.
d. Test the adequateness of the exponential model using the chi–square goodness of fit test and a confidence (or significance) level of 90%. Use the same intervals as in a, combining everything greater than 2000 hr into one group.
e. Suppose the test had been planned to stop upon the fifth failure, and hence ended at 112 hr. What would the estimated *MTTF* be for this experiment? Suppose instead the test had been planned to stop at 150 hr. What *MTTF* estimate would have resulted?

CONFIDENCE BOUNDS FOR λ AND THE *MTTF*

The estimate $\hat{\lambda}$ =(number of failures) / (total unit test hours) is a single number or point estimate of λ. Even though it may be the "best" estimate we can come up with, by itself it gives no measure of precision or risk. Can the true λ be as high as $10 \times \hat{\lambda}$? Or are we confident it is no worse than $1.2 \times \hat{\lambda}$? And how much better than $\hat{\lambda}$ might the true λ actually be?

These are very important questions if critical decisions must be made that depend on the true value of λ. No presentation of the test results nor a calculation of $\hat{\lambda}$ is complete without including an interval around that has a high degree of confidence of enclosing the true value of λ.

This kind of interval is called a confidence interval. A 90% confidence interval means that if the same experiment were repeated many times, and the same estimation method was used over and over again to construct an interval for λ, 90 percent of these intervals would contain the true λ. For the one time we actually do the experiment, our interval either does or does not contain λ. But if it is a 90% confidence interval, then we would give 9 to 1 odds in its favor. Alternatively, unless a 1 in 10 chance has occurred, the 90% confidence interval will contain the true (population) value for λ.

For Type I (time censored) or Type II (*r*th failure censored) data, factors based on the chi–square distribution can be derived and used as multipliers of $\hat{\lambda}$ to obtain the upper and lower ends of any size confidence interval. The remarkable thing about these multiplying factors is they depend only on the number of failures observed during the test.

Suppose we want a $100 \times (1 - \alpha)$ confidence interval for λ, where α is the risk we are willing to accept that our interval does not contain the true value of λ. For example, $\alpha = 0.1$ corresponds to a 90% interval. We will calculate a lower $100 \times \alpha/2$ bound for λ and an upper $100 \times (1 - \alpha/2)$ bound for λ. These two numbers give the desired confidence interval. When α is 0.1, this method sets a lower 5% bound and an upper 95% bound, having between them 90% chance of containing λ. The lower end of this interval, denoted by λ_5, is a 95% single–sided lower bound for λ. The upper end, or λ_{95}, is a single–sided 95% upper bound for λ.

The notation here can be confusing at first, and the reader should work out several examples until it is clear. As an illustration, suppose we want a 95% interval for λ. Then $\alpha = 0.05$ and $\alpha/2 = 0.025$. The interval will be $(\lambda_{2.5}, \lambda_{97.5})$. The lower limit is a 97.5% lower bound for λ, and the upper limit is a 97.5% upper bound.

Now that we have defined our notation, how do we calculate these lower and upper bounds? When we have Type II censored data, it can be shown that the lower $100 \times \alpha/2$ percentile of the chi–square distribution with $2r$ degrees of freedom (r is the number of failures), divided by $2r$, is a factor we can multiply $\hat{\lambda}$ by to get $\lambda_{100 \times \alpha/2}$. Similarly, the upper $100 \times (1 - \alpha/2)$ percentile of the same chi–square, divided by $2r$, is a factor we can multiply $\hat{\lambda}$ by to get $\lambda_{100 \times (1-\alpha/2)}$. The distribution statement corresponding to these factors is as follows:

$\chi^2 = 2r\lambda/\hat{\lambda}$ has a chi–square distribution with $2r$ degrees of freedom (Type II censoring without replacement of failed units).

$$P\left[\frac{\hat{\lambda}\left(\chi^2_{2r;100\alpha/2}\right)}{2r} \leq \lambda \leq \frac{\hat{\lambda}\left(\chi^2_{2r;100(1-\alpha/2)}\right)}{2r} \right] = 1 - \alpha$$

For the more common Type I censored data, intervals using the above factors are approximately correct. Exact intervals can be calculated only if failed units are replaced immediately, during the course of the test. In this case, the lower bound factor is exactly as above, while the upper factor uses the $100 \times (1-\alpha/2)$ percentile of a chi–square with $2 \times (r + 1)$ degrees of freedom, still divided by $2r$. Since this chi–square factor produces a slightly more conservative upper bound, we recommend using it for Type I censoring. The probability statement is as follows:

$$P\left[\frac{\hat{\lambda}\left(\chi^2_{2r;100\alpha/2}\right)}{2r} \leq \lambda \leq \frac{\hat{\lambda}\left(\chi^2_{2(r+1);100(1-\alpha/2)}\right)}{2r} \right] \geq 1 - \alpha$$

Using this for Type I censoring without replacement gives an upper bound equivalent to assuming that another fail occurred exactly at the end of the test and that we had Type II censoring with $r + 1$ fails.

The factors described above depend only on the number of failures, r. We will denote the factor that generates an upper or lower $100 \times (1 - \alpha)$ bound by $k_{r;1-\alpha}$. The key equation is

$$k_{r;(1-\alpha)} \times \hat{\lambda} = \lambda_{100(1-\alpha)}$$

for a Type I censored experiment with r failures. The factors are given in two tables: Table 3.5 has upper bound k factors, and Table 3.6 has lower bound factors. Use of the these tables is shown in Example 3.8. Note: since the tables give one-sided confidence bound factors, to obtain, for example, a 90% two–sided interval, one would use the factors corresponding to the 95% upper and lower columns.

Adjustment For Type II Censoring

If the data is Type II censored, the lower bound factors from Table 3.6 still apply. To get the proper upper bound factor from Table 3.5 the following trick must be used: look along the row for $r-1$ failures, instead of r. Find the k factor at the desired confidence and adjust it by multiplying by $(r-1)/r$. This adjusted factor is the correct Type II upper bound multiplier (see Example 3.8). For $r = 1$, use the constants given in Table 3.7, divided by the total POH instead of divided by nT.

Example 3.8   Confidence Bounds for λ

Compute 60% and 90% confidence bounds for λ using the data in Exercise 3.7, part b). Also calculate a 90% interval for λ for the two cases given in Exercise 3.7, part e).

Solution

A complete sample is "censoring Type II with $r = n$". Since $\hat{\lambda}$ =99.38%/K and r = 50, the upper bound factor for a 60% interval (an 80% upper bound factor from Table 3.5) is $(49/50) \times k_{49;0.8} = 0.98 \times 1.14 = 1.12$. Similarly, the upper bound factor for a 90% interval is $0.98 \times k_{49;0.95} = 0.98 \times 1.27 = 1.24$. The lower bound factors come directly from Table 3.6 for $r = 50$ and an 80% and 95% confidence level. They are 0.88 and 0.78, respectively. The 80% interval is (87.5, 111.3) and the 90% interval is (77.5, 123.2), both given in %/K.

$\hat{\lambda}$ for the first case in part e) is $1/1068 = 0.000936$ fails/hr and $r = 5$. The factors, after adjusting the upper bound for Type II censoring, are 0.39 and 1.83. The interval is (0.000365, 0.00171), given in fails/hr.

$\hat{\lambda}$ for the second case in part e) is $1/1173.8 = 0.000852$. The censoring is Type I with $r = 6$, so Table 3.5 applies without any adjustments. The factors are 0.44 and 1.97, and the interval is (0.000375, 0.001678) in fails/hr.

TABLE 3.5   Factors for One-Sided Exponential Upper Bound (Type I time censoring)

No. Fails	60%	80%	90%	95%	97.5%	99%	99.9%
1	2.02	2.99	3.89	4.74	5.57	6.64	9.25
2	1.55	2.14	2.66	3.15	3.61	4.21	5.63
3	1.39	1.84	2.23	2.58	2.92	3.35	4.33
4	1.31	1.68	2.00	2.29	2.56	2.90	3.67
5	1.26	1.58	1.86	2.10	2.33	2.62	3.30
6	1.22	1.51	1.76	1.97	2.18	2.43	2.99
7	1.20	1.46	1.68	1.88	2.06	2.29	2.79
8	1.18	1.42	1.62	1.80	1.97	2.18	2.64
9	1.16	1.39	1.58	1.74	1.90	2.09	2.53
10	1.15	1.37	1.54	1.70	1.84	2.02	2.41
11	1.14	1.34	1.51	1.66	1.79	1.95	2.32
12	1.13	1.33	1.48	1.62	1.75	1.90	2.25
13	1.12	1.31	1.46	1.59	1.71	1.86	2.19
14	1.12	1.29	1.44	1.56	1.68	1.82	2.14
15	1.11	1.28	1.42	1.54	1.65	1.78	2.08
16	1.11	1.27	1.40	1.52	1.62	1.75	2.04
17	1.10	1.26	1.39	1.50	1.60	1.72	2.00
18	1.10	1.25	1.38	1.48	1.58	1.70	1.96
19	1.09	1.24	1.36	1.47	1.56	1.68	1.94
20	1.09	1.24	1.35	1.45	1.54	1.65	1.90
21	1.09	1.23	1.34	1.44	1.53	1.64	1.87
22	1.09	1.22	1.33	1.43	1.51	1.62	1.85
23	1.08	1.22	1.32	1.42	1.50	1.60	1.83
24	1.08	1.21	1.32	1.41	1.49	1.59	1.81
25	1.08	1.21	1.31	1.40	1.48	1.57	1.79
26	1.08	1.20	1.30	1.39	1.47	1.56	1.77
27	1.07	1.20	1.30	1.38	1.46	1.55	1.75
28	1.07	1.19	1.29	1.37	1.45	1.54	1.73
29	1.07	1.19	1.28	1.36	1.44	1.52	1.71
30	1.07	1.19	1.28	1.36	1.43	1.51	1.71
31	1.07	1.18	1.27	1.35	1.42	1.50	1.69
32	1.07	1.18	1.27	1.34	1.41	1.49	1.68
33	1.07	1.18	1.26	1.34	1.40	1.49	1.66
34	1.06	1.17	1.26	1.33	1.40	1.48	1.66
35	1.06	1.17	1.25	1.33	1.39	1.47	1.64
36	1.06	1.17	1.25	1.32	1.38	1.46	1.63
37	1.06	1.16	1.24	1.31	1.38	1.45	1.62
38	1.06	1.16	1.24	1.31	1.37	1.45	1.61
39	1.06	1.16	1.24	1.31	1.37	1.44	1.60
40	1.06	1.16	1.23	1.30	1.36	1.43	1.59
41	1.06	1.15	1.23	1.30	1.36	1.43	1.58

1. Best F/R estimate: $\hat{\lambda} = \dfrac{\text{no. fails}}{\text{total POH}} \times 10^5 \; (\%/K)$. POH = Power-On Hours or total unit test hours.

2. Obtain upper bound at confidence level desired by multiplying by appropriate factor from above table.

3. Mean time to fail estimate: $\text{MTTF} = \dfrac{1}{0.01 \times \hat{\lambda}}$ in KPOH. Lower bound = $\dfrac{1}{0.01 \times \hat{\lambda}_{\text{upper bound}}}$ in KPOH. KPOH = Power-On Hours/1000.

4. For Type II censoring at the rth failure time, use the factor above corresponding to r − 1 fails, multiplied by $\dfrac{(r-1)}{r}$. For r = 1, use the constants given in Table 3.7, divided by the total POH instead of divided by nT.

TABLE 3.5   Factors for One-Sided Exponential Upper Bound (Type I time censoring) (continued)

No. Fails	60%	80%	90%	95%	97.5%	99%	99.9%
42	1.05	1.15	1.23	1.29	1.35	1.42	1.58
43	1.05	1.15	1.23	1.29	1.35	1.42	1.57
44	1.05	1.15	1.22	1.29	1.34	1.41	1.57
45	1.05	1.15	1.22	1.28	1.34	1.41	1.55
46	1.05	1.14	1.22	1.28	1.33	1.40	1.55
47	1.05	1.14	1.21	1.28	1.33	1.40	1.54
48	1.05	1.14	1.21	1.27	1.33	1.39	1.53
49	1.05	1.14	1.21	1.27	1.32	1.39	1.53
50	1.05	1.14	1.21	1.27	1.32	1.38	1.52
55	1.05	1.13	1.20	1.25	1.30	1.36	1.50
60	1.04	1.12	1.19	1.24	1.29	1.34	1.47
65	1.04	1.12	1.18	1.23	1.27	1.33	1.44
70	1.04	1.11	1.17	1.22	1.26	1.32	1.43
75	1.04	1.11	1.16	1.21	1.25	1.30	1.41
80	1.04	1.11	1.16	1.20	1.24	1.29	1.40
85	1.04	1.10	1.15	1.20	1.24	1.28	1.38
90	1.03	1.10	1.15	1.19	1.23	1.27	1.37
95	1.03	1.10	1.14	1.19	1.22	1.27	1.36
100	1.03	1.09	1.14	1.18	1.22	1.26	1.35
110	1.03	1.09	1.13	1.17	1.21	1.25	1.33
120	1.03	1.08	1.13	1.16	1.20	1.23	1.32
130	1.03	1.08	1.12	1.16	1.19	1.22	1.30
140	1.03	1.08	1.12	1.15	1.18	1.21	1.29
150	1.03	1.07	1.11	1.15	1.17	1.21	1.28
160	1.02	1.07	1.11	1.14	1.17	1.20	1.27
170	1.02	1.07	1.11	1.14	1.16	1.19	1.26
180	1.02	1.07	1.10	1.13	1.16	1.19	1.25
190	1.02	1.07	1.10	1.13	1.15	1.18	1.24
200	1.02	1.06	1.10	1.12	1.15	1.18	1.24
250	1.02	1.06	1.09	1.11	1.13	1.16	1.21
300	1.02	1.05	1.08	1.10	1.12	1.14	1.19
400	1.02	1.04	1.07	1.09	1.10	1.12	1.16
500	1.01	1.04	1.06	1.08	1.09	1.11	1.15
600	1.01	1.04	1.05	1.07	1.08	1.10	1.13
700	1.01	1.01	1.05	1.06	1.08	1.09	1.12
950	1.01	1.03	1.04	1.05	1.07	1.08	1.10
1500	1.01	1.02	1.03	1.04	1.05	1.06	1.08
3000	1.01	1.01	1.02	1.03	1.04	1.04	1.06
5000	1.00	1.01	1.02	1.02	1.03	1.03	1.04
20,000	1.00	1.01	1.01	1.01	1.01	1.02	1.02

1. Best F/R estimate: $\hat{\lambda} = \dfrac{\text{no. fails}}{\text{total POH}} \times 10^5$ $(\%/K)$. POH = Power-On Hours or total unit test hours.

2. Obtain upper bound at confidence level desired by multiplying by appropriate factor from above table.

3. Mean time to fail estimate: $\text{MTTF} = \dfrac{1}{0.01 \times \hat{\lambda}}$ in KPOH. Lower bound = $\dfrac{1}{0.01 \times \hat{\lambda}_{\text{upper bound}}}$ in KPOH. KPOH = Power-On Hours/1000.

4. For Type II censoring at the rth failure time, use the factor above corresponding to r − 1 fails, multiplied by $\dfrac{(r-1)}{r}$. For r = 1, use the constants given in Table 3.7, divided by the total POH instead of divided by nT.

TABLE 3.6   Factors for One-Sided Exponential Lower Bound (I and II censoring

No. Fails	60%	80%	90%	95%	97.5%	99%	99.9%
1	0.51	0.22	0.11	0.05	0.03	0.01	0.00
2	0.69	0.41	0.27	0.18	0.12	0.07	0.02
3	0.76	0.51	0.37	0.27	0.21	0.15	0.06
4	0.80	0.57	0.44	0.34	0.27	0.21	0.11
5	0.83	0.62	0.49	0.39	0.31	0.26	0.15
6	0.85	0.65	0.53	0.44	0.37	0.30	0.19
7	0.86	0.68	0.56	0.47	0.40	0.33	0.22
8	0.87	0.70	0.58	0.50	0.43	0.36	0.25
9	0.88	0.71	0.60	0.52	0.46	0.39	0.27
10	0.89	0.73	0.62	0.54	0.48	0.41	0.30
11	0.90	0.74	0.64	0.56	0.50	0.43	0.32
12	0.90	0.75	0.65	0.58	0.52	0.45	0.34
13	0.91	0.76	0.67	0.59	0.53	0.47	0.35
14	0.91	0.77	0.68	0.60	0.55	0.48	0.37
15	0.91	0.78	0.69	0.62	0.56	0.50	0.38
16	0.92	0.79	0.70	0.63	0.57	0.51	0.40
17	0.92	0.79	0.70	0.64	0.58	0.52	0.41
18	0.92	0.80	0.71	0.65	0.59	0.53	0.42
19	0.93	0.80	0.72	0.65	0.60	0.54	0.44
20	0.93	0.81	0.73	0.66	0.61	0.55	0.45
21	0.93	0.81	0.73	0.67	0.62	0.56	0.46
22	0.93	0.82	0.74	0.68	0.63	0.57	0.47
23	0.93	0.82	0.74	0.68	0.63	0.58	0.48
24	0.94	0.83	0.75	0.69	0.64	0.59	0.48
25	0.94	0.83	0.75	0.70	0.65	0.59	0.49
26	0.94	0.83	0.76	0.70	0.65	0.60	0.50
27	0.94	0.84	0.76	0.71	0.66	0.61	0.51
28	0.94	0.84	0.77	0.71	0.66	0.61	0.52
29	0.94	0.84	0.77	0.72	0.67	0.62	0.52
30	0.94	0.84	0.77	0.72	0.67	0.62	0.53
31	0.94	0.85	0.78	0.72	0.68	0.63	0.53
32	0.95	0.85	0.78	0.73	0.68	0.63	0.54
33	0.95	0.85	0.78	0.73	0.69	0.64	0.55
34	0.95	0.85	0.79	0.74	0.69	0.64	0.55
35	0.95	0.86	0.79	0.74	0.70	0.65	0.55
36	0.95	0.86	0.79	0.74	0.70	0.65	0.56
37	0.95	0.86	0.80	0.75	0.70	0.66	0.57
38	0.95	0.86	0.80	0.75	0.71	0.66	0.57
39	0.95	0.86	0.80	0.75	0.71	0.67	0.58
40	0.95	0.86	0.80	0.75	0.71	0.67	0.58
41	0.95	0.87	0.81	0.76	0.72	0.67	0.59

1. Best F/R estimate $\hat{\lambda} \dfrac{\text{no. fails}}{\text{total POH}} \times 10^3$ (%/K).

2. Obtain lower bound at confidence level desired multiplying by appropriate factor from above table.

TABLE 3.6   Factors for One-Sided Exponential Lower Bound (I and II censoring (continued)

No. Fails	60%	80%	90%	95%	97.5%	99%	99.9%
42	0.95	0.87	0.81	0.76	0.82	0.68	0.59
43	0.95	0.87	0.81	0.76	0.72	0.68	0.59
44	0.95	0.87	0.81	0.77	0.73	0.68	0.60
45	0.96	0.87	0.81	0.77	0.73	0.69	0.60
46	0.96	0.87	0.82	0.77	0.73	0.69	0.61
47	0.96	0.88	0.82	0.77	0.73	0.69	0.61
48	0.96	0.88	0.82	0.77	0.74	0.70	0.61
49	0.96	0.88	0.82	0.78	0.74	0.70	0.62
50	0.96	0.88	0.82	0.78	0.74	0.70	0.62
55	0.96	0.89	0.83	0.79	0.75	0.71	0.63
60	0.96	0.89	0.84	0.80	0.76	0.72	60.5
65	0.96	0.89	0.84	0.81	0.77	0.73	0.66
70	0.97	0.90	0.85	0.81	0.78	0.74	0.67
75	0.97	0.90	0.85	0.82	0.79	0.75	0.68
80	0.97	0.90	0.86	0.82	0.79	0.76	0.69
85	0.97	0.91	0.86	0.83	0.80	0.77	0.70
90	0.97	0.91	0.87	0.83	0.80	0.77	0.70
95	0.97	0.91	0.87	0.84	0.81	0.78	0.71
100	0.97	0.92	0.87	0.84	0.81	0.78	0.72
110	0.97	0.92	0.88	0.85	0.82	0.79	0.73
120	0.97	0.92	0.89	0.85	0.83	0.80	0.74
130	0.98	0.93	0.89	0.86	0.84	0.81	0.75
140	0.98	0.93	0.89	0.87	0.84	0.81	0.76
150	0.98	0.93	0.90	0.87	0.85	0.82	0.77
160	0.98	0.93	0.90	0.87	0.85	0.83	0.77
170	0.98	0.94	0.90	0.88	0.85	0.83	0.78
180	0.98	0.94	0.91	0.88	0.86	0.83	0.79
190	0.98	0.94	0.91	0.88	0.86	0.84	0.79
200	0.98	0.94	0.91	0.89	0.87	0.84	0.80
250	0.98	0.95	0.92	0.90	0.88	0.86	0.82
300	0.98	0.95	0.93	0.91	0.89	0.87	0.83
400	0.99	0.96	0.94	0.92	0.90	0.89	0.85
500	0.99	0.96	0.94	0.93	0.91	0.90	0.87
700	0.99	0.97	0.95	0.94	0.93	0.91	0.89
900	0.99	0.97	0.96	0.95	0.94	0.92	0.90
2000	0.99	0.98	0.97	0.96	0.96	0.92	0.93
3000	1.00	0.98	0.98	0.97	0.96	0.96	0.94
5000	1.00	0.99	0.98	0.98	0.97	0.97	0.96
20,000	1.00	0.99	0.99	0.99	0.99	0.98	0.98
50,000	1.00	1.00	0.99	0.99	0.99	0.99	0.99

1. Best F/R estimate $\hat{\lambda} \dfrac{\text{no. fails}}{\text{total POH}} \times 10^3$ (%/K).

2. Obtain lower bound at confidence level desired multiplying by appropriate factor from above table.

Exercise 3.8

Experimental results, analyzed based on an exponential distribution of failure times assumption, give $\hat{\lambda}$ = 200 FITs. Censoring was Type I and there were 3 failures during the test. Compute a 95% two–sided confidence interval for λ.

THE CASE OF ZERO FAILURES

When a test ends after T with none of the n test units having failed, the point estimate previously defined is zero. This value is not a realistic estimate, as it does not even take into account the number on test. An upper $100\,(1 - \alpha)$ confidence limit for λ is given by

$$\lambda_{100\,(1 - \alpha)} = \frac{\chi^2_{2;100\,(1 - \alpha)}}{2nT} = \frac{-\ln \alpha}{nT}$$

where $\chi^2_{2;100\,(1 - \alpha)}$ is the upper $100\,(1 - \alpha)$ percentile of the chi–square distribution with 2 degrees of freedom.

The 50% zero failures estimate is often used as a point estimate for λ. This should be interpreted very carefully. It is a value of λ that makes the likelihood of obtaining zero failures in the given experiment similar to the chance of getting a head when flipping a coin. We are not really 50% confident of anything; we have just picked a λ that will produce zero failures 50% of the time.

Table 3.7 gives zero fail $\lambda_{100\,(1 - \alpha)}$ formulas for several percentiles. As shown in the equation above, chi–square tables are not needed since the easy formula $\lambda_{100\,(1 - \alpha)} = (-\ln \alpha)/\,nT$ can be used to calculate any desired percentile.

Example 3.9   Zero Failures Estimation

Two hundred samples from a population of units believed to have a constant failure rate are put on test for 2,000 hr. At that time, having observed no failures, the

TABLE 3.7   Exponential Zero Fail Estimates

Percentile	Estimate
50	$0.6931/nT$
80	$1.6094/nT$
90	$2.3026/nT$
95	$2.9957/nT$
97.5	$3.6889/nT$
99	$4.6052/nT$

experimenter puts an additional 200 units on test. Three thousand hours later, there were still no failures, and all the units were removed from test. What is a 50% estimate of the failure rate? A 95% upper bound? Use the general formula to calculate a 70% upper bound on the true failure rate.

Solution

The total unit test is 200 × 5000 plus 200 × 3000 or 1,600,000 hr. The 50% estimate is 0.6931/1,600,000, which is 433 FITs. The 95% failure rate upper bound is 2.9957/1,600,000, or 1,872 FITs. For a 70% upper bound, $\alpha = 0.3$, and the general formula yields $\lambda_{100 \times 1-\alpha} = (-\ln 0.3)/1,600,000 = 752$ FITs.

So far, all upper and lower bounds have been for the failure rate and not the *MTTF*. But since the *MTTF* is $1/\lambda$, we can work with λ and bounds on λ, and then take reciprocals to convert to *MTTF* estimates with bounds. Note that the reciprocal of the upper bound for λ becomes the lower bound for the *MTTF*.

Example 3.10   Confidence Bounds on MTTF

Two hundred units were tested 5000 hr with 4 failures. Give 95% upper and lower bounds on the failure rate and the *MTTF*. What difference would it make if the test was designed to end at the fourth failure?

Solution

The $\hat{\lambda}$ estimate is 4/1,000,000. From the tables with $r = 4$, we find the upper bound factor of 2.29 and the lower bound factor of 0.34. After multiplying by these factors, we obtain $\lambda_5 = 1.36 \times 10^{-6} = 0.136$ %/K and $\lambda_{95} = 9.16 \times 10^{-6} = 0.916$%/K. These taken together form a 90% confidence interval for λ. The 95% lower bound on the *MTTF* is $1/\lambda_{95} = 109,170$ hr. The 95% upper bound on the *MTTF* is $1/\lambda_5 = 735,294$ hr.

If the data was Type II censored at the fourth fail, we would have the same $\hat{\lambda}$ and λ_5. The factor to obtain λ_{95} would be derived by taking the factor 2.58 from $r = 3$ in Table 3.5 and multiplying it by 3/4. This gives a result of 1.935. Thus $\lambda_{95} = 1.935 \times 4 \times 10^{-6}$ or $7.74 \times 10^{-6} = 0.774$%/K.

Exercise 3.9

Suppose 500 units are tested for 1000 hr with no fails. Estimate the 75% upper bound for λ in FITs using the general formula. Also calculate the 50% estimate. Interpret these two estimates.

Exercise 3.10

Using the 50% estimate for λ in Exercise 3.9, estimate the CDF for the exponential distribution at $t = 1,000$ hr. Also estimate the probability of zero failures among 500 units at 1,000 hr for an exponential distribution with this failure rate. What do you conclude?

PLANNING EXPERIMENTS USING THE EXPONENTIAL DISTRIBUTION

Proper experimental planning, or good experimental design, is acknowledged to be one of the most important ingredients of successful experimentation. Unfortunately, design is often neglected in the case of reliability testing and modeling. The complexity of censored data, as well as the difficult forms of many of the life distributions used in reliability analysis, present problems in experimental design that are hard to overcome in most typical applications.

This planning difficulty is not present when using the exponential distribution. Here we can, and should, give early consideration to the sample sizes and test durations of the experiment. If we carefully state our objectives, we can plan the right experiment to achieve them. In most cases, only a simple look–up of the right $k_{r,1-\alpha}$ factor from the preceding section is necessary.

Case I.   How Many Units Should Be Put On Test?

The following items must be specified before a sample size can be chosen:

1. a failure rate or *MTTF* objective
2. a confidence level for ensuring we meet this objective
3. a test duration
4. a somewhat arbitrary and predetermined number of failures that we want to allow to occur during the test and still be able to meet the objective

With all these items specified, we know all the terms in the equation

$$\frac{r}{nt} \times k_{r;1-\alpha} = \lambda_{obj}$$

except n (the numerator is only an approximation of the total unit test for planning purposes). Solving this equation for n gives the required sample size.

Example 3.11   Choosing Sample Sizes

We wish to be 90% confident of meeting a 0.2%/K specification (*MTTF* = 500K hr). We can run a test for 5000 hr, and we agree to allow up to 5 failures and still pass the product. What sample size is needed?

Solution

The $k_{5;0.90}$ factor from Table 3.5 is 1.86. The basic equation is

$$\frac{5}{5000n} \times 1.86 = 0.000002$$

$$n = 930$$

Exercise 3.11

Suppose we want to be 90% confident of meeting a *MTTF* of at least 2,000,000 hr (in other words, a failure rate of 500 FITs). We can test for 2000 hr, and we want to allow for up to two fails. What sample size do we need? What is the sample size if we only want 60% confidence? What if we allow only 1 fail and still want 60% confidence?

Variation: How Long Must The Test Run?

Here, we have a fixed number of test units and have specified the failure rate objective, confidence level, and number of allowable failures. The only unknown remaining in the basic equation is the test time.

Example 3.12   Choosing the Test Times

We have 100 units to test, and we want to be 95% confident that the *MTTF* is greater than 20,000 hr. We will allow up to 10 failures. How long must the test run?

Solution

The $k_{10;0.95}$ factor is 1.7. Solving for T in

$$\frac{10}{100 \times T} \times 1.7 = 0.00005$$

results in a test time of 3400 hr.

Exercise 3.12

We have 300 units to test, and we want to be 80% confident that the failure rate is less than 1000 FITs. If we to allow up to 4 fails, how long must we test? What if we allow only 1 fail and reduce our confidence level to 60%?

Case II.   With a Test Plan in Place, How Many Fails Can Be Allowed?

It is good practice to clearly state the pass/fail criteria of a test in advance and have all interested parties agree on it. After the experiment is run, it is much more difficult to obtain such agreements.

As before, a failure rate or *MTTF* objective must be stated, along with the confidence level. The test plan will fix the sample size and test length. From the basic equation, we can solve for $r \times k_{r;1-\alpha}$. Since r is the only unknown, we can't pick the $k_{r;1-\alpha}$ from the table. We can, however, move down the $1-\alpha$ column of Table 3.5, multiplying each number by the corresponding r. The first value of r that produces a $r \times k_{r;1-\alpha}$ greater than our target value is 1 higher than the desired pass criteria.

Example 3.13   Choosing Pass/Fail Criteria

What is the maximum number of failures we can allow in order to be 80% confident of a failure rate no higher than 5%/K, if 50 units are to be tested for 2000 hr?

Solution

The basic equation is $[r/(50 \times 2000)] \times k_{r;1-\alpha} = 0.00005$. From this we obtain $r \times k_{r;1-\alpha} = 5$. Looking down the 80% column of Table 3.5, we calculate $1 \times 2.99 = 2.99$, $2 \times 2.14 = 4.28$, and $3 \times 1.84 = 5.52$. The last product, for $r = 3$, is greater than 5. Therefore, we subtract 1 from 3 and come up with a pass criteria of up to 2 failures on the test.

Exercise 3.13

What is the maximum number of failures that can be allowed in order to be 95% confident that the failure rate is no more than 20,000 FITs (*MTTF* = 50,000 hr) if we have 500 units to test for 2000 hr? What if we run the test for 3000 hr? What if we use 60% confidence and 3000 hr?

Case III.   What Are Minimum Test Sample Sizes We Can Use?

As before, we have to specify a failure rate objective, a confidence level, and a test duration. For a minimum sample size, we anticipate the best possible outcome; namely, 0 failures. This means we are prepared to state the product has not demonstrated the specified failure rate at the required confidence level if we see even one fail.

Minimum sample sizes are derived by setting $k_{0;1-\alpha}/nT = \lambda_{obj}$, where $k_{0;1-\alpha}$ is the zero failures factor from Table 3.7. Solving for n gives $n = k_{0;1-\alpha}/(\lambda_{obj} \times T)$. An equivalent formula that works for any αlevel without needing tables is

$$n = \frac{-\ln \alpha}{\lambda_{obj} T}$$

Example 3.14   Minimum Sample Sizes

We want a minimum sample size that will allow us to verify a 40,000 hr $MTTF$ with 90% confidence, given the test can last 8,000 hr.

Solution

The $k_{0;0.9}$ factor is 2.3026 (from Table 3.7). Therefore, $n = (40,000 \times 2.3026)/8000 = 12$ (after rounding up).

Exercise 3.14

What is the minimum sample size that will allow us to verify a 500,000 hr $MTTF$ with 85% confidence, given that the test can run for 2500 hr?

Variation: How do we determine minimum testing times?

The number of test units, as well as the failure rate objective and the confidence level, are fixed in advance. The choice of T then becomes

$$T = \frac{k_{0;1-\alpha}}{n\lambda_{obj}} = -\frac{\ln \alpha}{n\lambda_{obj}}$$

As before, if one fail occurs when the test is run, the failure rate objective will not be confirmed at the desired confidence level.

Example 3.15   Minimum Test Times

The failure rate objective is the very low number 10 PPM/K. We want to confirm this at an 80% confidence level. The component is an inexpensive resistor, and we plan to test 10,000 of them. How long should the test run?

Solution

By substitution, $T = 1.6094/(10,000 \times 10 \times 10^{-9}) = 16,094$ hr.

This period is about two years of continuous testing and might be much too long to be practical, even though it is a "minimum" test time. This example illustrates the difficulties inherent in verifying very high levels of reliability. Since the trend is toward more and more failure rate objectives in the PPM/K range, reliability analysts will face this problem with increasing frequency. The better we make our components, the harder it becomes to assess the actual performance.

One way out of this quandary is to test at high levels of stress, accelerating failure times as compared to what would happen at actual use conditions. We already saw this concept employed (but not mathematically explained) with the light bulb equivalent month of failure data (Example 3.3). A full discussion of acceleration modeling is in Chapter 7 of this book. At this point, we just note that if the test in Example 3.15 could be carried out in a test condition that accelerates failure times by a factor of 10 (i.e. a 10 × acceleration factor), then only 1,604 test hours would be required, or under 10 weeks.

Exercise 3.15

Suppose we want to confirm a 100 FITs rate at 90% confidence. We have 2,000 components to test. What is the minimum test time? What if a 60% confidence limit is used?

Using a minimum sampling plan will generally save around 40% in sample size or test duration, as compared to allowing just one fail. On the other hand, deciding a product will not meet its objective based on a single fail leaves no margin for the one odd, defective unit that might have slipped into the sample. It may turn out better to allow for a few failures and pay the extra sample size price rather than have a test result that many will not accept as valid. Use of minimum sample sizes make sense if we are very confident the product is much better than the objective, or we think it is so much worse that we will see many failures, even with the minimum sample.

SIMULATING EXPONENTIAL RANDOM VARIABLES

A general method of simulating any random variable starting with uniformly distributed random variables was given in Chapter 1. First the inverse CDF F^{-1} is derived. Then, a uniform random variable U is substituted for F in this inverse equation. For the exponential, we have

$$F(t) = 1 - e^{-\lambda t} = U$$

$$F^{-1}(U) = t = \frac{-\ln(1 - U)}{\lambda}$$

Since both U and $1-U$ have the same uniform distribution over the unit interval $(0,1)$, for purposes of simulation this expression is equivalent to

$$t = \frac{-\ln U}{\lambda}$$

By substituting n random uniformly distributed numbers into the last equation, a random sample of size n from the exponential is generated.

Example 3.16   Simulating Exponential Data

We will simulate 25 random failure times from an exponential distribution with $MTTF = 3000$ hr. The first step is to use either a random number table or a hand calculator with a built in random number generator or a computer random number generator to generate uniform random numbers from the unit interval $(0,1)$. We used a computer program to obtain:

0.49099 0.00794 0.4535   0.03049 0.45408 0.77058 0.12525 0.21737 0.30474

0.76521 0.82923 0.9044   0.14665 0.77664 0.91914 0.87884 0.57361 0.63097

0.82923 0.75458 0.17498 0.92835 0.73708 0.14085 0.3722   0.82507

The next step is to put these in descending order starting with 0.92835, 0.91914, 0.9044, etc. Next, transform each of these to random exponential data by the formula $t = (-\ln U) \times MTTF$. The first failure time is $t_1 = (-\ln 0.92835) \times 3000 = 223$ hr. The second is 253, and so on (it is left to the reader to calculate the remaining 23 failure times).

Exercise 3.16

Take the first 20 random failure times generated in Example 3.16 and treat them as if they were obtained from a censored Type II life test with 25 units on test and $r = 20$. Estimate the $MTTF$ and give a 90% confidence interval for it.

SUMMARY

The exponential life distribution is defined by $F(t) = 1 - e^{-\lambda t}$. It is characterized by being the only distribution to have a constant failure rate. This constant failure rate has the same value as the one unknown parameter λ, which is also the reciprocal of the mean time to fail ($MTTF$).

Another important characteristic of the exponential is its "lack of memory." When failures seem to occur randomly at a steady rate, with no significant wearout or degradation mechanisms, then the exponential will be a good model. It is appropriate for the long flat portion of the widespread "bathtub" curve for failure rates.it is also useful when we want to verify an average failure rate over an interval, and the time pattern of fails over the interval is of little concern.

The key formulas and properties of the exponential distribution are summarized in Table 3.8 for easy reference.

The best estimate of λ from censored type I or type II data is the number of fails divided by the total hours on test of all units. For readout data, the denominator may not be known exactly. Even in this case, however, the total unit test hours can often be estimated with little loss of precision.

When λ has been estimated from censored type I or II data, there are factors that depend only on the number of fails which can be used to obtain upper and lower bounds on λ. These factors are given in Tables 3.5 and 3.6. Table 3.7 has formulas for upper bounds when there are no fails on the test.

The factors in Tables 3.5 and 3.7 can also be very useful in the planning stages of an experiment. Sample sizes, or test durations, can be calculated once the failure rate objective and a confidence level are specified. The zero fail formulas yield minimum sample sizes. Even when the unit to be tested does not have an

TABLE 1.8   Exponential Distribution Formulas and Properties

Name	Value or Definition
CDF $F(t)$	$1-e^{-\lambda t}$
Reliability $R(t)$	$e^{-\lambda t}$
PDF $f(t)$	$\lambda e^{-\lambda t}$
Failure rate $h(t)$ or renewal rate	λ
AFR (t_1, t_2)	λ
Mean $E(t)$ or $MTTF$ (also $MTBF$)	$1/\lambda$
Variance $V(t)$	$1/\lambda^2$
Median	$\dfrac{0.693}{\lambda}$
Mode	0
Lack of memory property	The probability a component fails in the next t_2 hours, after operating successfully for t_1 hours, is the same as the probability of a new component failing in its first t_2 hour.
Closure property	A system of n independent components, each exponentially distributed with parameters $\lambda_1, \lambda_2, \ldots, \lambda_s$, respectively, has an exponential distribution for the time to first failure, with parameter $\lambda_s = \lambda_1 + \lambda_2 + \ldots + \lambda_s$.

exponential distribution, the numbers obtained from these tables are useful for rough planning.

When data is available, a histogram plotted on the same graph as the estimated PDF $f(t)$ (properly scaled up), gives a good visual check on the adequacy of using an exponential model.In addition, the chi–square goodness of fit test provides an analytic check.

PROBLEMS

3.1 How many units do we need to verify a 500,000 hr *MTTF* with 80% confidence, given that the test can run for 2500 hr and 2 failures are allowed?

3.2 Suppose we want to confirm a 500 FITs rate at 90% confidence. We have 2500 components to test and we are willing to allow up to 2 fails. How long should the test run? If we reduce the confidence level to 60%, how long must we test? What is the minimum test time (no fails allowed) at 60% confidence? At 75% confidence?

3.3 We want to check for a failure rate of 1600 FITs at 60% confidence. We have 1000 units we can test for 2000 hr. How many fails can we allow?

3.4 Suppose a sampling plan designed to screen product lots requires a burn-in done on 300 pieces for 168 hr, with no more than 1 fail allowed for the lot to be accepted. Assuming a 60% confidence level, how many hours of burn-in are needed to get the same level of protection if we switch to only accepting lots with zero fails? How any units must we burn in if we shorten the time to 48 hr but keep 60% confidence and zero fails allowed and the same level of failure rate protection?

3.5 For the product screen discussed in problem 3.5, what is the danger of shortening the burn-in time, even if larger sample sizes are used? (Hint: remember the assumptions underlying the exponential model, such as the "Lack of Memory" property.) Does 100% burn-in of a lot make sense if the exponential distribution really applies to all the units in the lot? What is the point of burn–in applied to a small sample from such a lot?

3.6 Assume a population is known to follow the life distribution model

$$F(t) = 1 - e^{-\lambda t^2}$$

with λ unknown. An experiment (censored, Type I) is run and 10 units fail out of 25 on test during the 1000 hr test time. Describe how you could estimate λ and construct confidence bounds for λ using the methods and Tables given in Chapter 3. Hint: Consider the new random variable given by $X = t^2$, where t is a random failure time from the given population.

3.7   Simulate the failure times of 50 units randomly selected from an exponential population with *MTTF* = 10,000 hr. Estimate λ and the *MTTF*. Next, "censor" the simulated failure times at 2000 hr as if that was the end of a censored Type I test and re–estimate λ and the *MTTF*.

Chapter 4

The Weibull Distribution

In the last chapter, we saw simple, yet powerful methods for analyzing exponential data and planning life-test experiments. Questions about sample size selection, test duration and confidence bounds could all be answered using a few tables. However, these methods apply only under the constant failure rate assumption or the equivalent "lack of memory" property. As long as this assumption is nearly valid over the range of failure times we are concerned with, we can use the methods given. But what do we do when the failure rate is clearly decreasing (typical of early failure mechanisms), or increasing (typical of later life wearout mechanisms)?

This problem was tackled by Waloddi Weibull (1951). He derived the generalization of the exponential distribution that now bears his name. Since that time, the Weibull distribution has proved to be a successful model for many product failure mechanisms because it is a flexible distribution with a wide variety of possible failure rate curve shapes. In addition, the Weibull distribution also has a derivation as a so-called "extreme value" distribution, which suggests its theoretical applicability when failure is due to a "weakest link" of many possible failure points.

First we will derive the Weibull as an extension of the exponential. Then we will discuss the extreme value theory. We'll learn that not only does the Weibull appear to "work" in many practical applications, but there is also an explanation to tell us why it works and in what areas it is likely to be most successful.

EMPIRICAL DERIVATION OF THE WEIBULL DISTRIBUTION

The goal is to find a CDF that has a wide variety of failure rate shapes, with the constant $h(t) = \lambda$ as just one possibility. Allowing any polynomial form of the type $h(t) = at^b$ for a failure rate function achieves this objective.

81

To derive $F(t)$, it is easier to start with the cumulative failure rate or hazard function $H(t)$. Setting

$$H(t) = (\lambda t)^m$$

gives us the exponential constant failure rate when $m = 1$, and a polynomial failure rate for other values of m; that is,

$$h(t) = \frac{dH(t)}{dt} = m\lambda (\lambda t)^{m-1}$$

Now we use the basic identity relating and $F(t)$ and $H(t)$

$$F(t) = 1 - e^{-H(t)} = 1 - e^{-(\lambda t)^m}$$

We obtain the form we shall use for the Weibull CDF by making a substitution of $c = 1/\lambda$ in the above equation.

$$F(t) = 1 - e^{-(t/c)^m}$$

The parameter c is called the _characteristic life_. The parameter m is known as the _shape parameter_. Both c and m must be greater than zero, and the distribution is a life distribution defined only for positive times t.

The PDF and failure rate $h(t)$ and AFR for the Weibull are given by

$$f(t) = \frac{m}{t}\left(\frac{t}{c}\right)^m e^{-(t/c)^m}$$

$$h(t) = \frac{m}{c}\left(\frac{t}{c}\right)^{m-1} = \frac{m}{t}\left(\frac{t}{c}\right)^m$$

$$AFR(t_1, t_2) = \frac{(t_2/c)^m - (t_1/c)^m}{t_2 - t_1}$$

$$AFR(T) = \frac{1}{c}\left(\frac{T}{c}\right)^{m-1}$$

There is, unfortunately, no consistent convention used throughout the literature when naming the Weibull parameters. Often the shape parameter is known as β. The characteristic life parameter c may be designated by α or η and is sometimes called the *scale parameter*. There is also an alternative form of the Weibull often encountered where the scale parameter $\theta = c^m$ is used, resulting in

$$F(t) = 1 - e^{-\left(\frac{t^m}{\theta}\right)}$$

Because of this confusion of terminology and the meanings of the parameters, the reader should be careful about the definitions used when reading literature about the Weibull.

Example 4.1   Weibull Properties

A population of capacitors is known to fail according to a Weibull distribution with characteristic life $c = 20,000$ power-on hours. Evaluate the probability a new capacitor will fail by 100, 1000, 20,000, and 30,000 hr, for the cases where the shape parameter $m = 0.5$ or 1.0 or 2.0. Also calculate the failure rates at these times for these three shape parameters.

Solution

Table 4.1 gives the various values requested. Several results are worth noting. In particular, observe that when the shape value is 0.5, the CDF at 100 hours is much higher than when $= 1$ or $= 2$. Also, the failure rate values for $m = 0.5$ are highest at the early time, and decrease with each later time. Exactly the opposite is true for the failure rate values when $m = 2$.

Table 4.1 also shows, as the Weibull derivation given in this section indicates, that the failure rate for $m = 1$ is a constant. Thus, the Weibull reduces to an exponential with $\lambda = 0.00005 = 1/c$.

Finally, note that the percent fail values at the characteristic lifetime value of 20,000 hr were uniformly 63.2, for all the choices of m.

All these observations based on Table 4.1 illustrate general points about the Weibull distribution that will be discussed in the next section.

TABLE 4.1   Solution to Example 4.1

	CDF in %			Failure Rate in %/K		
Time	$m = 0.5$	$m = 1.0$	$m = 2.0$	$m = 0.5$	$m = 1.0$	$m = 2.0$
100	7.0	0.5	0.002	35.4	5.0	0.05
1,000	20.0	4.9	0.2	11.2	5.0	0.5
20,000	63.2	63.2	63.2	2.5	5.0	10.0
30,000	70.6	77.7	89.5	2.0	5.0	15.0

The Weibull CDF equation has four quantities that may be known, assumed, or estimated from data. These are: the cumulative fraction failed, $F(t)$; the time, t; the shape parameter, m; and the characteristic life parameter, c. If any three of these are known the fourth can be calculated by one of the equations below.

$$F(t) = 1 - e^{-(t/c)^m}$$

$$m = \frac{\ln[-\ln(1-F)]}{\ln(t/c)}$$

$$t = c[-\ln(1-F)]^{1/m}$$

$$c = \frac{t}{[-\ln(1-F)]^{1/m}}$$

Weibull Calculations: Given that the population distribution is Weibull:

Exercise 4.1

Find the characteristic life necessary for 10% failures by 168 hours, given shape parameter of 2.0.

Exercise 4.2

Find the expected cumulative percent fallout at 1,000 hours, given a characteristic life of 1,000,000 hours and a shape parameter of 0.5.

Exercise 4.3

Find the time to achieve 20% failures, given a characteristic life of 50,000 hours and a shape parameter of 1.0.

Exercise 4.4

Find the shape parameter necessary for 5% failures in the first 2,000 hours, given a characteristic life of 30,000 hours.

Exercise 4.5

Find the characteristic life necessary to have an average failure rate of 100 FITs over the first 40,000 hours, given shape parameter of 3.0.

PROPERTIES OF THE WEIBULL DISTRIBUTION

The strength of the Weibull lies in its flexible shape as a model for many different kinds of data. The shape parameter m plays the major role in determining how the Weibull will look. For $0 < m < 1$, the PDF approaches infinity as time approaches zero and is always decreasing rapidly towards zero as time increases. The failure rate behaves the same way, making this type of Weibull an useful model for an early failure mechanism typical of the front end of the "bathtub" curve. When $m = 1$, the Weibull reduces to a standard exponential with constant failure rate $\lambda = 1/c$.

For $m > 1$, the PDF starts at zero and increases to a peak at $c\,[\,1 - (\,1/m\,)\,]^{1/m}$. From then on it decreases towards zero as time increases. The shape is skewed to the right. The failure rate also starts at zero, but then increases monotonically throughout life. The rate of increase depends on the size of m. For example, if m is 2, the failure rate increases linearly (and the distribution is also known as the Rayleigh distribution). When m is 3, the failure rate has a quadratic rate of increase, and so on. This type of Weibull is a useful model for wearout failure mechanisms typical of the back end of the "bathtub" curve.

Figure 4.1 shows several examples of Weibull CDFs, Figure 4.2 illustrates the Weibull PDFs, and Figure 4.3 is a graph of a variety of Weibull failure rate (hazard) curves. Table 4.2 summarizes the way the Weibull varies according to the value of its shape parameter m.

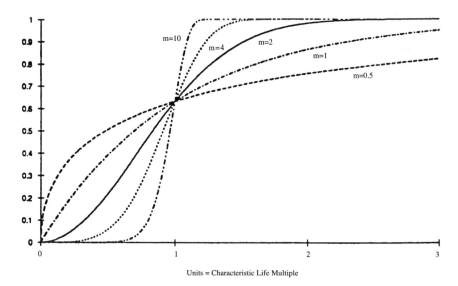

Units = Characteristic Life Multiple

Figure 4.1   Weibull Cumulative Distribution Function

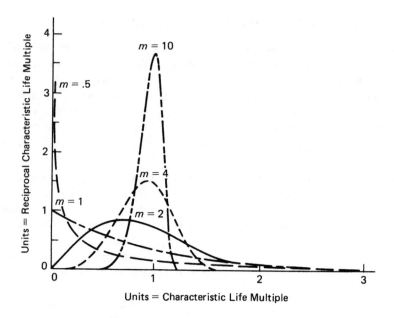

Figure 4.2   Weibull Probability Density

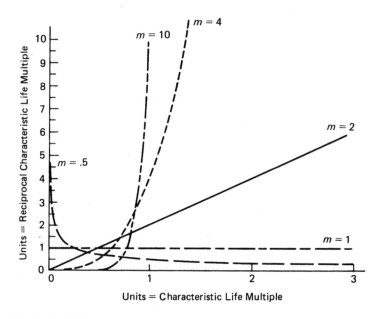

Figure 4.3   Weibull Failure Rate (Hazard)

TABLE 4.2   Weibull Distribution Properties

Shape Parameter, m	PDF	Failure rate, h(t)
$0 < m < 1$	Exponentially decreasing from infinity	Same
$m = 1$	Exponentially decreasing from $1/c$	Constant
$m > 1$	Rises to peak and then decreases	Increasing
$m = 2$	Rayleigh distribution	Linearly increasing
$3 \le m \le 4$	Has "normal" bell-shape appearance	Rapidly increasing
$m > 10$	Has shape very similar to Type I extreme value distribution	Very rapidly increasing

The second parameter of the Weibull, c, is a scale parameter that fixes one point of the CDF: the 63.2 percentile or characteristic life point. If we substitute c for time in the CDF, we obtain

$$F(c) = 1 - e^{-(c/c)^m} = 1 - e^{-1} = 0.632$$

In other words, 63.2% of the population fails by the characteristic life point, independent of the value of the shape parameter m.

PDF and Hazard Rate Plots

The median, or T_{50} point for the Weibull is found by letting $F(T_{50}) = 0.5$ to yield

$$T_{50} = c (\ln 2)^{1/m}$$

Exercise 4.6

A manufacturer produces parts that have a median width of 2.0 cm. The population distribution is Weibull with a shape parameter of 2.0. If the specification for acceptable parts is 1.0 cm to 3.0 cm, what fraction of the parts produced are rejectable.

To give the mean and the variance of the Weibull distribution, it is necessary to introduce a mathematical function known as the Gamma function. This function is defined by

$$\Gamma(v) = \int_0^\infty y^{v-1} e^{-y} dy$$

and is tabulated in many places; for example, see Abramovitz and Stegun (1964). In particular, when v is an integer, $\Gamma(v) = (v-1)!$. When $v = 0.5$, $\Gamma(0.5) = \sqrt{\pi}$. In general, for $v > 0$, $\Gamma(v) = (v-1)\Gamma(v-1)$.

The mean of the Weibull is $c\Gamma(1 + 1/m)$, and the variance is

$$c^2\Gamma(1 + 2/m) - [c\Gamma(1 + 1/m)]^2$$

For example, for $m = 0.5$, 1 and 2, the means are $2c$, c, and $c\sqrt{\pi}/2$, respectively. Note that this mean or *MTTF* no longer has any direct relationship to the failure rate (unless $m = 1$). So while the Weibull has a *MTTF*, it is not that useful or meaningful a number as compared to a graph of the failure rate or the average failure rate calculated over an interval of interest.

Exercise 4.7

A Weibull distribution has $m = 0.5$ and $c = 1,000$ hours. Determine the mean, median, and standard deviation. Sketch the PDF curve and indicate the positions of the mean and median.

The Weibull also has a closure or reproductive property, similar to the exponential. If a system is composed of n parts, each having an independent Weibull distribution with the same shape parameter but not necessarily the same characteristic life, and the system fails when the first component fails, then the time to the first system fail also follows a Weibull distribution. If the characteristic life parameters are $c_1, c_2, \ldots, c_n$, and the shape parameter is m, then the system failure distribution also has shape parameter m. The system characteristic life is given by

$$c_s = \left(\sum_{i=1}^{n} \frac{1}{c_i^m} \right)^{-\frac{1}{m}}$$

Example 4.2   Weibull Closure Property

A car manufacturer uses five different hoses as a part of the engine cooling system for one of its models. The hose manufacturer specifies that each hose has a lifetime modeled adequately by a Weibull distribution with shape parameter 1.8. The five hoses have characteristic lives, in months of average car use, of 95, 110, 130, 130, and 150. What is the life distribution for time to first car hose failure? What is the *MTTF*? What is the median or T_{50}? How likely is it no hose fails in the first year of car life? By four years?

Solution

Applying the closure property relationship, we have that the hose system (first) failure distribution is a Weibull with shape parameter $m = 1.8$ and characteristic life

$$c_s = \left(\frac{1}{95^{1.8}} + \frac{1}{110^{1.8}} + \frac{1}{130^{1.8}} + \frac{1}{130^{1.8}} + \frac{1}{150^{1.8}} \right)^{-1/1.8} = 48.6$$

The MTTF is $c\Gamma(1 + 1/1.8) = 43.2$ months. The T_{50} is $c(\ln 2)^{1/m} = 39.6$ months. The reliability at 12 months, or $[1 - F(12)]$, is 92 percent. The reliability at 48 months is approximately $(100 -$ the characteristic life percent), or 37 percent.

Exercise 4.8

A system consists of 4 identical, independent components. If one component fails, the system fails. The components failure times have a Weibull distribution with characteristic life $c = 10,000$ hours and shape parameter $m = 0.5$. Determine the system failure distribution. Find the system MTTF and median life. How likely is it that no system failures occur in the first 100 hours?

The parameter definitions and key Weibull formulas given in this section are summarized in Table 4.3.

EXTREME VALUE DISTRIBUTION RELATIONSHIP

In general, if we are interested in the minimum of a large number of similar independent random variables, as for example, the first time of failure reached by many competing similar defect sites located within a material, then the resulting life distribution can converge to only one of three types. These types, known as smallest extreme value distributions, depend on whether we define the random variables on the entire x-axis, or just the positive or negative halves of the axis.

The extreme value distributions were categorized by Gumbel in 1958 and are

$$F(x) = 1 - e^{-e^x} \qquad -\infty < x < \infty \qquad \text{(Gumbel Type 1)}$$

$$F(x) = 1 - e^{-(-x)^{-a}} \qquad x \le 0$$

$$F(x) = 1 - e^{-x^a} \qquad x \ge 0 \qquad \text{(Weibull)}$$

In the above formulas, the random variables have been suitably shifted by any location parameters and divided by scale parameters so as to appear "normal-

TABLE 4.3 Weibull Formulas

Name	Value or Definition
CDF $F(t)$	$1 - e^{-(t/c)^m}$
Reliability $R(t)$	$e^{-(t/c)^m}$
PDF $f(t)$	$(m/t)\,(t/c)^m\,e^{-(t/c)^m}$
Characteristic life c	$F(c) = 63.2$
Shape parameter m	Determines shape of Weibull
Failure rate $h(t)$	$m/c(t/c)^{m-1}$
AFR (t_1, t_2)	$\dfrac{\left[(t_2/c)^m - (t_1/c)^m\right]}{t_2 - t_1}$
Mean $E(t)$	$c\Gamma\left(1 + \dfrac{1}{m}\right)$
Variance $V(t)$	$c^2\Gamma\left(1 + \dfrac{2}{m}\right) - \left[c\Gamma\left(1 + \dfrac{1}{m}\right)\right]^2$
Median T_{50}	$c(\ln 2)^{1/m}$
Mode	$c(1 - 1/m)^{1/m}$
System characteristic life c_s when components are independent Weibull with parameters (m, c_1)	$c_s = \left(\displaystyle\sum_{i=1}^{n} \dfrac{1}{c_i^m}\right)^{-\frac{1}{m}}$

ized." Of the three possible types, the only one that is a life distribution (i.e., defined only for non-negative values) turns out to be the Weibull. Since failure can often be modeled as a weakest link of many competing failure processes, the wide applicability of the Weibull is not surprising.

The justification for using the Weibull based on extreme value theory is an important and useful one. It can also be abused and lead to vigorously defended misapplications of the Weibull. For this reason, the strengths and weaknesses of this derivation should be carefully noted. If there are *many identical and independent competing processes* leading to failure, and the *first* to reach a critical stage determines the failure time, then—*provided "many" is large enough*—we can derive a Weibull distribution. In a typical modeling application, we might suspect or hope all the italicized assumptions apply, but we are not likely to be certain.

How do we pick a life distribution model from a practical viewpoint? The approach we recommend is as follows: Use a life distribution model primarily because it works; that is, *fits the data well and leads to reasonable projections when extrapolating beyond the range of the data*. Look for a new model when the one previously used no longer "works." Select models by researching which models have been used in the literature successfully for similar failure mechanisms, or by using theoretical arguments applied to the physical models of failure. For example, extreme value theory, the multiplicative degradation model theory of the next chapter on the lognormal distribution, or the "lack of memory" exponential model of the preceding chapter, are properties to consider.

Another situation where theoretical derivations prove very useful is when several distributions all seem to work well with the available data, but give significantly different results when projected into critical tail regions that are beyond the range of the data. In this case, pick the model that has a theoretical derivation most closely matching the cause of failure.

There is another very interesting mathematical relationship between the Weibull distribution and the Type I extreme value distribution. It turns out that the natural logarithms of a population of Weibull failure times form a population following the Type I distribution. In other words, the natural logarithm of a Weibull random variable is a random variable that has the Type I extreme value CDF. This relationship, as we shall see in the next chapter, is exactly the same as exists between the lognormal life distribution and the normal distribution. So, if the Weibull had not been named in honor of its chief advocate, it probably would have been called the log-extreme value distribution.

The exact statement of this relationship is as follows: let t_f be a Weibull random variable with CDF

$$F(t) = 1 - e^{-(t/c)^m}$$

Then the random variable $X = \ln t_f$ has the Type I extreme value CDF given by

$$1 - e^{-e^{(x-a)/b}}$$

with $a = \ln c$ and $b = 1/m$.

A reliability analyst who has computer programs or graph paper designed to handle Type I extreme value data can also analyze Weibull data after first transforming it into extreme value data (by taking natural logarithms). Later on the scale, parameter b can be used to estimate the Weibull shape via $m = 1/b$. The location parameter a is transformed into the Weibull characteristic life via $c = e^a$.

AREAS OF APPLICATION

After the introduction of the Weibull distribution, its use spread across a wide variety of applications, running the range from vacuum tubes and capacitors to ball bearings and relays and material strengths. The primary justification for its use has always been its flexible ability to match a wide range of phenomena. There are few, if any, observed failure rates that cannot be accurately described over a significant range of time by a polynomial or Weibull hazard function.

Some particular applications, such as modeling capacitor dielectric breakdown, fit nicely into the "worst link" (or first of many flaws to produce a failure) extreme value theory. Dielectric materials contain many flaws, all "competing" to be the eventual catastrophic failure site. In many cases, the failures occur mostly early in life and a Weibull with a shape parameter less than 1 works best.

On the other hand, there is less reason to expect a Weibull to apply when failure is due to a chemical reaction or a degradation process such as corrosion or migration or diffusion (although even here the many competing sites argument might still possibly apply). It is in precisely such applications, typical of many semiconductor failure mechanisms, that the lognormal distribution (Chapter 5) has replaced the Weibull as the most popular distribution.

One particular form of Weibull deserves special mention. When $m = 2$, as noted in Table 4.2, the distribution is called the Rayleigh. The failure rate increases linearly with $h(t) = (2/c^2)t$ and the CDF is given by

$$F(t) = 1 - e^{-(t/c)^2}$$

There is an interesting measurement error problem that also leads to this same CDF.

Assume you are measuring, or trying to locate, a particular point on the plane. A reasonable model that is often used measures independent x and y coordinates. Each measurement has a random amount of error, modeled as usual by the Normal distribution (Chapter 5). Assume each error distribution has zero mean and the same standard deviation, σ. If the error in the x direction is the random variable X, and the y direction error is Y, then the total radial error (or distance from the correct location) is

$$R = \sqrt{X^2 + Y^2}$$

Using standard calculus methods, the CDF of R can be derived. It turns out to be

$$F(r) = 1 - e^{-\left(r^2/2\sigma^2\right)}$$

which is the Rayleigh distribution with $c = \sqrt{2}\sigma$.

Example 4.3   Rayleigh Radial Error

An appliance manufacturer wants to purchase a robot arm to automate a particular assembly operation. The quality organization has been asked to evaluate the reliability of an arm under consideration. A key point in the evaluation is whether the arm can repeatedly go to specified points in its operating range, within a tolerable margin of positioning error. The literature on the arm says that it will repeatedly arrive at programmed points with an accuracy of plus or minus 0.3 cm in either the x coordinate or the y coordinate direction.

Tests have determined that the operation will succeed as long as the arm arrives no further than 0.4 cm from the designated point. If an error rate of less than 1 in a 1000 is required, will the arm under consideration be adequate?

Solution

This example is instructive, not only of the Rayleigh distribution, but also of the kind of detective work a statistician or reliability analyst must often carry out. What does an accuracy of plus or minus 0.3 cm really mean? Often a phone call to the supplier will not produce an immediately satisfactory answer. In the meantime, an analyst can make an evaluation based on making typical assumptions. These will involve a knowledge of the normal distribution, discussed in the next chapter.

Assume that plus or minus 0.3 cm refers to plus or minus three standard deviations (or sigmas) of the typical normal error distribution. Sigma is then 0.1 cm. If we also assume that the placement errors in each coordinate are independent with an average value of zero, the point the robot arm arrives at will have a random distance from the objective point with a CDF given by the Rayleigh distribution

$$F(r) = 1 - e^{-r^2/2(0.1)^2} = 1 - e^{-50r^2}$$

Substituting 0.4 for r yields $F(0.4) = 0.9997$. This result means the arm will be more than 0.4 cm off only about 3 times every 10,000 operations, which meets the less than 1 in a 1000 objective.

Exercise 4.9

An expert dart thrower claims he can throw a dart repeatedly at a target and end up within 0.5 cm of the origin of the target, in any direction. The bull's-eye is centered at the origin and has a diameter of 0.7 cm. If he throws a dart at the target 50 times, what's his expected number of hits within the bull's-eye? What assumptions are you making? What's the expected number of hits inside the bull's-eye if the distribution of darts hitting the target is uniform within the radius of 0.5 cm?

WEIBULL PARAMETER ESTIMATION

Two analytic methods for estimating m and c from data (either complete or censored or grouped samples) will be described in this section. The most recommended procedure is called the _method of maximum likelihood_. It is a standard, well known technique, described fully in most statistics textbooks (for example, see Wilks, 1962). Its use for censored reliability data is described in detail in Nelson (1982) and also in Mann et al. (1974). Unfortunately, its implementation for censored or grouped data is generally not practical unless appropriate computer programs are available to do the calculations. These programs may be purchased by the reader as a part of statistical analysis packages such as CENSOR (described by Meeker and Duke (1981) or Statpac (developed by W. Nelson and described by Strauss, 1980). The reader with programming experience can write his own procedures from the equations given in Nelson (1982). The second estimation method is easier to implement. It is an analytic version of the graphical estimation techniques described in Chapter 6. Only the common least squares or regression programs found in many hand held calculators are needed.

At this point, it is natural to inquire why a complicated method requiring fairly rare computer programs is recommended when simple, intuitive graphical techniques, combined with a curve-fitting routine that eliminates subjective judgment, can be used. Not only that, but the graphical approach (as we shall see in the next section) offers an immediate visual test of whether the Weibull distribution fits the data or not (based on whether the data points line up in an approximate line on special Weibull graph paper). The reason for the recommendation has to do with the concept of accuracy, in a statistical sense.

In a real-life problem, when we estimate parameters from data we can come up with many often widely differing results, depending on the estimation method we use. Important business decisions may depend on which estimate we choose. Obviously, it's important to have objective criteria which tell us which method is best in a particular situation. Accuracy, in a statistical sense, starts with a definition of desirable properties an estimation method may have and continues with an investigation into which of these "good" properties the methods available actually have. Note that we are talking about the method of estimation—not the estimates that can be derived from the use of these methods on a specific set of data. Due to the random nature of sample data, an almost arbitrary guess based on looking at a few of the data points might, on one given day, turn out to be closer than the estimate obtained from the most highly recommended computer program. Statistical theory describes how well various methods compare to each other in the long run—over many, many applications. We have to accept the logic that it makes sense to use the best long run method on any single set of data we want to analyze.

The most desirable attributes defined for estimation methods are the following:

1. *Lack of bias*. The expected value of the estimate equals the true parameter (or on the average, you're centered "on target").
2. *Minimum variance*. A minimum variance estimator has less variability on the average than any other estimator and, if also unbiased, is likely to be closer to the true value than another estimator.
3. *Sufficiency*. The estimate makes use of all the statistical information available in the data.
4. *Consistency*. The estimate tends to get closer to true value with larger size samples ("infinite" samples yield perfect estimates).

In addition, we want our estimates to have a known distribution which we can utilize for forming confidence intervals and carrying out tests of hypotheses.

In general, no known method provides all of the attributes mentioned. Indeed, it may be difficult to find any method with a lack of bias or minimum variance when dealing with life distributions and censored data. The maximum likelihood method can be shown to possess all the above properties as sample sizes (or numbers of failures) become large enough. This property, called asymptotic behavior, assures us that, for reasonable amounts of data, no other estimation technique is "better." Asymptotic theory does not tell us "how large" is "large" but practical experience and simulation experiments indicate that more than 20 failures is "large" and, typically, if there are over 10 failures the maximum likelihood estimates (MLEs) are accurate. For smaller amounts of data, the unbiased minimum variance property can not be claimed, but better techniques are hard to come by.

A loose but useful description of the MLE technique is as follows: the "probability" of the sample is written by multiplying the density function evaluated at each data point. This product, containing the data points and the unknown parameters, is called the likelihood function. By finding parameter values that maximize this expression we make the set of data observed "more likely." In other words, we choose parameter values which are most consistent with our data by maximizing the likelihood of the sample.

The MLE technique is therefore equivalent to maximizing an equation of several variables. In general, the standard calculus approach of taking partial derivatives with respect to each of the unknown parameters and setting them equal to zero, will yield equations that have the MLEs as solutions. In most cases, by first taking natural logarithms of the likelihood equation, and then taking partial derivatives to solve for a maximum, the calculations are simplified. The same parameter values that maximize the log likelihood will, of course, maximize the likelihood. However, for censored or grouped data, these equations are non-linear and complicated to set up and solve. Consequently, appropriate computer programs are needed.

When the life distribution is the exponential, the MLE equations are easy to derive and solve, even for censored data. This procedure is shown in the next example, which illustrates how the MLE method works.

Example 4.4   MLE for the Exponential

Show that the estimate of the exponential parameter λ given in Chapter 3 (i.e., the number of failures divided by the total unit test hours) is the MLE estimate for complete or censored Type I (time censored) or censored Type II (r^{th} fail censored) data.

Solution

The likelihood equation is given by

$$LIK = k \left(\prod_{i=1}^{r} f(t_i) \right) [1 - F(T)]^{n-r}$$

or

$$LIK = k\lambda^r \left(e^{-\lambda \sum_{i=1}^{r} t_i} \right) \left(e^{-\lambda T} \right)^{n-r}$$

where k is a constant independent of λ and not important for the maximizing problem. The last term in LIK is the probability of $n - r$ sample units surviving past the time T. If T is fixed in advance, we have Type I censoring. If T is the time of the rth fail, we have Type II censoring. If $r = n$, the sample is complete or uncensored.

If we let L denote the log likelihood (without any constant term), then

$$L = r \ln\lambda - \lambda \sum_{i=1}^{r} t_i + (n-r)(-\lambda T)$$

To find the value of λ that maximizes L, we take the derivative with respect to λ and set it equal to 0.

$$\frac{dL}{d\lambda} = \frac{r}{\lambda} - \sum_{i=1}^{r} t_i - (n-r)T = 0$$

Solving for λ, we have the estimate given in Chapter 3:

$$\hat{\lambda} = \frac{r}{\sum_{i=1}^{r} t_i + (n-r)T}$$

If the data is readout or grouped, the MLE, even in the simple exponential case, is not easy to obtain. The likelihood equation is

$$LIK = kF(t_1)^{r_1} \left\{ \prod_{i=2}^{m} [F(t_i) - F(t_{i-1})]^{r_i} \right\} [1 - F(t_m)]^{n-r}$$

where

$t_1, t_2, \ldots, t_m$ = the readout times

$r_1, r_2, \ldots, r_m$ = the failures first observed at those times

F = the CDF for the assumed life distribution

$$r = \sum_{i=1}^{m} r_i = \text{the total number of failures out of the } n \text{ on test}$$

The partial derivative equations that result in the Weibull case may be found in Chace (1976). For multicensored data, with l_i units removed from test at time L_i, we include a term of the form $[1 - F(L_i)]^{l_i}$ to the LIK equation. This term reflects the fact that we know only that those units have survived to time L_i.

Let's prescribe a general method for writing the LIK equation corresponding to a given set of data. Each observation that is an exact failure time contributes a term of the form $f(t_i)$, where t_i is the time of failure. Each observation that indicates a unit failed somewhere between two times, say t_{i-1} and t_i, needs an expression of the type $[F(t_i) - F(t_{i-1})]$. Each observation for a unit still surviving at t_i hours, with no further information known about that unit after t_i hours, generates a statement of the form $[1-F(t_i)]$. In the last situation, t_i is sometimes called a *run-time*. All of these terms, and constant multipler term that does not affect the MLE solutions, are multiplied together to derive the LIK equation.

To elicit all the information needed to write the LIK equation, we ask the following set of questions:

1. What are the exact times of failure, if any?
2. What are the starting and ending times of all intervals that contain failures whose exact failure times are not known, and what are the corresponding numbers of failures for each interval?
3. What are the censoring or run times when unfailed units are no longer observed, and what are the associated numbers of censored units?

With the answers to these questions and an assumed model for $F(t)$, then we can set up the LIK equation and solve the MLEs for unknown parameters. In most cases, the complexity of the LIK equation necessitates the use of a computer program employing numerical algorithms for solution.

A very useful part of the large sample theory for MLE estimates is that they have an asymptotic Normal distribution. The mean is the true parameter value and the standard deviation can be estimated from equations based on partial derivatives of the log likelihood equations. This theory will not be given here: it is described in detail by Nelson (1982), and the calculations should be part of any program that obtains MLEs from reliability data.

One more use of MLEs deserves mention. As stated in Chapter 3, when doing a goodness of fit test to check a distribution assumption, the type of estimates to use in place of unknown parameters are MLEs.

The second estimation method for Weibull parameters is based on a procedure called "linear rectification." The idea is to put the Weibull CDF equation into a form that, with the proper substitution of variables, is linear. The equations are:

$$F(t) = 1 - e^{-(t/c)^m}$$

$$\ln[1 - F(t)] = -(t/c)^m$$

$$\ln\{-\ln[1 - F(t)]\} = m\ln t - m\ln c$$

$$Y = mX + b$$

In the final linear form, Y is the estimated value of at time t, and X is the natural log of t. To estimate Y, we have to estimate $F(t)$. The estimate to use, as described in Chapter 6, depends on whether exact times of failure are available (in that case, use $F(t_i) = (i - 0.3)/(n + 0.4)$ for the ith failure time), or whether the data is readout; use $F(t_i) = $ (total number of failures up to time t_i).

When the calculations are completed, there is an (X,Y) pair for each data point or readout time. A least squares fit, or regression of Y on X, yields estimates for the slope m and the intercept $b = -m \ln c$. Estimate c by $\hat{c} = e^{-\hat{b}/\hat{m}}$. Programs or calculators that will accept the (X,Y) pairs as inputs and give least squares estimates of m and b as outputs are common. (An elementary discussion of using the least squares method to fit a line to data points, and linear rectification, is given in Chapter 6).

Weibull graph paper, discussed in Chapter 6, has scales adjusted in the same way as the transformations that obtained (X,Y) from the CDF estimate and the time of fail. Therefore, a plot of the CDF estimates versus time on this specially constructed paper will yield an approximate straight line with slope m and intercept $m \ln c$, provided the data follow a Weibull model. The computer least-squares procedure based on linear rectification is an objective way to put a line on Weibull graph paper that minimizes the squared deviations in the cumulative percent failure scale direction.

In many statistical applications, the estimates obtained via least-squares or regression methods can be shown to have very desirable properties similar to those described for MLEs. However, there are several key assumptions about the (X,Y) points that must be made in order for the least squares "optimality" properties to hold. Basically, the random errors in Y at the X points must be uncorrelated and have zero average value and the same variance. All three of these assumptions are known *not* to hold in the application of least squares described above. This method of estimating Weibull parameters gives convenient analytic estimates, which will be good for large amounts of data (consistency property). Little else can be said for them in terms of desirable properties. Also, any confidence bounds on m or b given in the output of a regression program are not valid for this application.

Example 4.5   Weibull Parameter Estimation

Capacitors, believed to have operating lifetimes modeled adequately by a Weibull distribution, were tested at high stress to obtain failure data. Fifty units were run and 25 failed by the end of the test. First assume the failure times were continuously monitored and the times of fail were, in hours to the nearest tenth, 0.7, 52.7, 129.4, 187.8, 264.4, 272.8, 304.2, 305.1, 309.8, 310.5, 404.8, 434.6, 434.9, 479.2, 525.3, 620.3, 782.8, 1122, 1200.8, 1224.1, 1322.7, 1945, 2419.5,

2894.5, and 2920.1. Assume the test ended at the last fail and estimate the
Weibull parameters. Next, assume only readout data were taken. The readouts
were made at 24, 168, 200, 400, 600, 1000, 1500, 2000, 2500 and 3000 hr. The
new failures observed corresponding to these readout times were 1, 2, 1, 6, 5, 2,
4, 1, 1, and 2. Again, estimate m and c.

Solution

An MLE program yields $\hat{m}$ = 0.62 and $\hat{c}$ = 5078 for the exact fail time data.
Using the regression method, the Y and X values shown below were input into a
least squares line fitting program, yielding $\hat{m}$ = 0.55 and $\hat{b}$ = –4.8277. These
results give $\hat{c}$ = $e^{4.8277/0.55}$ = 6488.

Y	X	Y	X
–4.27	–0.36	–1.06	6.17
–3.37	3.96	–0.99	6.26
–2.90	4.86	–0.91	6.43
–2.57	5.24	–0.84	6.66
–2.12	5.58	–0.77	7.02
–1.95	5.61	–0.70	0.09
–1.80	5.72	–0.64	0.11
–1.66	5.72	–0.57	0.19
–1.54	5.74	–0.51	7.57
–1.43	5.74	–0.45	7.79
–1.33	6.00	–0.40	7.97
–1.24	6.07	–0.36	7.98
–1.15	6.08		

For the readout data, MLE estimates of m and c are $\hat{m}$ = 0.64 and $\hat{c}$ = 5025.
The least squares method uses the X and Y values below to obtain $\hat{m}$ = 0.78 and
$\hat{c}$ = 3759.

Y	X
–3.90	3.18
–2.78	5.12
–2.48	5330
–1.50	5.99
–1.03	6.40
–0.88	6.91
–0.61	7.31
–0.55	7.60
–0.48	7.82
–0.37	8.01

Since the data in Example 4.5 was simulated, we can compare the various estimates obtained to the true Weibull parameter values: $m = 0.6$ and $c = 4000$. The closest estimates for the critical m parameter were given by the MLE method. The least-squares method with interval data happens, in this example, to come closest to the correct c value.

CDF values calculated using either of the above estimation methods will seldom differ significantly for times within the range where experimental failures occurred. However, we are often concerned with extrapolating back to very early times in the front tail of the life distribution—percentiles much smaller than experimental sample sizes allow us to observe directly. Here, small changes in estimated parameter values (especially in the shape parameter) can make orders of magnitude difference in the CDF estimate. Hence, it is important to use the best technique available for final parameter estimates.

Chapter 6 will describe techniques for plotting Weibull data and obtaining quick parameter estimates. Even though these estimates are not recommended for use in critical applications, the value obtained from looking at your data on the appropriate graph paper can not be overemphasized. Sometimes a strange pattern of points on graph paper may lead an analyst to ask questions that lead to a valuable insight that would have been lost had the entire analysis been done by computer programs. Even if the plot only serves to confirm the model chosen, it is useful for presentation and validation purposes.

Bain and Engelhardt (1991) provide tables for an another, alternative approach, using simple estimators, for Weibull parameter estimation. Cohen and Whitten (1988) provide additional information on MLE procedures.

Exercise 4.10

Using the exact failure time data from Example 4.5, plot the estimated CDF [that is, the plotting position for the ith failure time is $(i - 0.3)/(n + 0.4)$] at each failure time on a linear-by-linear graph. Plot the Y and X values for the exact failure times on a log-by-log graph. Comment on the results.

Exercise 4.11

Using the readout data from Example 4.5, plot the estimated CDF (that is, the plotting position for i total failures is i/n) at each readout time on a linear-by-linear graph. Plot the Y and X values for the readout failure times on a log-by-log graph. Comment on the results.

SIMULATING WEIBULL RANDOM VARIABLES

As we discussed in Chapter 1, random variables may be simulated through the use of the uniform random variable defined in the interval (0,1) and inverse expressions for the desired CDFs. Specifically, if F is any CDF with an inverse F^{-1}, then by substituting the unit uniform random variable U for F in the inverse expression denoted by F^{-1}, we generate a random variable distributed according to F.

For the Weibull distribution with shape parameter m and characteristic life c, the inverse expression is:

$$F^{-1} = t = c\left[-\ln\left(1 - F\right)\right]^{1/m}$$

Thus, substitution of the unit uniform random variable U gives:

$$F^{-1}(U) = t = c\left[-\ln\left(1 - U\right)\right]^{1/m}$$

Since $1 - U$ and U both represent a uniformly distributed variate in the interval $(0,1)$, for purposes of simulation we may write the inverse expression as:

$$F^{-1}(U) = t = c = c\left[-\ln U\right]^{1/m}$$

Class Project 4.1: *Simulating Weibull Random Variables.* A method to generate the unit uniform random variables is needed. Calculators with such a capability may be used. However, if a supply of 10-sided dice of various colors is obtainable, rolling the dice can be fun. Each student takes three different colored ten-sided dice and designates one color as the first digit, a second color as the second digit, and the third color as the last digit. Thus, the roll of the three dice will simulate a uniform random variable in the interval 0 to 999. Adding 1 to the number so obtained and dividing by 1000 will make the interval 0.001 to 1.000.

Each student rolls the 3 dice 10 times to simulate 10 uniform pseudo-random variables. Add 1 to each number and divide by 1000 to get 10 uniform unit random variables in the interval $(0.001, 1.000)$. Substitute each value for U so obtained into the Weibull inverse expression. One-third of the class will use the shape parameter $m = 0.5$, one-third will use $m = 2$, and the remaining one-third will use $m = 10$. The characteristic life c will be equal to 100 hr in all cases.

Thus, each student generates 10 pseudo-random, unordered, Weibull variables based on specific parameters. Those students with the same m value will combine results and draw a histogram on the blackboard or on an overhead transparency for other class members to view. Each students may wish to calculate the median time to failure and compare with the median expected from the formula

$$c\left(\ln 2\right)^{1/m} = c\left(0.6931\right)^{1/m}$$

Exercise 4.12

Simulate data from a stress test on twenty units assuming the failure times follow a Weibull distribution with $c = 5,000$ hr and $m = 1.5$. Estimate the *MTTF*, median, and standard deviation. Order the observations from smallest to largest. How many failures occur by 2,000 hours? By 4,000 hours? By 6,000 hours?

SUMMARY

The Weibull distribution, with CDF $F(t) = 1 - e^{-(t/c)^m}$, is a flexible, convenient life distribution model. It has a family of polynomial shaped failure rate functions, depending on the value of the shape parameter m, including decreasing, constant, and increasing failure rates. Many components and systems have reliability properties that are successfully modeled by the Weibull.

The Weibull also has a theoretical derivation as an extreme value distribution applying to the smallest random time of failure out of many independent competing times. Consequently, one expects the Weibull will apply when failure occurs at a defect site within a material and the are many such sites competing with each other to be the first to produce a failure. Capacitor dielectric material is an example, and the Weibull distribution has proved very successful as a model for capacitor lifetime.

The best way to analyze Weibull data and estimate m and c is by the technique of maximum likelihood estimation (MLE). For censored and readout data, special computer programs are needed. A less optimal estimation procedure uses least squares and is equivalent to fitting a line to the data on Weibull graph paper using an objective regression procedure. Whatever method is used for the final parameter estimates, Weibull data plots, as shown in Chapter 6, should be part of the analysis.

PROBLEMS

4.1 A population of devices are known to fail according to a Weibull distribution with characteristic life $c = 200,000$ hr. Evaluate the probability a new device will fail by 1000, 4000, 10,000, 40,000, and 100,000 hr for cases where the shape parameter is 0.5, 1.0, and 2.5. Calculate the failure rate (in FITS) at these times for the three shape parameters.

4.2 A radio manufacturer uses four different resistors as part of a tuner system. The resistor manufacturer specifies that each resistor has a lifetime modeled adequately by a Weibull distribution with shape parameter 2.0. The four resistors have characteristic lives, in hours of average radio usage, of 1000, 1200, 1500, and 2000 hr, respectively. What is the life distribution for time to first resistor failure? What is the radio *MTTF*? (Hint: $\Gamma(3/2) =$

0.88623.) What is the median failure time for the radio? How likely is it that no resistor fails in the first 168 hr? In the first 1000 hr?

4.3 Left truncated data means that all population failure times prior to a given time, say T, are not observed. This situation might occur when a vendor does a burn-in equivalent to time T use hours but does not inform customers about burn-in fallout. Show that the likelihood function LIK can be obtained by ignoring the fact that the population observed was truncated, and then multiplying this expression by a term equivalent to n units having run-times of T, where n is the total sample size. (Hint: All n units under observation are know survivors to at least T hr, so use conditional probabilities.)

4.4 An experiment to estimate Weibull parameters produced the following data: 10 exact times of failure at $t_1, t_2, \ldots, t_{10}$; 5 more failures in the readout interval $(0, T_1)$, and 7 more fails between T_1 and T_2. The mixture of exact times and readout times came about because two test chambers were used, only one of which had continuous monitoring capability. The chamber with exact failure times ended test at time T_2, with 10 unfailed units. The other chamber ended test also at time T_2 with 8 unfailed units. Write the equation for LIK. How would LIK change if you learned that the failure at time t_7 was caused by a chamber mishap, and not by the failure mechanism under investigation? (Hint: See discussion in Chapter 2 on failure mode separation.)

4.5 Simulate data from a stress test on 100 units assuming the failure times follow a Weibull distribution with $c = 10{,}000$ hours and $m = 0.75$. Do 100 iterations. For each iteration, estimate the characteristic life c using the time to the 63rd failure. Estimate also the *MTTF*, median, and standard deviation. Using the estimated characteristic life and the median, estimate the shape parameter via the expression

$$m = \frac{c \ln (\ln 2)}{T_{50}}$$

Generate histograms from the 100 estimates for both c and m. Comment on the results.

Chapter 5

The Normal and Lognormal Distributions

The lognormal distribution has become the most popular life distribution model for many high technology applications. In particular, it is very suitable for semi-conductor degradation failure mechanisms. It has also been used successfully for modeling material fatigue failures and failures due to crack propagation. Some of its success comes from its theoretical properties; in other cases, it "works" because it is flexible and fairly easy to use.

The primary purpose of this chapter is to discuss the properties and areas of application of the lognormal distribution. Many of these properties come directly from the properties of the normal distribution, because a simple logarithmic transformation changes lognormal data into normal data. Anything we know how to do with the normal distribution and normal data, we can therefore do for the lognormal distribution and lognormal data. For that reason, we will start this chapter with a review of the normal distribution.

NORMAL DISTRIBUTION BASICS

Figure 5.1 shows the familiar bell shaped curve that is the normal PDF. This curve is defined over the entire x-axis, from $-\infty$ to $+\infty$. This domain of definition differs from the life distributions in the previous chapters, which were defined only on the positive x-axis. The equation for the normal PDF is

$$f(x) = \frac{1}{\sigma\sqrt{2\pi}} e^{-(x-\mu)^2/2\sigma^2}$$

with the two parameters designated by μ (mu, the mean), and σ (sigma, the stand-ard deviation). As Figure 5.1 indicates, the distribution is symmetrical about its

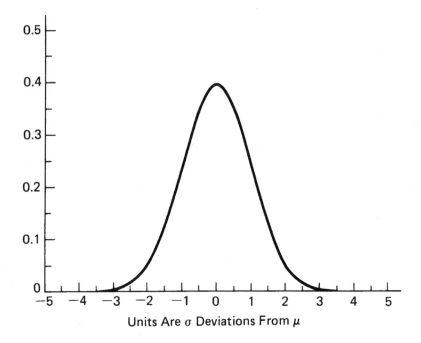

Units Are σ Deviations From μ

Figure 5.1   The Normal Distribution PDF

center μ, and σ is a scale parameter which tells how close to the center the area under the curve is packed. A range of ±1 σ from the center contains about 68 percent of the area or population, ±2 σ contains about 95 percent, and ±3 σ covers 99.7 percent of any normal population.

The parameters μ and σ are natural in the sense that μ = E(X), the mean, and σ² = E[(X − μ)²], the variance for a normal random variable X. The standard deviation, σ, is the square root of the variance.

The CDF for the normal distribution is obtained by integrating the PDF and has an S shape as shown in Figure 5.2. The hazard function is defined for all real x, but it has limited value in a failure rate context, since the normal distribution is seldom used as a life distribution model.

If we start with a normal random variable X with parameters μ and σ, a simple transformation yields

$$Z = \frac{X - \mu}{\sigma}$$

This transformation is known as normalization, and Z has the "standard" normal distribution, which is a normal distribution with mean 0 and standard devia-

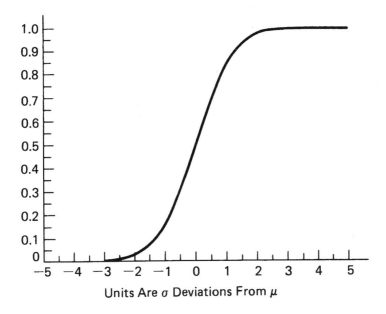

Units Are σ Deviations From μ

Figure 5.2   The Normal Distribution CDF

tion 1. The symbols $\phi(z)$ and $\Phi(z)$ are commonly used to represent the standard normal PDF and CDF, respectively. The respective equations are

$$\phi(z) = \frac{1}{\sqrt{2\pi}} e^{-z^2/2}$$

$$\Phi(z) = \int_{-\infty}^{z} \phi(z)\, dz$$

The CDF for the normal distribution cannot be obtained in closed form by integration of the PDF. Thus, tables obtained by numerical integration procedures are used. The fact that any normal variable can easily be transformed into a standard normal by subtracting its mean and dividing by its standard deviation, means that one set of tabled values for $\Phi(z)$ has wide applicability. Since, by symmetry, $\Phi(z)$ = $1 - \Phi(-z)$, many authors provide values of the CDF only for positive z, that is, starting at $\Phi(0) = 0.5$. For ease of calculation, Table 5.1 gives values of $\Phi(z)$ from $z = -4.09$ to $z = 4.09$, in steps of 0.01. This table will handle most applications. Alternatively, hand-held calculators and personal computer software programs are available which provide CDF values directly for many distributions, including the normal.

TABLE 5.1   Standard Normal CDF

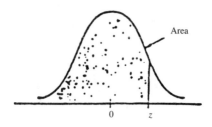

Area

0   z

Standard Normal Curve Area

Z	0.00	0.01	0.02	0.03	0.04	0.05	0.06	0.07	0.08	0.09
-4.0	0.00003	0.00003	0.00003	0.00003	0.00003	0.00003	0.00002	0.00002	0.00002	0.00002
-3.9	0.00005	0.00005	0.00004	0.00004	0.00004	0.00004	0.00004	0.00004	0.00003	0.00003
-3.8	0.00007	0.00007	0.00007	0.00006	0.00006	0.00006	0.00006	0.00005	0.00005	0.00005
-3.7	0.00011	0.00010	0.00010	0.00010	0.00009	0.00009	0.00008	0.00008	0.00008	0.00008
-3.6	0.00016	0.00015	0.00015	0.00014	0.00014	0.00013	0.00013	0.00012	0.00012	0.00011
-3.5	0.00023	0.00022	0.00022	0.00021	0.00020	0.00019	0.00019	0.00018	0.00017	0.00017
-3.4	0.00034	0.00032	0.00031	0.00030	0.00029	0.00028	0.00027	0.00026	0.00025	0.00024
-3.3	0.00048	0.00047	0.00045	0.00043	0.00042	0.00040	0.00039	0.00038	0.00036	0.00035
-3.2	0.00069	0.00066	0.00064	0.00062	0.00060	0.00058	0.00056	0.00054	0.00052	0.00050
-3.1	0.00097	0.00094	0.00090	0.00087	0.00084	0.00082	0.00079	0.00076	0.00074	0.00071
-3.0	0.00135	0.00131	0.00126	0.00122	0.00118	0.00114	0.00111	0.00107	0.00104	0.00100
-2.9	0.00187	0.00181	0.00175	0.00169	0.00164	0.00159	0.00154	0.00149	0.00144	0.00139
-2.8	0.00256	0.00248	0.00240	0.00233	0.00226	0.00219	0.00212	0.00205	0.00199	0.00193
-2.7	0.00347	0.00336	0.00326	0.00317	0.00307	0.00298	0.00289	0.00280	0.00272	0.00264
-2.6	0.00466	0.00453	0.00440	0.00427	0.00415	0.00402	0.00391	0.00379	0.00368	0.00357
-2.5	0.00621	0.00604	0.00587	0.00570	0.00554	0.00539	0.00523	0.00508	0.00494	0.00480
-2.4	0.00820	0.00798	0.00776	0.00755	0.00734	0.00714	0.00695	0.00676	0.00657	0.00639
-2.3	0.01072	0.01044	0.01017	0.00990	0.00964	0.00939	0.00914	0.00889	0.00866	0.00842
-2.2	0.01390	0.01355	0.01321	0.01287	0.01255	0.01222	0.01191	0.01160	0.01130	0.01101
-2.1	0.01786	0.01743	0.01700	0.01659	0.01618	0.01578	0.01539	0.01500	0.01463	0.01426
-2.0	0.02275	0.02222	0.02169	0.02118	0.02068	0.02018	0.01970	0.01923	0.01876	0.01831
-1.9	0.02872	0.02807	0.02743	0.02680	0.02619	0.02559	0.02500	0.02442	0.02385	0.02330
-1.8	0.03593	0.03515	0.03438	0.03362	0.03288	0.03216	0.03144	0.03074	0.03005	0.02938
-1.7	0.04457	0.04363	0.04272	0.04182	0.04093	0.04006	0.03920	0.03836	0.03754	0.03673
-1.6	0.05480	0.05370	0.05262	0.05155	0.05050	0.04947	0.04846	0.04746	0.04648	0.04551
-1.5	0.06681	0.06552	0.06426	0.06301	0.06178	0.06057	0.05938	0.05821	0.05705	0.05592
-1.4	0.08076	0.07927	0.07780	0.07636	0.07493	0.07353	0.07215	0.07078	0.06944	0.06811
-1.3	0.09680	0.09510	0.09342	0.09176	0.09012	0.08851	0.08692	0.08534	0.08379	0.08226
-1.2	0.11507	0.11314	0.11123	0.10935	0.10749	0.10565	0.10383	0.10204	0.10027	0.09853
-1.1	0.13567	0.13350	0.13136	0.12924	0.12714	0.12507	0.12302	0.12100	0.11900	0.11702
-1.0	0.15866	0.15625	0.15386	0.15151	0.14917	0.14686	0.14457	0.14231	0.14007	0.13786
-0.9	0.18406	0.18141	0.17879	0.17619	0.17361	0.17106	0.16853	0.16602	0.16354	0.16109
-0.8	0.21186	0.20897	0.20611	0.20327	0.20045	0.19766	0.19489	0.19215	0.18943	0.18673
-0.7	0.24196	0.23885	0.23576	0.23270	0.22965	0.22663	0.22363	0.22065	0.21770	0.21476
-0.6	0.27425	0.27093	0.26763	0.26435	0.26109	0.25785	0.25463	0.25143	0.24825	0.24510
-0.5	0.30854	0.30503	0.30153	0.29806	0.29460	0.29116	0.28774	0.28434	0.28096	0.27760
-0.4	0.34458	0.34090	0.33724	0.33360	0.32997	0.32636	0.32276	0.31918	0.31561	0.31207
-0.3	0.38209	0.37828	0.37448	0.37070	0.36693	0.36317	0.35942	0.35569	0.35197	0.34827
-0.2	0.42074	0.41683	0.41294	0.40905	0.40517	0.40129	0.39743	0.39358	0.38974	0.38591
-0.1	0.46017	0.45620	0.45224	0.44828	0.44433	0.44038	0.43644	0.43251	0.42858	0.42465
-0.0	0.50000	0.49601	0.49202	0.48803	0.48405	0.48006	0.47608	0.47210	0.46812	0.46414

TABLE 5.1 Standard Normal CDF (continued)

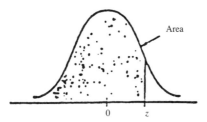

Area

0   z

Z	0.00	0.01	0.02	*Standard Normal Curve Area* 0.03	0.04	0.05	0.06	0.07	0.08	0.09
0.0	0.50000	0.50399	0.50798	0.51197	0.51595	0.51994	0.52392	0.52790	0.53188	0.53586
0.1	0.53983	0.54380	0.54776	0.55172	0.55567	0.55962	0.56356	0.56749	0.57142	0.57535
0.2	0.57926	0.58317	0.58706	0.59095	0.59483	0.59871	0.60257	0.60642	0.61026	0.61409
0.3	0.61791	0.62172	0.62552	0.62930	0.63307	0.63683	0.64058	0.64431	0.64803	0.65173
0.4	0.65542	0.65910	0.66276	0.66640	0.67003	0.67364	0.67724	0.68082	0.68439	0.68793
0.5	0.69146	0.69497	0.65847	0.70194	0.70540	0.70884	0.71226	0.71566	0.71904	0.72240
0.6	0.72575	0.72907	0.73237	0.73565	0.73891	0.74215	0.74537	0.74857	0.75175	0.75490
0.7	0.75804	0.76115	0.76424	0.76730	0.77035	0.77337	0.77637	0.77935	0.78230	0.78524
0.8	0.78814	0.79103	0.75389	0.79673	0.79955	0.80234	0.80511	0.80785	0.81057	0.81327
0.9	0.81594	0.81859	0.82121	0.82381	0.82639	0.82894	0.83147	0.83398	0.83646	0.83891
1.0	0.84134	0.84375	0.84614	0.84849	0.85083	0.85314	0.85543	0.85769	0.85993	0.86214
1.1	0.86433	0.86650	0.86864	0.87076	0.87286	0.87493	0.87698	0.87900	0.88100	0.88298
1.2	0.88493	0.88686	0.88877	0.89065	0.89251	0.89435	0.89617	0.89796	0.89973	0.90147
1.3	0.90320	0.90490	0.90658	0.90824	0.90988	0.91149	0.91308	0.91466	0.91621	0.91774
1.4	0.91924	0.92073	0.92220	0.92364	0.92507	0.92647	0.92785	0.92922	0.93056	0.93189
1.5	0.93319	0.93448	0.93574	0.93699	0.93822	0.93943	0.94062	0.94179	0.94295	0.94408
1.6	0.94520	0.94630	0.94738	0.94845	0.94950	0.95053	0.95154	0.95254	0.95352	0.95449
1.7	0.95543	0.95637	0.95728	0.95818	0.95907	0.95994	0.96080	0.96164	0.96246	0.96327
1.8	0.96407	0.96485	0.96562	0.96638	0.96712	0.96784	0.96856	0.96926	0.96995	0.97062
1.9	0.97128	0.97193	0.97257	0.97320	0.97381	0.97441	0.97500	0.97558	0.97615	0.97670
2.0	0.97725	0.97778	0.97831	0.97882	0.97932	0.97982	0.98030	0.98077	0.98124	0.98169
2.1	0.98214	0.98257	0.98300	0.98341	0.98382	0.98422	0.98461	0.98500	0.98537	0.98574
2.2	0.98610	0.9865	0.98679	0.98713	0.98745	0.98778	0.98809	0.98840	0.50870	0.98899
2.3	0.98928	0.98956	0.98983	0.99010	0.99036	0.99061	0.99086	0.99111	0.99134	0.99158
2.4	0.99180	0.99202	0.99224	0.99245	0.99266	0.99286	0.99305	0.99324	0.99343	0.99361
2.5	0.99379	0.99396	0.99413	0.99430	0.99446	0.99461	0.99477	0.99492	0.99506	0.99520
2.6	0.99534	0.99547	0.99560	0.99573	0.99585	0.99598	0.99609	0.99621	0.99632	0.99643
2.7	0.99653	0.99664	0.99674	0.99683	0.99693	0.99702	0.99711	0.99720	0.99728	0.99736
2.8	0.99744	0.99752	0.99760	0.99767	0.99774	0.99781	0.99788	0.99795	0.99801	0.99807
2.9	0.99813	0.95819	0.99825	0.99831	0.99836	0.99841	0.99846	0.99851	0.99856	0.99861
3.0	0.99865	0.99869	0.99874	0.99878	0.99882	0.99886	0.99889	0.99893	0.99896	0.99900
3.1	0.99903	0.99906	0.99910	0.99913	0.99916	0.99918	0.99921	0.99924	0.99926	0.99929
3.2	0.99931	0.99934	0.99936	0.99938	0.99940	0.99942	0.99944	0.99946	0.99948	0.99950
3.3	0.99952	0.99953	0.99955	0.99957	0.99958	0.99960	0.99961	0.99962	0.99964	0.99965
3.4	0.99966	0.99968	0.99969	0.99970	0.99971	0.99972	0.99973	0.99974	0.99975	0.99976
3.5	0.99977	0.99978	0.99978	0.99979	0.99980	0.99981	0.99981	0.99982	0.99983	0.99983
3.6	0.99984	0.99985	0.99985	0.99986	0.99936	0.99987	0.99987	0.99988	0.99988	0.99989
3.7	0.55555	0.99990	0.99990	0.99990	0.99991	0.99991	0.99992	0.99992	0.99992	0.99992
3.8	0.99993	0.99993	0.99993	0.99994	0.99994	0.99994	0.99994	0.99995	0.99995	0.99995
3.9	0.99995	0.99995	0.99996	0.99996	0.99996	0.99996	0.99996	0.99996	0.99997	0.99997
4.0	0.99997	0.99997	0.99997	0.99997	0.99997	0.99997	0.99998	0.99998	0.99998	0.99998

Any area under the normal PDF curve, or any probability or population proportion, can be calculated from Table 5.1. An example will illustrate the techniques involved.

Example 5.1   Normal Distribution Calculations

An electronics manufacturer uses interconnection wire which has a nominal strength of 11 grams (i.e., it takes an average pull force of 11 grams to break the wire). If the population distribution about this average value is normal with a standard deviation of 1.2 grams, find the following:

1. the proportion of wires that will survive a pull of 13 grams
2. the probability a wire breaks under a load of 8 grams
3. the proportion of wires that will survive a pull of at least 8.5 grams but not over 13.2 grams

Solution

Before we can use tables based on the standard normal distribution, we have to standardize the numbers given in the problem. The proportion greater than 13 for a normal with $\mu = 11$ and $\sigma = 1.2$ is the same as the area to the right of $(13 - 11)/1.2 = 1.67$ for a standard normal distribution. This area is $1 - \Phi(1.67)$. Table 5.1 gives $\Phi(1.67) = 0.95254$, and so $1 - 0.95254 = 0.04746$.

The probability that a wire will break under an eight-gram load is just $[(8 - 11)/1.2] = (2.5)$. We look up $\Phi(-2.5)$ in Table 5.1 and obtain $\Phi(-2.5) = 0.00621$.

The final question asks for the area between the z values of $(8.5 - 11)/1.2$ and $(13.2 - 11)/1.2$. This region is the area to the right of -2.08 but to the left of 1.83. That area is given by $\Phi(1.83) - \Phi(-2.08)$. From Table 5.1, this proportion is $0.96638 - 0.01876 = 0.94762$.

Exercise 5.1

A soft drink vending machine dispenses liquid into an 8 oz cup. The target value is 8.0 oz, and the cup will take up to 8.72 oz before overflowing. The lower limit before customer dissatisfaction is 7.5 oz. Assume the distribution of fill is normal with mean on target at 8.0 oz and standard deviation of 0.25 oz. What is the probability of overflow? What is the probability of a cup being filled below the lower limit? What is the amount of liquid corresponding to the 0.5 percent point of the fill distribution?

Some additional properties of normal random variables extend the usefulness of the standard normal CDF values given in Table 5.1 even further. These properties are:

1. If X and Y are independent, normal random variables with X having mean μ_x and standard deviation σ_x, and Y having mean μ_y and standard deviation σ_y, then the sum $W = X + Y$ is also distributed normally with mean $\mu_w = \mu_x + \mu_y$ and standard deviation

$$\sigma_w = \sqrt{\sigma_x^2 + \sigma_x^2}$$

This property can be extended to the sum of any number of independent normal variables by adding more mean terms to get the mean of the sum, and adding more variance terms under the square root sign to get the standard deviation of the sum.

2. Linear combinations of independent normal random variables are also normal. For X and Y as in property 1 above, $Z = aX + bY$ is normal with mean $(a\mu_x + b\mu_y)$ and standard deviation

$$\sqrt{a^2 \sigma_x^2 + b^2 \sigma_y^2}$$

3. If $x_1, x_2, \ldots, x_n$ are a random sample of observations from a normal population with mean μ and standard deviation σ, then the sample mean

$$\bar{x} = \sum_{i=1}^{n} x_i / n$$

also has a normal distribution with mean $\mu_{\bar{x}} = \mu$ and standard deviation $\sigma_{\bar{x}} = \sigma / \sqrt{n}$.

Example 5.2   Root-Mean-Square Example

A computer designer must carefully control the delay time it takes signals to travel from one location within a system to another, so that they arrive at their destination on time according to the system "clock." Each component within a given path has a nominal delay specified, as well as a 3-sigma upper and lower bound to use in testing the safety of the design. Assume a given path has four stages, two of which have nominal delays of 10 ns, with a 3-sigma upper bound of 13 ns, and the other two have nominal delays of 8 ns, with a 3-sigma upper bound of 11 ns. Characterize the nominal and 3-sigma upper bound of the total path and estimate the probability a delay is longer than 42.5 ns.

Solution

The rules given for calculating the mean and standard deviation of a sum of independent normal random variables are valid even if the assumption of normality is dropped. The average or nominal total path delay is the sum of all the nominal delays (independence is not even required for this property to hold). Thus the nominal delay is $10 + 10 + 8 + 8 = 36$. If the delays at each stage *are independent*, the value of sigma for the total delay is the square root of the sum of all the individual variances. Three sigma for the first two stages is $13 - 10 = 3$, so sigma for each of these is 1. Three sigma for the next two stages is $11 - 8 = 3$, and again sigma is 1. The total path sigma is the square root of $1^2 + 1^2 + 1^2 + 1^2$, or $\sqrt{4} = 2$.

Since normality is usually a reasonable assumption for delay lengths, the probability of any delay exceeding 42.5 ns can be calculated by finding $1 - \Phi[(42.5 - 36)/2] = 1 - \Phi(3.25) = 0.0006$, from Table 5.1.

Note that adding up 3-sigma limits for each stage (i. e., adding $13 + 13 + 11 + 11 = 48$) would give an upper limit on the path delay equivalent to six sigma units ($36 + 6 \times 2 = 48$). This limit would be unrealistic.

Exercise 5.2

An electronic component acts to transform an input signal of level X into an output signal of level $Y = 3X + 2$. If X is normally distributed with mean 10 and standard deviation 2, estimate the average and standard deviation of Y. Find the probability that Y exceeds 40. Find the probability that the average of 10 randomly selected output signals exceeds 35.

APPLICATIONS OF THE NORMAL DISTRIBUTION

As the last example showed, the normal distribution is frequently used to model measurement errors of almost any kind. In addition, many of the populations frequently encountered in industrial (or other) applications generate sample results that give bell shaped symmetrical histograms that can be fitted adequately by the normal PDF.

Even if a population has a skewed histogram which cannot be modeled by a normal distribution, the mean of a sample from this population (or $\bar{x}$), considered as a random variable in its own right, will tend to have a more nearly normal shape. (This property is a consequence of the Central Limit Theorem, described in the next section.) Therefore, we can use the normal distribution to model populations of sample means, with nearly universal success. This feature may not seem very useful, but in fact, it is the basis for the wide applicability of control chart techniques for monitoring industrial processes.

For any process we want to maintain at its best operating level, we can take regular samples of a key process parameter and treat the sample averages as if they were normally distributed. A control chart has upper and lower lines usually located 3-sigma units (the sigma we speak of here is that of the sample averages, or $\sigma/(\sqrt{n})$, where σ is the population standard deviation for the individual values and n is the number of measurements included in each average) above and below the overall process parameter average. Based on properties of the normal distribution, only about three sample averages in a thousand should fall outside of the region bounded by the 3-sigma lines. In other words, plus or minus 3-sigma limits from the center of the population contain 99.7 percent of the population values, as can be calculated from Table 5.1 by looking at the area to the left of three and to the right of minus three. So any point that does fall outside the control limits is highly unusual and unlikely, unless something abnormal happened to change the process. Control charts are discussed in more detail in Chapter 9.

If we investigate the process every time an "out of control point" outside of the control limits occurs, we will be chasing false alarms only an average of three times in a thousand, or once a year if we take one sample every day, seven days per week. This rare false alarm is a small price to pay for an alert system that tells us when some unusual variability has probably been introduced into our process.

Exercise 5.3

A process has established a stable mean at 10.5 microns. The population standard deviation for the individual values is 0.5. If sampling consists of four measurements for each consecutive time period, calculate the 3-sigma upper and lower control limits for this process.

THE CENTRAL LIMIT THEOREM

The Central Limit Theorem, or CLT, which insures the approximate normality of $\bar{x}$, has many versions. Most statements involve intricate mathematical conditions. (See Gnedenko, et al., 1969, page 33, for example.) For our purposes, a heuristic verbal statement will suffice.

Under fairly general conditions, the CLT shows that the sum of a large number of random variables, each of which contributes only a small amount to the total, will have an approximately normal distribution. As the number of contributing factors increases, with the share of each growing smaller, the approximation becomes more exact.

The sample average, as the sample size n increases, is a perfect example of a situation satisfying the conditions of the CLT. Each term in the sum $\bar{x} = (x_1/n) + (x_2/n) + \ldots + (x_n/n)$ grows smaller as n increases, and the distribution of approaches normality *no matter what underlying distribution the x_i come from* (as long as this distribution has a finite mean and variance). In terms of practical application, when n is as little as 4 or 5, $\bar{x}$ typically shows good normal characteristics.

Another excellent example is the variability introduced whenever anything is measured. Measurement error comes from a multiplicity of small random factors, combining to produce a deviation from the "true" value of the quantity under measurement. Repeated measurements produce a histogram with a normal shape, as one would expect from the CLT. In fact, the great nineteenth century mathematician Karl Gauss derived the normal distribution as the appropriate model for measurement errors, and it is often called the Gaussian error distribution.

An example similar to measurement error would be the variability when a mechanical operation is repeated over and over again. (This property formed the basis of Example 4.3 of the preceding chapter, where we saw that normal errors in two coordinate directions led to a Rayleigh distribution for the radial error). The rule of thumb of designing so that 3-sigma deviations from nominal are within specifications comes from assuming that manufacturing variation and repeatability variation leads to a normal shaped variability about the nominal value, and plus or minus 3-sigma limits will therefore capture 99.7 percent of the population.

The CLT justifies the empirical observation that the normal model works for many types of data, since it makes sense to model many important random phenomena as the sum of a large number of small contributing factors. Since fitting a normal model to data means picking a $\hat{\mu}$ estimate of μ and a $\hat{\sigma}$ estimate of σ, methods of estimation will be looked at next.

Class Project 5.1: *Illustrating the Central Limit Theorem.* A supply of dice is needed. Divide class into three person teams. One individual on each team will roll dice, one will count, and the third will record. Teams get two, five, or ten dice. Each team determines possible outcomes for the average of the particular number of dice, rolls the collection of dice 50 times, and constructs frequency tables and histograms. The class then compares sampling distributions, noting in particular, the overall sample mean and the spread in the average roll. What do we expect under the CLT? What should the distribution of 50 rolls of 10 dice look like if we record the outcome of each die and not the averages ?

NORMAL DISTRIBUTION PARAMETER ESTIMATION

The standard estimates derived from complete samples of observations from a normal distribution are well known and the best that can be used. For a sample of size n, they are

$$\hat{\mu} = \bar{x} = \sum_{i=1}^{n} x_i / n, \qquad \hat{\sigma} = s = \sqrt{\sum_{i=1}^{n} \frac{(x_i - \bar{x})^2}{n-1}}$$

These are also the MLEs (that is, maximum likelihood estimates), except that the MLE of σ has a divisor of n, instead of $(n - 1)$, which is a negligible difference for large sample sizes.

If the data is censored or multi–censored or grouped by readout intervals, the standard estimates no longer apply. Either graphical estimation (see Chapter 6) or computer programs using the maximum likelihood method can be used. This latter approach is treated extensively in Cohen and Whitten (1988). See also Bain and Engelhardt (1991) and Nelson (1982).

The graphical method is equivalent to the fitting of a straight line through the data points plotted on normal probability paper, and using the slope of the line and the 50th percentile interception value to calculate σ and μ estimates, respectively. The graphical approach starts by setting the CDF estimate at any of the data points, or readout times, approximately equal to the standard normal CDF Φ evaluated at the appropriate normalized point. The inverse of Φ is then applied to derive an equation that is linear in the unknown parameters μ and σ. The equations are

$$\Phi\left(\frac{x - \mu}{\sigma}\right) = F(x)$$

$$\Phi^{-1}\Phi\left(\frac{x - \mu}{\sigma}\right) = \Phi^{-1}F(x)$$

$$\frac{x - \mu}{\sigma} = \Phi^{-1}F(x)$$

$$x = \mu + \sigma\Phi^{-1}F(x)$$

The last equation is the key one, since it links an observable variable x to a quantity $\Phi^{-1}F(x)$ that can be calculated from the CDF estimate corresponding to x. Indeed, this linear equation is the basis of probability plotting, discussed in Chapter 6.

For r exact observations censored out of a sample of n, first order the points so that x_i is the ith smallest (x_1 is the smallest), and then estimate the CDF $F(x_i)$ by $(i - 0.3)/(n - 0.4)$. However, if the data comes in grouped intervals with endpoints $x_1, x_2, \ldots, x_k$, then estimate $F(x_i)$ by the number of observations less than or equal to x_i, divided by n. (The rationale for these CDF estimates is given in Chapter 6.)

The quantity $\Phi^{-1}(F)$ can be obtained from Table 5.1. For example, if $F = 0.62$, look up this value on the probability scale and find that the z value that has $\Phi(z) = 0.62$ is approximately 0.31. Therefore, $\Phi^{-1}(0.62) = 0.31$. If we want $\Phi^{-1}(0.33)$, we look in Table 5.1 and obtain $z = -0.44$. This shows $\Phi^{-1}(0.33) = -0.44$.

Example 5.3   Censored Normal Data

The electronics manufacturer in Example 5.1 is considering buying interconnection wires from a new vendor. A reliability engineer is asked to estimate the pull strength distribution from a sample of 500 wires. His test apparatus increases the number of grams of pull in steps of one, from one gram to 13. His procedure is to test each wire and record at what step it breaks, or whether it went all the way up to a pull of 13 grams without breaking. He observes one failure at 9 grams, 10 at 10 grams, 79 at 11 grams, 164 at 12 grams and 170 at 13 grams. There were 76 wires that survived the test. Calculate least squares estimates of μ and σ.

Solution

The data is grouped by readout intervals so the proper CDF estimate to use at a pull strength step is the total number of failures up to that point divided by the sample size of 500. Table 5.2 shows a worksheet for setting up the inputs to a least squares routine. The inverse CDF column was calculated using Table 5.1. The plot of the data is shown in Figure 5.3.

Often, a regression program, in which we input the first column from a table similar to Table 5.2 as the "Y" or dependent variable and input the third column as the "X" or independent variable, is used to estimate the slope and intercept. However, we caution that such a least squares procedure treats all data points as equal, when in fact, the sampling uncertainties (that is, the standard errors) associated with the tails of the distribution are greater than those near the center of the distribution. Hence, tail points should be weighted less than center values. Usually, the fitting of the data by eye yields acceptable estimates.

Here, the least squares regression gave an intercept estimate of $\hat{\mu} = 11.97$, and a slope estimate of $\hat{\sigma} = 1.01$. (A maximum likelihood method program for readout data gives almost exactly the same estimates.) The least squares line is shown in Figure 5.3.

Exercise 5.4

Using Figure 5.3, manually estimate the intercept and slope of the straight line. Compare to the least squares regression results.

TABLE 5.2   Example 5.3 Worksheet

Readout Point	CDF Estimate F	$\Phi^{-1}(F)$
9	1/500 = 0.002	−2.88
10	11/500 = 0.022	−2.01
11	90/500 = 0.180	−0.92
12	254/500 = 0.508	0.02
13	424/500 = 0.848	1.03

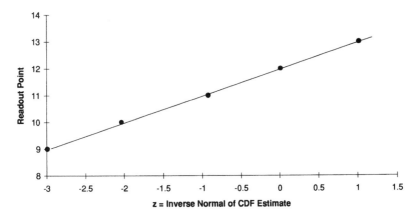

Figure 5.3   Plot of Data from Table 5.2

This survey of some of the properties of the normal distribution has been included primarily for what will be needed when we discuss the lognormal life distribution. Major topics, such as confidence intervals for parameter estimates and tests of hypotheses concerning normal populations, have not even been mentioned. The reader interested in reviewing these areas of basic statistics should consult a text such as Dixon and Massey (1969), Ostle and Mensing (1975), or Ott (1993).

SIMULATING NORMAL RANDOM VARIABLES

As we discussed in Chapter 1, random variables may be simulated through the use of the uniform random variable defined in the interval (0,1) and inverse expressions for the desired CDFs. Specifically, if F is any CDF with an inverse F^{-1}, then by substituting the unit uniform random variable U for F in the inverse expression denoted by F^{-1}, we generate a random variable distributed according to F. However, for a normal distribution, a closed–formed expression for the inverse does not exist, although there are programs available that numerically approximate the inverse normal integral transform. Alternatively, we can enlist other methods based on the results of probability theory.

The simplest method for the simulation of the normal distribution would involve the use of tables of random normal numbers such as those found in Dixon and Massey (1969). Equivalently, we can simulate the normal distribution using Table 5.1. A unit uniform random variable U is selected, and reference to Table 5.1 allows us to chose a standard normal variate Z corresponding to U. For example, if we randomly draw the value $u = 0.5322$, Table 5.1 provides us with the $z = 0.0808$. However, such procedures do not lend themselves to automation, unless the inverse normal program mentioned previously is available.

An alternative method for simulating the normal distribution employs the Central Limit Theorem (CLT). The CLT states that for any population with a finite variance σ^2 and mean μ, the distribution of the sample averages (for n items randomly selected) will approach the normal distribution with variance σ^2/n and mean μ as the sample size n increases. The sample average is just the sum of the observations divided by the number of observations n.

Let the random variable U have a uniform distribution defined in the interval $(0,1)$. U is called a *unit rectangular variate*. Many hand calculators contain programs to generate the uniform random variate. U can be shown to have mean $1/2$ and variance $1/12$. According to the Central Limit Theorem, the distribution of averages

$$\overline{U} = \frac{\displaystyle\sum_{i=1}^{n} U_i}{n}$$

approaches the normal distribution as n increases. Hence, the quantity

$$\frac{\overline{U} - \dfrac{1}{2}}{\sqrt{\dfrac{1}{12n}}}$$

approaches the standard normal variate Z for increasing n. Generally, it is sufficient (See Hastings and Peacock, 1974) to take $n = 12$, resulting in the expression

$$\left(\sum_{i=1}^{12} U_i\right) - 6$$

having an approximate standard normal distribution.
    Let

$$Z = \frac{\overline{U} - \dfrac{1}{2}}{\sqrt{\dfrac{1}{12n}}}$$

Then any normal variate X with mean μ and standard deviation σ can be obtained from the equation

$$X = \mu + \sigma Z$$

This procedure for generating a normal variate is thus quite straightforward but requires considerable computation. Note that Z can be written as

$$Z = \left(\sum_{i=1}^{n} U_i - \frac{n}{2} \right) \sqrt{\frac{12}{n}}$$

To get a pseudo–random sample of N normally distributed numbers, we first generate $n \times N$ random numbers uniformly distributed in the unit interval (0,1). Next, these numbers are added together in groups of size n to get N sums. After subtracting $n/2$ from each sum and multiplying that quantity by $\sqrt{12/n}$, we get a set of N pseudo–random numbers from a standard normal distribution. Multiply these numbers by the desired standard deviation and add the desired mean to obtain a pseudo–random sample of N unordered numbers from a normal distribution with mean μ and standard deviation σ.

Another approach is called the Box–Muller technique. (See Ross, 1990.) Two independent and standardized, normally distributed random variables X and Y can be generated through the expressions

$$X = \sqrt{-2\ln U_1}\cos(2\pi U_1)$$

$$Y = \sqrt{-2\ln U_2}\sin(2\pi U_2)$$

where U_1 and U_2 are random, uniform variates in the interval (0,1). However, this procedure is computationally time–consuming because logs, square roots, and trigonometric functions must be evaluated for each pair of random numbers. Ross (1993) shows a more efficient, computational approach to evaluate the expressions for X and Y.

Example 5.4   Simulation of C_{pk} Distribution

Let's consider a quality control example involving a stable process with measured values modeled by a normal distribution. As we discussed earlier in this chapter, 99.7 percent of normally distributed values are contained within $\pm 3\sigma$ units, and so 6σ is commonly used to represent the process width or spread. We are interested in how closely the center of the distribution comes to the nearest— either upper or lower—specification limit. We define the process capability index, C_{pk} (see Montgomery, 1991), as the minimum of the distances of the process center to either specification limit divided by 3σ, which is half the process spread. Thus,

$$C_{pk} = min\left(\frac{USL - \mu}{3\sigma}, \frac{\mu - LSL}{3\sigma} \right)$$

where USL and LSL represent the upper and lower specification limits, respectively, and μ is the process mean. Since the process standard deviation is estimated by the sample standard deviation s, which will vary from sample to sample, C_{pk} is a statistic whose value will also vary; that is, C_{pk} will have its own sampling distribution. In the absence of applicable theory, simulation can provide distributional properties of C_{pk} for specified conditions.

Let's assume a quality control document states the following for a sampling scheme: *Take a random sample of 30 units from a process lot and estimate C_{pk}. If C_{pk} so determined is less than 1.6, hold the product for disposition. Otherwise, ship the product.* What is the probability a lot is held? In this case, simulation involves the generation of 30 pseudo–random variables from a normal population. With the standard deviation equal to 1, the mean may be set so that its distance to the nearest specification limit—in units of sigma—achieves a desired population process capability index, C_{pk}. For each group of 30 samples, corresponding to a lot, we estimate σ and C_{pk}. For each possible C_{pk}, the simulated lot sampling process is repeated, for example 100 times, to obtain 100 simulated lot estimates for C_{pk}. A sufficient number of C_{pk}s are thus produced to delineate the sampling distribution of C_{pk} under the defined conditions.

We ran the simulation, using a PC spreadsheet program with random number generation capability, for C_{pk} values from 1.5 to 2.0, in steps of 0.1. Based on 100 iterations for each C_{pk} value, the ratio of the number of C_{pk} estimates that were less than 1.6 (causing lot hold) to 100 provided the basis for estimating the probability of lot hold. Table 5.3 displays the results.

Exercise 5.5

Redo the simulation of Example 5.4 based on a random sample of 50 units instead of 30 per lot. Do 100 iterations for various C_{pk} values.

THE LOGNORMAL LIFE DISTRIBUTION

The simple relationship between a normal random variable X and its derived lognormal random variable t_f is the following: if X has mean μ and standard devia-

TABLE 5.3   Results of Simulation Example (100 lots per C_{pk})

Population C_{pk}	1.5	1.6	1.7	1.8	1.9	2.0
Probability of Lot Hold	0.6	0.55	0.31	0.17	0.06	0.01

tion σ, then $t_f = e^X$ has a lognormal distribution with parameters $T_{50} = e^\mu$ and σ. Alternatively, if we start with a population of random failure times t_f modeled by a lognormal distribution with median parameter T_{50} and shape parameter σ, then the population of logarithmic failure times $X = \ln t_f$ is normal with mean $\mu = \ln T_{50}$ and standard deviation σ.

The logarithm of a lognormal is normal. In a mathematical sense, we never have to deal with the lognormal as a separate distribution; we can take logarithms (natural) of all the data points and analyze the transformed data as we would analyze normal data. This procedure is the basis for almost all lognormal analysis routines. After completing a normal analysis on the logarithmic time scale, the results are displayed in terms of the lognormal distribution and real time. The T_{50} parameter of the lognormal distribution is a "natural" parameter in the sense that it is the median time to fail of the population of lognormal lifetimes. The parameter σ, on the other hand, causes much confusion, as it is not a "natural" quantity describing the population of times to fail. σ *should be thought of only as a shape parameter for the lognormal distribution. It is not the standard deviation of the population of lifetimes.* σ is a standard deviation, in units of logarithmic time, for the normal distribution describing the population of logarithmic times to failure. In a sense, σ is a "borrowed" parameter, only used for the lognormal distribution because of mathematical convenience.

Exercise 5.6

A manufacturer produces parts that have a median width of 2.0 cm. The population distribution is lognormal with a shape parameter (sigma) of 0.2. If the specified width for acceptable parts is 1.0 to 3.0 cm, what fraction of the parts are rejectable?

PROPERTIES OF THE LOGNORMAL DISTRIBUTION

The PDF for the lognormal distribution is given by

$$f(t) = \frac{1}{\sigma t \sqrt{2\pi}} e^{-\left(1/2\sigma^2\right)(\ln t - \ln T_{50})^2}$$

The CDF $F(t)$ is the integral of the PDF from 0 to time t. It can also be expressed in terms of the standard normal CDF as

$$F(t) = \Phi\left[\frac{\ln(t/T_{50})}{\sigma}\right]$$

This last equation allows us to use table 5.1 to evaluate probabilities of failure and survival for the lognormal distribution.

Example 5.5   Lognormal Properties

A population of components, when tested at high laboratory stresses, fail according to a lognormal distribution with $T_{50} = 5000$ and $\sigma = 0.7$. What percent of failures are expected on a 2000 hour test?

Solution

Direct substitution gives

$$F(2000) = \Phi\left[\frac{\ln(2000/5000)}{0.7}\right] = \Phi(-1.309)$$

Using Table 5.1, we get 0.095 or 9.5 percent.

Exercise 5.7

For the population in Example 5.5, estimate the percent of failures expected on 4000 hour and 6000 hour tests.

The lognormal PDF has a wide variety of appearances, depending on the critical shape parameter σ. As shown in Figure 5.4, the variety of shapes lognormal data can have resembles the shapes taken on by the Weibull distribution (see Chapter 4, Figure 4.1). This flexibility makes the lognormal an empirically useful model for right skewed data. The similarity between Figure 5.4 and the Weibull PDF shapes shown in Figure 4.1 also suggests that both models will often fit the same set of experimental data equally well. One trick that occasionally helps choose whether a lognormal or a Weibull will work better for a given set of data is to look at a histogram of the logarithm of the data. If this is symmetrical and bell–shaped, the lognormal will fit the original data well. If, on the other hand, the histogram now has a left skewed appearance, a Weibull fit to the original data might work better.

Some examples of lognormal CDF curves are plotted in Figure 5.5. These curves show a long right hand tail (right skew) for large values of σ. For a small σ such as 0.2, the PDF and CDF have a normal shape.

The lognormal failure rate function $h(t)$ has to be calculated using the basic definition $h(t) = f(t)/[1 - F(t)]$.

The various shapes that can be taken by the lognormal failure rate are shown in Figure 5.6. These are also similar to the variety of shapes taken on by the Weibull failure rate (as shown in Figure 4.2). Large values of σ behave like small values

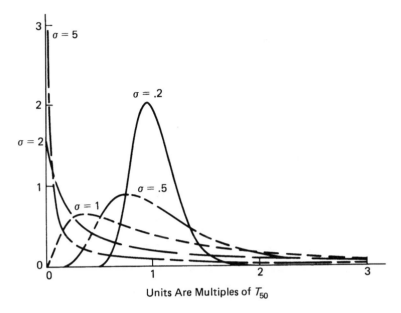

Figure 5.4   Lognormal Distribution of PDF

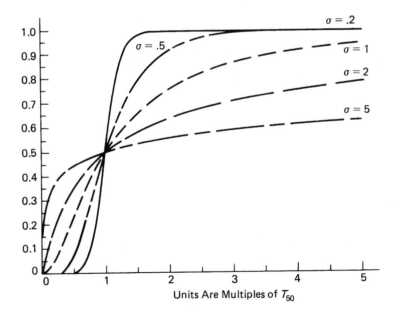

Figure 5.5   Lognormal Distribution of CDF

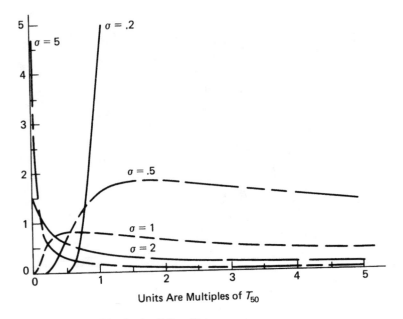

Figure 5.6 Lognormal Distribution Failure Rate

of m for the Weibull; if σ is larger than 2, the failure rate has a decreasing shape. (However, it always starts at 0 and rises to a maximum before decreasing. For large σ the rise is so quick that, for practical purposes, the failure rate appears to decrease throughout life). Values of σ around 1 give failure rate functions that first rise quickly, then have a fairly constant failure rate for a long period of time. Low values of σ correspond to wearout failure rates that are eventually rapidly increasing.

The mean, or *MTTF*, for a lognormal distribution is given by

$$E(t_f) = MTTF = T_{50}e^{\sigma^2/2}$$

and the variance is

$$Var(t_f) = T_{50}e^{\sigma^2}\left(e^{\sigma^2} - 1\right)$$

The true standard deviation is, of course, the square root of the variance.

Exercise 5.8

A lognormal distribution has median time to failure equal to 40,000 hours and shape parameter sigma equal to 0.9. Calculate the mean and standard deviation of the distribution.

The key formulas and properties of the lognormal distribution are summarized in Table 5.4.

LOGNORMAL DISTRIBUTION AREAS OF APPLICATION

The preceding section showed empirical justification for the suitability of the lognormal distribution as a life distribution model: the lognormal has a very flexible PDF and failure rate function, and its relationship to the normal distribution makes it convenient to work with.

Even though both the lognormal model and the Weibull model will fit most sets of life test data equally well, there is a major difference between them that can have critical importance. That difference comes about when we use the fitted model to extrapolate beyond the range of the sample data.

TABLE 5.4   Lognormal Formulas and Properties

Name	Formula or Property
PDF	$f(t) = \dfrac{1}{\sigma t \sqrt{2\pi}} e^{-(1/2\sigma^2)(\ln t - \ln T_{50})^2}$
CDF	$F(t) = \int_0^t f(t)\,dt = \Phi\left\{ \dfrac{\ln \dfrac{t}{T_{50}}}{\sigma} \right\}$
Reliability	$R(t) = 1 - F(t)$
Failure rate	$h(t) = \dfrac{f(t)}{R(t)}$
T_{50}	Median lifetime of 50% failure point
σ or sigma	Shape parameter. Large σ (≥ 2) means high early failure rate decreasing with time. Low σ (≤ 0.5) means increasing (wearout) type failure rate and a PDF with a "normal" shape. For σ close to 1, the failure rate is fairly flat.
Relation to normal	If t_f is lognormal with parameters (T_{50}, σ), then $X = \ln t_f$ is normal with mean $\mu = \ln T_{50}$ and a standard deviation σ.
Mean	$E(t) = T_{50} e^{\sigma^2/2}$
Variance	$Var(t) = T_{50} e^{\sigma^2}\left(e^{\sigma^2} - 1\right)$

If we have 100 units on test, the smallest percentile we can actually observe is the 1 percent point. We need a parametric model to project to the earlier time when 0.001 percent of the population might fail, or to estimate a proportion of failures for a time much smaller than the time of the first observed test failure. If we use a lognormal model, the projection to smaller percentiles is usually optimistic, as compared to a projection based on a Weibull distribution fit. In other words, the lognormal will extrapolate lower AFRs at early times. Sometimes the difference can be several orders of magnitude.

In Chapter 7 we will see that the use failure rates we are concerned with often must be estimated from the very early percentiles of a laboratory test conducted at high stress. How, then, can we decide whether it is appropriate to use the more optimistic lognormal model or the Weibull model?

A reasonable answer to this is to look for a theoretical justification for one model or the other, based on the failure mechanism under investigation. In Chapter 4, we learned that the Weibull can be derived as the extreme value distribution that applies when many small defect sites compete with each other to be the one that causes the earliest time of failure. Similarly, there is a derivation that leads to the lognormal as a model for processes that degrade over time, eventually reaching a failure state.

The precise model that leads to a lognormal distribution is called a multiplicative (or proportional) growth model. At any instant of time, the process undergoes a random increase of degradation that is proportional to its present state. The multiplicative effect of all these random (and assumed independent) growths builds up to failure.

This model was used with great success by Kolmogorov (1941) to describe the dimensions of particles (such as grains of sand by the ocean) constantly undergoing a pulverizing process. References describing the multiplicative model include Mann et al. (1974) and Gnedenko et al. (1969).

The derivation depends on viewing the process at many discrete points of time over the interval $(0,t)$. Let $x_0, x_1, x_2, \ldots, x_n$ be random variables measuring the state of degradation of the process $\{x_t\}$ as we move along in time. The multiplicative model says that

$$ x_i = (1 + \delta) x_{i-1} $$

where δ is the small proportional growth that takes the process from x_{i-1} to x_i. Then it follows that

$$ x_n = \left[\prod_{i=1}^{n} (1 + \delta_i) \right] x_0 $$

$$\ln x_n = \sum_{i=1}^{n} \ln(1 + \delta_i) + \ln x_0$$

$$\ln x_n \approx \sum_{i=1}^{n} \delta_i + \ln x_0$$

Now we invoke the Central Limit Theorem on the sum of the small random quantities δ_i, and obtain that $\ln x_n$ has an approximate normal distribution. Therefore, x_n, and hence x_t, has a lognormal distribution for any t. This argument shows that the probability the process has degraded to a failure state by time t is approximately given by the lognormal distribution.

This derivation gives us a theoretical reason to prefer a lognormal model when we can hypothesize a multiplicative degradation process is going on. Many semiconductor failure mechanisms could fit this model. Some examples are chemical reactions such as corrosion, or material movement because of diffusion or migration or even crack growth propagation. It makes more sense to use a lognormal model for these kinds of failures, than a Weibull (extreme value) model.

LOGNORMAL PARAMETER ESTIMATION

If we have a complete sample of exact times to failure, then the best way to estimate T_{50} and σ is to take natural logarithms of all times of failure and then calculate $\bar{x}$ and s for the sample of logarithmic data. The estimate of T_{50} is $e^{\bar{x}}$, and the estimate of σ is s. The formulas are

$$\bar{x} = \frac{\sum\limits_{i=1}^{n} \ln t_i}{n}$$

$$s = \sqrt{\frac{\sum\limits_{i=1}^{n} (\ln t_i - \bar{x})^2}{n-1}}$$

$$\hat{T}_{50} = e^{\bar{x}}$$

$$\hat{\sigma} = s$$

If the data is censored or grouped by readout intervals, these estimates cannot be calculated. Instead, a computer program that calculates MLEs should be used (see the description of MLEs in Chapter 4). Commercial programs like STATPAC (see Strauss, 1980) and CENSOR (see Meeker and Duke, 1981) are available.

Graphical methods can also be used to obtain (less precise) estimates. These procedures are described in Chapter 6. Here we will show a least squares technique for putting a straight line through the data points one would plot on lognormal probability paper. This analytic method can be used for censored or readout data to estimate lognormal parameters. The estimates do not have the large sample optimality properties of MLEs, but when MLE computer programs are not available, it is a reasonable method to use.

The least squares technique is similar to that described earlier for the normal distribution. By using the relationship between the lognormal and the normal, the linear equation involving the estimated sample CDF given in the section on normal parameter estimation becomes

$$\ln t_f = \ln T_{50} + \sigma \Phi^{-1} F(t)$$

Least squares estimates of $\ln T_{50}$ and σ are obtained from any standard regression program by inputting the $\ln t_i$ values as the dependent or Y variable, and the CDF estimates as the independent or X variable (similar to Example 5.3). After the least squares fit is completed, the slope estimate is $\hat{\sigma}$, and the antilogarithm of the intercept is $\hat{T}_{50}$.

Example 5.6   Lognormal Parameter Estimation

A 4000–hr life test of 20 components yields 9 failures. Table 5.5 shows the exact times of failure data, along with the corresponding CDF estimates. (The formula used for the CDF estimates when exact times of failure are recorded is explained in Chapter 6.) Estimate T_{50} and σ for this data.

TABLE 5.5   Life Test Failure Data (20 units on test)

Order of Failure	Time of Failure	CDF Estimate $(i - 0.3)/20.4$
1	1317	0.034
2	2243	0.083
3	2248	0.132
4	2282	0.181
5	2362	0.230
6	2773	0.279
7	2797	0.328
8	3104	0.377
9	3600	0.426

Solution

The natural logarithm of the time of failure column is the dependent variable, and the standard normal inverse of the CDF column is the independent variable. The dependent variable values are formed from the natural logs of the times of failure and are, respectively: 7.183, 7.716, 7.718, 7.733, 7.767, 7.928, 7.936, 8.04, 8.188. The independent variable values are the standard normal variates (that is, the inverse normal transforms) corresponding to the estimated CDFs and are, respectively: −1.825, −1.385, −1.117, −0.911, −0.739, −0.585, −0.445, −0.313, −0.186. The lognormal plot is shown as Figure 5.7.

Least squares estimates of the intercept and slope are 8.219 and 0.5, respectively. The fitted least squares line is shown in Figure 5.7. The T_{50} estimate is $e^{intercept}$ or 3710. The sigma estimate is the slope or $\hat{\sigma} = 0.5$. MLEs for the same data are $\hat{T}_{50} = 4164$ and $\hat{\sigma} = 0.58$. (This example used simulated data from a population with true $T_{50} = 5000$ and true $\sigma = 0.7$. Therefore, in this example, the MLEs were more accurate.)

Exercise 5.9

Add 3000 hours to the time of failure shown in Table 5.4. Construct a lognormal plot and estimate the lognormal parameters (median life and sigma).

SOME USEFUL LOGNORMAL EQUATIONS

There are equations for the lognormal distribution that can facilitate estimation of parameters. We may be interested in solving for T_{50}, assuming σ and $F(t)$ are

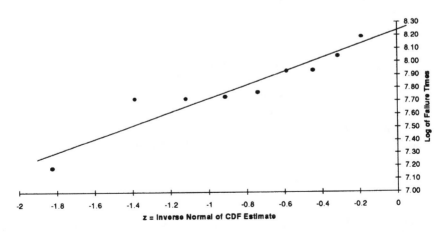

Figure 5.7   Plot of Failure Data from Table 5.5

known, or solving for σ, assuming T_{50} and $F(t)$ are known, or even solving for t, assuming the other quantities are known. These alternatives are often useful, and the equations are

$$T_{50} = te^{-\sigma\Phi^{-1}F(t)}$$

$$\sigma = \frac{\ln{(t/T_{50})}}{\Phi^{-1}F(t)}$$

$$T = T_{50}e^{\sigma\Phi^{-1}F(t)}$$

Example 5.7   Lognormal Calculations

What must the T_{50} of a lognormal distribution be to have 0.1 percent cumulative failures by 60,000 hours, assuming σ is known to be 0.8? What σ is needed to have an AFR over the first 40,000 hours of life of 0.05%/K, given the T_{50} is 300,000? For this σ, how many hours does it take to reach one percent cumulative failures?

Solution

We have

$$T_{50} = 60,000e^{-0.8\Phi^{-1}(0.001)}$$

and we use Table 5.1 to find $\Phi^{-1}(0.001) = -3.09$. Thus, $T_{50} = 60,000e^{2.472} = 710,767$ hours.

To answer the second question, we first have to convert the AFR given to a CDF value. But AFR(40,000) = 10^5 × {–ln R(40,000)}/40,000, where the factor of 10^5 is needed to convert the units to %/K. Setting this AFR equal to 0.05, we obtain R(40,000) = 0.9802 and F(40,000) = 0.0198. Using the formula σ = [ln (40,000/300,000)]/Φ^{-1}(0.0198), and the Table 5.1 value for Φ^{-1}(0.0198) = 2.06, we obtain σ = 0.98. The time point when the CDF is one percent is given by

$$t = 300,000e^{0.98\Phi^{-1}(0.01)} = 300,000e^{-0.98 \times 2.33} = 30,581$$

Lognormal Calculations: Given that the population distribution is lognormal, perform the following exercises.

Exercise 5.10

Find the median life T_{50} necessary for 5 percent failures by 96 hours, given shape parameter of 1.5.

Exercise 5.11

Find the expected cumulative percent fallout at 2,000 hours, given a median life T_{50} of 500,000 hours and a shape parameter of 5.0.

Exercise 5.12

Find the time to achieve 10 percent failures, given T_{50} of 100,000 hours and a shape parameter of 0.75.

Exercise 5.13

Find the shape parameter necessary for 1 percent failures in the first 2,000 hours, given a T_{50} of 60,000 hours.

Exercise 5.14

Find the T_{50} necessary to have an average failure rate of 50 FITS over the first 30,000 hours, given shape parameter of 2.5.

SIMULATING LOGNORMAL RANDOM VARIABLES

We invoke the relationship of the lognormal distribution to the standard normal distribution to generate lognormal pseudo–random numbers. Thus, if Z has the standard normal distribution (mean 0, standard deviation 1), then, the random variable T defined by

$$ T = \mu e^{\sigma Z} $$

has a lognormal distribution with median μ and shape parameter σ. Thus, to obtain random lognormal values, we first generate the random, standardized normal values, and use the expression for T above to get the desired lognormally distributed random variables.

Exercise 5.15

Generate 30 lognormal random variables from a population with median life = 5,000 hours and $\sigma = 4.0$. Estimate the population mean, median, standard deviation, and shape parameter from the complete sample data. Order the observations from smallest to largest. How many failures would have been observed if the testing had stopped at 2,000 hours? Describe how you might estimate the lognormal distribution parameters from the censored data.

SUMMARY

The normal distribution, while not a suitable model for population lifetimes, finds many applications in reliability analysis because of its relationship to the lognormal distribution and its use as an error model and a model for control chart applications (see Chapter 9). Since any normal variable can be transformed into a standard normal variable by subtracting its mean and dividing by its standard deviation, the standard normal table of CDF values given as Table 5.1 are frequently used.

Estimates of the normal distribution parameters μ and σ are $\bar{x}$ and s for complete samples. For censored and grouped data, graphical methods (Chapter 6), or the method of maximum likelihood, can be used. A simple regression procedure that objectively finds the best line to fit the data points to on normal probability paper, can also be used to obtain estimates.

The lognormal distribution is a very useful and flexible model for reliability data. It is closely related to the normal distribution since the logarithm of a lognormal random variable has a normal distribution. The parameters of the lognormal distribution are its median, T_{50}, and a shape parameter σ. If we work in a logarithmic time scale, by taking natural logarithms of all the data points, the resulting normal population has mean parameter $\mu = e^{T_{50}}$ and standard deviation σ. This feature allows us to use Table 5.1 to calculate probabilities associated with a lognormal distribution.

The lognormal distribution can be derived from a model for degradation processes. The main requirement is that the change in the degradation process at any time be a small random proportion of the accumulated degradation up to that time. This derivation may explain why the lognormal has been so successful modeling failures due to chemical reactions or molecular diffusion or migration. Some types of crack growth might also be expected to have a lognormal distribution.

Formulas and properties associated with the lognormal distribution are shown in Table 5.4. Estimation methods are identical to those mentioned for the normal distribution, because of the close relationship between the lognormal and the normal.

PROBLEMS

5.1   A population of devices are known to fail according to a lognormal distribution with median life $T_{50} = 200,000$ power-on hours (POHs). Evaluate the probability a new device will fail by 1,000, 4,000, 10,000, 40,000, and 1,000,000 POH for the cases where the shape parameter sigma is 0.5 or 1.0 or 2.5. Also, calculate the failure rates (in FITS) at these times for these three shape parameters.

5.2   Given that the population distribution is lognormal,

   a. Find the median life T_{50} necessary for 10 percent failures by 168 hours, given shape parameter of 2.

   b. Find the expected cumulative percent fallout at 1,000 hours, given a median life T_{50} of 1,000,000 hours and a shape parameter of 10.

   c. Find the time to achieve 20 percent failures, given T_{50} of 50,000 hours and a shape parameter of 1.

   d. Find the shape parameter necessary for 5 percent failures in the first 2,000 hours, given a T_{50} of 50,000 hours.

   e. Find the T_{50} necessary to have an average failure rate of 100 FITS over the first 40,000 hours, given shape parameter of 3.

5.3   A manufacturer produces parts with nominal width of 2.0 cm. The population distribution about this average value is normal with a standard deviation of 0.1 cm. The specification for the part is 1.8 to 2.2 cm. What fraction of the parts produced end up being scrapped? What is the answer if the standard deviation is 0.2? What if the standard deviation is 0.5? Calculate the C_{pk} for each case.

5.4   A manufacturer makes three kinds of parts: A, B, and C. A has an average length of 10 meters, and a 3-sigma upper limit of 11.5 meters. Part B has a nominal length of 2 meters and a 3-sigma upper limit of 2.75 meters. For part C, the mean value is 7.0 meters, with a standard deviation of 0.4 meters. If the final unit consists of one part A, one part B, and one part C joined end to end, what is the average length so formed? What is the 3-sigma upper limit for the final part? What assumptions have you made?

5.5   Use a maximum likelihood program to calculate estimates of T_{50} and σ for the readout pull strength data in Example 5.3 and verify that they are almost the same as the least squares estimates.

5.6   Use one of the simulation methods described in the text to generate 20 random failure times from a lognormal distribution with $T_{50} = 10,000$ and σ =1.0. Order these times and assume the first 10 are data from a censored life test that ended at the 10th failure time. Use both the least squares estimate method and the maximum likelihood method to estimate T_{50} and σ. How close to the true values are these estimates? What do you think would happen if you did this process many times and averaged the answers to get new estimates.

5.7   Assume the exact times of failure in Example 4.5, Chapter 4, came from a lognormal population and estimate T_{50} and σ by the least squares and maximum likelihood methods. Now assume that the high stress is so accelerated over normal use conditions that one second of high stress is equivalent to one day of normal use. Use both the Weibull model estimated parameters and the lognormal estimated parameters to calculate the cumulative percent failure total at one second. Which of the two models gives more optimistic (by more than five orders of magnitude, in the MLE case) results?

Chapter 6

Reliability Data Plotting

We have discussed in previous chapters analytical techniques for various distributions. However, it is useful to have simple graphical procedures that allow checking the applicability or quickly estimating certain parameters of the assumed distribution model. The methods we show in this chapter will permit rapid analysis of either exact time of failure data or interval, that is, readout type data. Complete, single-censored, or multicensored data will be treated. We will illustrate the procedures for the exponential, Weibull, normal, and lognormal distributions. We begin this chapter by first reviewing the properties of straight lines. Then the concepts of least squares regression analysis and linear rectification are described. This introductory material will prepare us for the graphical examples that follow.

PROPERTIES OF STRAIGHT LINES

Consider the straight line drawn through two points P_1 at (x_1, y_1) and P_2 at (x_2, y_2) in Figure 6.1. The slope of the line is defined as the ratio of the change in the y coordinates to the change in the x coordinates, for points on the line. Designating the slope by the letter m, we have:

$$m = \frac{rise}{run} = \frac{y_2 - y_1}{x_2 - x_1} = \frac{\Delta y}{\Delta x}$$

Note the calculation of the slope is independent of the location of the points on the line, but the best precision will be obtained by selecting the two points as far apart as possible. We can write the above equation in the form

$$\Delta y = m\Delta x$$

135

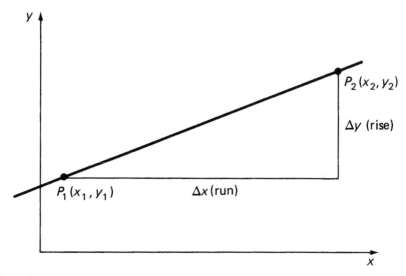

Figure 6.1   Straight Line Plot

which says that the change in y is proportional to the change in x, and the slope m is the proportionality factor. This property is called a linear relation.

Parallel line have equal slopes. It is easy to show (Thomas, 1960) that perpendicular lines have the product of the slopes equal to -1.

Consider the expression

$$y_2 - y_1 = m(x_2 - x_1)$$

Fix the point (x_1, y_1) on the line. For any (x, y) on the line,

$$y - y_1 = m(x - x_1)$$

or

$$y = mx + (y_1 - mx_1)$$

Since $y_1 - mx_1$ is fixed, we can set this quantity equal to a constant b and write:

$$y = mx + b$$

Constant b is called the *intercept* since at $x = 0$, we get $y = b$, indicating that b is the y coordinate at the point where the straight line crosses the y-axis.

An alternative expression for a straight line is

$$Ax + By + C = 0$$

where A, B, and C are constants. Any equation in this form (for example, $3x - 4y + 2 = 0$) describes a straight line.

Exercise 6.1

Consider the two equivalent equations for a straight line: $y = mx + b$ and $Ax + By + C = 0$. Express m and b in terms of A, B, and C. Express A, B, and C in terms of m and b.

Example 6.1   Linear Equations

Let C and F denote, respectively, corresponding Celsius and Fahrenheit temperature readings. Given that there is a linear relation between the two readings, F and C, find the equation from the data: $F = 32$ at $C = 0$, and $F = 212$ at $C = 100$. Is there a temperature at which $F = C$?

Solution

Use

$$F_2 - F_1 = m(C_2 - C_1)$$

Then,

$$212 - 32 = m(100 - 0)$$

or

$$m = \frac{180}{100} = \frac{9}{5}$$

At $C = 0$, $F = b = 32$. So the equation is $F = (9/5)C + 32$. If we want the temperature at which $F = C$, we solve $F = (9/5)F + 32$, or $F = -40$.

Exercise 6.2

For calculating electronic component reliability at use conditions, it is a common practice to select, often with little justification, $55°\,C$ as the ambient, operating, system temperature. What is the Fahrenheit equivalent?

LEAST SQUARES FIT (REGRESSION ANALYSIS)

Because of variation that will occur whenever measurements are made in experimentation, even if a linear relationship is expected between two variables, there will be a scatter of data points around the expected line.

Is there a "best fitting" line for a given set of data points? What do we mean by best fit? Intuitively, we would like to be able to accurately predict the y value from a given x value measured without error. We assume y is the dependent variable and x is the independent variable. Then, one definition of best fit would be to choose a line such that the sum of the squares of the deviations of the fitted y values from the observed y values is a minimum. In statistics such a line is called a *regression line* (see Figure 6.2).

It can be shown (Neter, Wasserman, and Kutner, 1990) that the equation of the best fitting (that is, least squares) line is

$$\hat{y} = mx + (\bar{y} - m\bar{x})$$

where

$$\bar{x} = \text{mean of } x = \frac{\sum\limits_{i=1}^{n} x_i}{n}, \qquad \bar{y} = \text{mean of } y = \frac{\sum\limits_{i=1}^{n} y_i}{n}$$

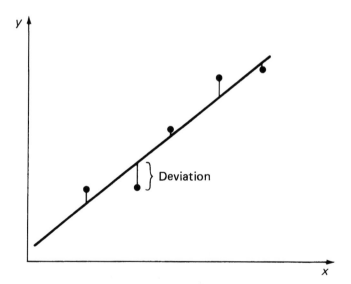

Figure 6.2   Regression Line Example

and the slope of the regression line is

$$
m = \frac{n \sum\limits_{i=1}^{n} x_i y_i - \left(\sum\limits_{i=1}^{n} x_i \right) \left(\sum\limits_{i=1}^{n} y_i \right)}{n \sum\limits_{i=1}^{n} x_i^2 - \left(\sum\limits_{i=1}^{n} x_i \right)^2}
$$

The intercept is $b = \bar{y} - m\bar{x}$. Note in the equation for $\hat{y}$, when $x = \bar{x}$, we have $y = \bar{y}$; that is, the least squares line always goes through the point $(\bar{x}, \bar{y})$.

While this formula appears a bit tedious to evaluate, it is expected that most individuals working in the field of reliability will have access to statistical packages that will perform the calculations. Many inexpensive hand calculators are designed to do least squares regression calculations.

If certain assumptions (for example, the deviations about the best fitting line are independent, normally distributed with expected value of zero, and have constant variance across all x values) are made concerning the linear model relating the dependent variable to the independent variable, then further statistical properties of the slope, intercept, and predicted values can be derived. The reader is referred to a text like that by Neter, Wasserman, and Kutner (1990) for a full discussion of this topic.

Example 6.2   Regression Line

Find the regression line for the x, y pairs: $(1, 3.1)$, $(2, 4.0)$, $(3, 5.1)$, $(4, 5.7)$, and $(5, 7.1)$.

Solution

Set up a simple table.

	x	y	x^2	xy
	1	3.1	1	3.1
	2	4.0	4	8.0
	3	5.1	9	15.3
	4	5.7	16	22.8
	5	7.1	25	35.5
Sums	15	25.0	55	84.7

$$\bar{x} = 3, \quad \bar{y} = 5, \quad \left(\sum_{i=1}^{n} x_i \right)^2 = 225$$

$$m = \frac{5 \times 84.7 - 15 \times 25}{5 \times 55 - 225} = 0.97, \quad \text{and } b = \bar{y} - m\bar{x} = 2.09$$

Thus, the regression line is $\hat{y} = 0.97x + 2.09$.

The difference between the observed value, y, and the fitted value, $\hat{y}$ (that is, $y - \hat{y}$) is called a *residual*. Plots of residuals are very useful for checking model assumptions. For further discussion of this topic, see Neter, Wasserman, and Kutner (1990).

Exercise 6.3

Using the data from Example 6.2, plot y versus x. Calculate the fitted values, $\hat{y}$, at each x and draw the least squares line. Does the fit appear reasonable? Plot the residuals, $y - \hat{y}$, versus the fitted values, $\hat{y}$. Comment on the residual plot.

RECTIFICATION

The techniques of linear regression can be used to estimate the constants and parameters of many equations, either directly or via a procedure called linear rectification. The simplest example is the problem of determining a resistance value from current and voltage pairs (I, V). By Ohm's Law, $V = IR$, where I is the current, V is the voltage, and R is the resistance. This equation is already in the form of $y = m x + b$, with $y = V, x = I, m = R$, and $b = 0$. So a plot of V versus I should approximate a straight line, going through the origin $(0, 0)$, with slope R.

Next, consider the Gas Law given by $pv = RT$, where p is the pressure, v is the volume, T is the temperature, and R is the gas constant. Suppose we wish to determine R from an experiment in which the volume is varied at a constant temperature (adiabatically) and the pressure is measured. A plot of p versus v would appear as Figure 6.3.

If, however, p is plotted versus the reciprocal of v (that is, $y = p, x = 1/v$), then we should obtain a straight line with slope RT. Since T is known and fixed, R is simply determined. See Figure 6.4.

Example 6.3   Linear Rectification

A model states the variable Q is proportional to some power of the variable s (that is, $Q = \alpha s^k$), where k is unknown and α is the unknown constant of proportionality. How do we determine k and α?

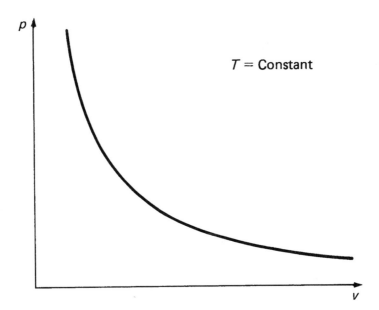

$T = $ Constant

Figure 6.3   Gas Law Plot

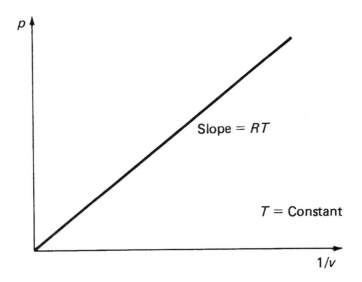

Slope $= RT$

$T = $ Constant

Figure 6.4   Gas Law Plot Using Rectification

Solution

Taking logs of both sides of the equation, we get

$$\ln Q = \ln \alpha + \ln s^k$$

or

$$\ln Q = \ln \alpha + k \ln s$$

If we now plot Q versus s on log-log paper (or, equivalently, plot $\ln Q$ versus $\ln s$ on linear-linear paper) and the model applies, we should get a straight line with slope k and intercept α (or $\ln \alpha$ on linear-linear paper). Note on log-log paper, the intercept is read at $s = 1$, since $\ln 1 = 0$.

The concept of linear rectification (or doing what is necessary to convert an equation into a linear form) is very important because it may allow us to easily determine estimates of the parameters of a model from experiments. This utility shall become very evident in Chapter 7 when acceleration models are introduced. Now, however, we shall see the application of rectification to probability plotting.

Exercise 6.4

A model is proposed that states the variable $Y^2 = Ae^{(B/T)S}$, where A and B are unknown constants, T is the temperature (°K), and S is the stress applied. We run an experiment in which Y is measured at various S levels, with the temperature T fixed. Using the principle of rectification, transform the model into a form suitable for linear plotting and describe how to estimate the unknown constants A and B.

Exercise 6.5

Based on the model in Exercise 6.4, the experiment, run at $T = 300°$ K, produces the following (Y, S) pairs: $(2.0, 1), (4.4, 2), (7.7, 3), (15.2, 4), (35.0, 5)$. Plot Y versus S on linear-linear paper. Using rectification, plot the transformed data, draw a line by eye, and estimate the unknown constants A and B. Check your visual estimates using least squares regression.

PROBABILITY PLOTTING FOR THE EXPONENTIAL DISTRIBUTION

Graphical analysis of reliability data is based simply on the concept of rectifying the data in such a way that approximately straight lines can be generated when the data is plotted. Then, the graph can be quickly checked to determine if a straight line can reasonably fit the data. If not, the assumed distribution is rejected and another may be tried. The distributions we discuss in this text can all be analyzed in a graphical manner, so if one form doesn't fit, we can experiment with another representation.

The techniques we present here will be illustrated first for the exponential distribution. However, the procedures are general, and the later sections contain examples involving the Weibull, normal, and lognormal distributions.

We will show the techniques for the exponential distribution using several variations. Two methods for handling exact times to failure will be discussed. The third variation involves readout data. These examples will also show how to estimate the failure rate, λ, by using the slope of the straight line drawn through the data.

Exact Failure Times

Method 1: Cumulative Hazard Method

An extensive development of this topic is given by Nelson (1972). We will illustrate the cumulative hazard method on the times-to-failure data shown in Table 6.1. Of the 20 components in the experiment, all eventually failed.

Recall from Chapter 2 that the cumulative hazard function $H(t)$ of any distribution is related to the CDF, $F(t)$, by

$$H(t) = -\ln[1 - F(t)]$$

For the exponential distribution,

$$F(t) = 1 - e^{-\lambda t}$$

Hence,

$$H(t) = -\ln\left(e^{-\lambda t}\right) = \lambda t$$

TABLE 6.1   Failure Times of 20 Components under Normal Operating Conditions (time in hours)

3.04	76.6	114.6	245.6
4.45	76.7	121.2	314.8
6.25	103.9	130.2	407.9
37.1	107.7	220.0	499.2
42.7	110.8	236.8	627.4

So the cumulative hazard function of the exponential distribution varies linearly with time and the proportionality constant is λ. Thus, if the exponential distribution holds for the data under analysis, a plot of the estimated $H(t)$ versus time should yield a linear fit with slope λ and intercept zero.

First, however, we need a method for estimating $H(t)$. At each time t, $H(t)$ is the sum of the individual hazard terms found by dividing the number of failures by the number of units surviving just before time t. For exact, unique times to failure, the number of failures at each failure time is always 1. Thus, a simple procedure is to order the failure times from lowest to highest, and then associate with each failure time the reverse rank starting with the initial sample size. For example, with 20 units, we have the first failure matched with 20, the second with 19, and so forth. Then for each time, the cumulative hazard function estimate is just the sum of the reciprocals of the reverse ranks, since there is one failure for each of the reverse ranks. Each reciprocal calculated this way is called a "hazard value." The calculations are illustrated in Table 6.2. The data is plotted on linear paper in Figure 6.5.

TABLE 6.2   Cumulative Hazard Calculation, Individual Failure Times (n = 20)

Failure Time	Failure No.	Reverse Rank (k)	Hazard Value [100 × (1/k)]	Cumulative Hazard
3.04	1	20	5.00	5.0
4.45	2	19	5.26	10.3
6.25	3	18	5.56	15.8
37.1	4	17	5.88	21.7
42.7	5	16	6.25	28.0
76.6	6	15	6.67	34.6
76.7	7	14	7.14	41.8
103.9	8	13	7.69	49.5
107.7	9	12	8.33	57.8
110.8	10	11	9.09	66.9
114.6	11	10	10.00	76.9
121.2	12	9	11.1	88.0
130.2	13	8	12.5	100.5
220.0	14	7	14.3	114.8
236.8	15	6	16.7	131.4
245.6	16	5	20.0	151.4
314.8	17	4	25.0	176.4
407.9	18	3	33.3	210.0
499.2	19	2	50.0	260.0
627.4	20	1	100.0	360.0

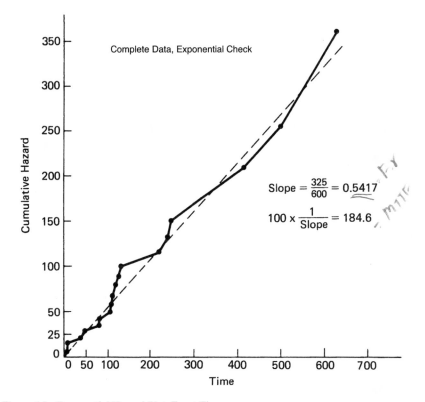

Figure 6.5   Exponential Hazard Plot, Exact Times

The fit to a straight line appears reasonable. The placement of the line to the data may be done by eye. It is possible to apply regression analysis to the data to estimate the slope of the line, but generally an eyeball fit is adequate. From the slope of the line, we obtain an estimate of the failure rate of $\hat{\lambda}$ = 542 %/k hr. The reciprocal of this slope estimates the *MTTF*, which is 185 hours.

Method 2: CDF Procedure

Recall that the CDF of the exponential distribution is given by

$$F(t) = 1 - e^{-\lambda t}$$

Rewriting this equation and taking natural logarithms, we get

$$1 - F(t) = e^{\lambda t}$$

$$-\ln\left[1 - F(t)\right] = \lambda t$$

or

$$\ln\left[\frac{1}{1 - F(t)}\right] = \lambda t$$

So if $\ln\left\{1 / \left[1 - F(t)\right]\right\}$ is plotted against the time on linear by linear graph paper or, equivalently, $1 / \left[1 - F(T)\right]$ is plotted versus the time on semi-log (i.e., log-linear) paper, then data should approximately fall on a straight line if the exponential distribution applies.

The main issue is now to estimate the CDF $F(t)$ from the data of exact times. A simple approach for a starting sample of size n would be to assign $1/n$ for the estimate of $F(t)$ at the first ordered failure time; $2/n$, at the second ordered time; and so forth. In general,

$$F(\hat{t}_i) = \frac{1}{n} \qquad i = 1, 2, 3, \ldots, n$$

However, this procedure has obvious problems. If we have only one unit, and it fails at time t_1, we do not expect the observed time to failure to represent the 100th percentile (that is, $F(t_1) = 1$) of the population distribution. $F(t_1) = 0.5$ would appear to be a more reasonable estimate. Similarly, if we have only two units and both fail, the simple approach assigns the unlikely values $F(t_1) = 0.5$ and $F(t_2) = 1$ to the first and second times to failure, respectively.

To address this issue, other authors (Hahn and Shapiro, 1967) suggest using

$$F(\hat{t}_i) = \frac{i}{n + 1}$$

or

$$F(\hat{t}_i) = \frac{i - \frac{1}{2}}{n}$$

as preferred estimators of the population CDF $F(t)$. Our choice, because of desirable statistical properties discussed in the paper by Johnson (1951), is to use what are called "median ranks," especially when the formula can be conveniently programmed on a computer.

The following equation can be derived to estimate the CDF $F(t)$:

$$F(\hat{t}_i) = \frac{0.5^{1/n}(2i - n - 1) + n - i}{n - 1} \qquad i = 1, 2, 3, \ldots$$

Fortunately there is a simple approximation to this formula given by

$$\left[F(\hat{t}_i) = \frac{i - 0.3}{n + 0.4} \right] \qquad i = 1, 2, 3, \ldots$$

We recommend that above formula be used for estimating the population CDF in all reliability plotting.

Median ranks can be described conceptually. Suppose we perform repeated sampling from a population with CDF F(t) and consider the time to failure corresponding to a specific order or rank from each repeated sample. For example, we repeat an experiment in which we stress 20 units from the same population an infinite number of times and generate the sampling distribution of the 5th time to failure across all stresses. When the population CDF is evaluated at the median of this sampling distribution of the 5th time to failure (that is, F[median of t_5]), this CDF value equals the median rank value for the 5th failure out of 20 units or approximately (5 − 0.3)/(20 + 0.4) = 0.23. Thus, we associate any time to failure with the estimate F(t) provided by the median rank.

Using the same data as before, we present the necessary calculations as Table 6.3. Figure 6.6 shows the plot on semilog paper. Again a straight line is reasonable. The failure rate is estimated from the slope of the line, and 530%/k hr is obtained.

Note that there is an equivalence between the CDF and cumulative hazard procedures since we have shown,

$$H(t) = -\ln[1 - F(t)]$$

However, the CDF procedure is often preferred by reliability engineers since the results follow from a simple transformation of the data. Unlike the CDF, the cumulative hazard has no simple interpretation, and so it becomes more difficult for an engineer to explain his analysis to other workers. Nevertheless, the cumulative hazard is a very useful mathematical concept, especially for handling multicensored data.

We plotted complete data (all units fail), but time-censored data can similarly be plotted up to the last available exact time of failure. The analysis procedure is identical to that applied to complete data.

TABLE 6.3   CDF Estimation by Median Ranks, Individual Failure Times (n = 20)

Failure Time t	Failure count i	Median Rank $F = (i - 0.3)/(n + 0.4)$	Transformation $1/(1 - F)$
3.04	1	0.034	1.035
4.45	2	0.083	1.091
6.25	3	0.132	1.152
37.1	4	0.181	1.221
42.7	5	0.230	1.299
76.6	6	0.279	1.388
76.7	7	0.328	1.489
103.9	8	0.377	1.606
107.7	9	0.426	1.743
110.8	10	0.476	1.906
114.6	11	0.524	2.103
121.2	12	0.574	2.345
130.2	13	0.623	2.650
220.0	14	0.672	3.046
236.8	15	0.721	3.580
245.6	16	0.770	4.343
314.8	17	0.819	5.519
407.9	18	0.868	7.567
449.2	19	0.917	12.030
627.4	20	0.966	29.360

Exercise 6.6

Ten units are stressed to failure. Exact failure times are measured at 1, 2, 4, 6, 7, 10, 15, 30, 55, and 96 hr. Assuming an exponential model for the parent population, analyze the data using the cumulative hazard and CDF procedures. Estimate the *MTTF* for each method.

Interval Data

When data occurs in the form of readout intervals, the CDF plotting method is applied more frequently than the cumulative hazard procedure because of the more direct interpretation of the CDF. However, the concept of median ranks, used to provide accurate plotting positions when exact failure times are known, is no longer needed. The best CDF estimate at a given readout time is just the cumulative number of failures up to that time divided by the starting sample size.

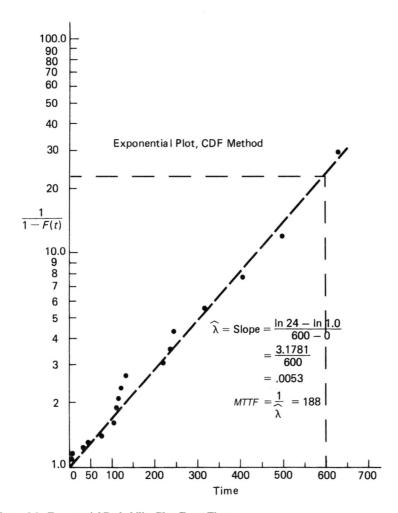

Figure 6.6   Exponential Probability Plot, Exact Times

We show in Table 6.4 some readout times for a group of one hundred components stressed in a study that ran to 600 hr. Here we have censored data with two components still surviving at the end of the experiment. The calculations are also shown in Table 6.4. Note the CDF is estimated by

$$F(t) = \frac{\text{total failures by time } t}{\text{starting sample size}}$$

The graph is shown as Figure 6.7. The failure rate is estimated again by the slope as 618%/k hr.

TABLE 6.4   Readout Data (n = 100)

Interval (hours)	Frequency of Failures	Cumulative Frequency	Estimated CDF F(t)	Transformation 1/(1 − F)
0–50	29	29	0.29	1.409
50–100	21	50	0.50	2.000
00–150	11	61	0.61	2.564
150–200	10	71	0.71	3.448
200–250	4	75	0.75	4.000
250–300	8	83	0.83	5.882
300–350	6	89	0.89	9.091
350–400	3	92	0.92	12.50
400–450	1	93	0.93	−14.29
450–500	3	96	0.96	25.00
500–550	1	97	0.97	33.33
550–600	1	98	0.98	50.00

Alternative Estimate of the Failure Rate and Mean Life

We note that at $t = 1/\lambda$, $F(1/\lambda) = 1 - e^{-1} = 0.63212\ldots$, so an alternative estimator of the $MTTF = 1/\lambda$ can be found by noting the time where the plotted line intersects the horizontal line obtained at $F(t) = 0.63212$ or, equivalently, where the transformed variable $1/[1 - F(MTTF)] = e = 2.7183\ldots$. Such a procedure is a quick way to get the $MTTF$ without calculating the slope. It may be difficult to read off the $MTTF$ value accurately using this procedure, but in many cases it will suffice. Here, we estimate from the Figure 6.7 that the $MTTF = 160$ hr or $\lambda = 625\%/K$.

Exercise 6.7

Consider the failure data in Table 6.3. Suppose, instead of exact failure times, the experimenter had obtained readouts on the number of failures at the running times of 24, 48, 96, 168, 336, and 672 hr. Analyze the interval data so obtained using graphical procedures, assuming of an exponential distribution of failure times. Estimate the $MTTF$.

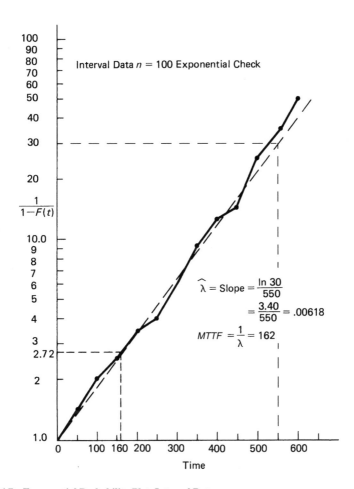

Figure 6.7   Exponential Probability Plot, Interval Data

PROBABILITY PLOTTING FOR THE WEIBULL DISTRIBUTION

The equation for the Weibull distribution is as follows:

$$F(t) - 1 - e^{-(t/c)^m}$$

where c, m, and $t > 0$. As was done in Chapter 4, we rewrite this equation in the form

$$1 - F(t) = e^{-(t/c)^m}$$

and take natural logarithms of both sides twice to get

$$\ln \{-\ln [1 - F(t)]\} = m \ln t - m \ln c$$

If we rename the variable $y = \{-\ln [1 - F(t)]\}$ and x = ln t, then it is obvious that the above transformed equation represents a straight line with slope m and intercept $-m \ln c$. Now we use the fact from Chapter 2 that the cumulative hazard $H(t)$ is related to the CDF $F(t)$ by the expression

$$H(t) = -\ln [1 - F(t)]$$

So the transformed Weibull equation can be written in terms of the cumulative hazard as

$$\ln H(t) = m \ln t - m \ln c$$

Thus, a plot of the log of the cumulative hazard versus the log of time should approximate a straight line with slope m and intercept $-m \ln c$ if the assumed Weibull model applies.

The previous discussion shows that several options are available for analyzing Weibull data graphically:

1. Plot $H(t)$ versus t on log-log paper using the reverse rank procedure for estimating $H(t)$.
2. Plot $-\ln[1 - F(t)]$ versus t on log-log paper, using the median ranks to estimate $F(t)$.

In addition, special-purpose Weibull graph paper is available that permits direct plotting of the estimated CDF versus time. These papers also have attached scales that allow quick parameter estimation. Since the first two methods are basically equivalent, we shall illustrate Weibull plotting using the second procedure, and then show how to use a special purpose paper on the same data.

Weibull Plotting, Exact Failure Times, CDF Method

Example 6.4   Weibull CDF Plotting—Exact Times

Twenty test vehicles to investigate dielectric breakdown strength are placed under stress. The test is concluded after 600 hr at which time 18 units have failed. The data is shown in Table 6.5. Plot the results to determine if the Weibull distri-

TABLE 6.5   Weibull Example (n = 20)

Failure Time t	Failure Count	Median Rank $F = (i - 0.3)/(n + 0.4)$	Transformation $-\ln(1 - F)$
0.69	1	0.034	0.035
0.94	2	0.083	0.087
1.12	3	0.132	0.142
6.79	4	0.181	0.200
9.28	5	0.230	0.262
9.31	6	0.279	0.328
9.95	7	0.328	0.398
12.90	8	0.377	0.474
12.93	9	0.426	0.556
21.33	10	0.476	0.645
64.56	11	0.524	0.743
69.66	12	0.574	0.852
108.38	13	0.623	0.974
124.88	14	0.672	1.114
157.02	15	0.721	1.275
190.19	16	0.770	1.469
250.55	17	0.819	1.708
552.87	18	0.868	2.024

bution is a reasonable fit to the data. Estimate from the graph the parameters of the distribution: shape factor m and characteristic life c.

Solution

For exact, unique failures times, the estimation of $F(t)$ is done by the same procedure used in the exponential case, that via the median ranks as shown in Table 6.3. The Weibull plot is shown as Figure 6.8. The fit appears reasonable to the model.

To estimate the slope from the log-log paper plot, we first pick two points on the log time axis, say $t_1 = 0.1$ and $t_2 = 1000$ hr, and find the corresponding $Y_1 = -\ln[1 - F(t)]$ on the vertical log axis to be $Y_1 = 0.03$ and $Y_2 = 3.2$. The slope is then evaluated from the equation

$$m = \frac{\ln(Y_2/Y_1)}{\ln(t_2/t_1)} = \frac{\ln(3.2/0.03)}{\ln(1000/0.1)} \approx 0.507$$

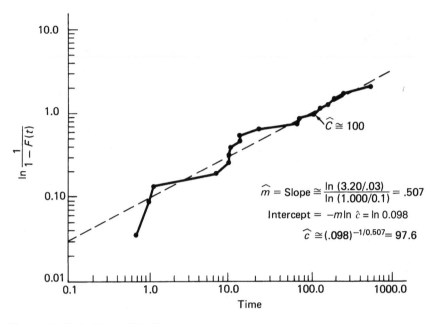

Figure 6.8   Weibull Probability Plot, Exact Times

The intercept $-m \ln c$ is found at $t = 1$ (since $\ln 1 = 0$) and equals about $\ln 0.098$ in the logarithmic scale, implying $c = (0.098)^{-1/0.507} \approx 98$.

Note that an alternative quick estimator of the characteristic life c is found at the CDF value for which $F(c) = 1 - e^{-1} = 0.63212$ or $-\ln[1 - F(c)] = 1.00$. From Figure 6.8, we get $\hat{c} \approx 100$.

Instead of plotting the data on log-log paper, special purpose charts are available (TEAM, of Tamworth, New Hampshire is one source) to facilitate the Weibull analysis. Using the same data in Table 6.5, we plot the CDF median rank estimate versus time on one type of Weibull probability paper in Figure 6.9. A straight line has been visually drawn through the points. The fit appears acceptable. To estimate the shape factor m, we now draw a line, starting at the small circle marked origin in the upper left hand corner of the graph, parallel to the line through the data points. The intersection of that line with the scale on the left gives a direct estimate of m. Here, $\hat{m} \approx 52$. The characteristic life c is estimated by finding the intersection of the line through the points with the line at the percent failure level of 63.2, marked as a small circle in the figure. Here, $\hat{c} \approx 96$ hr.

We used one form of Weibull paper. Other types involving direct plotting of the estimated cumulative hazard values are also available. However, we emphasize that for all distributions we treat in this text, one can make probability plots

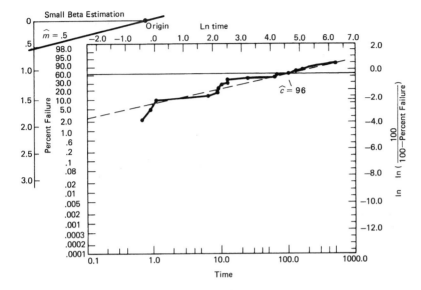

Figure 6.9   Weibull Plot on Special Chart, Exact Times

from common linear or logarithmic papers by applying the transformations we show and possibly using some common tables.

Exercise 6.8

Ten drill bits are randomly selected from a process. The times to failure (that is, loss of acceptable sharpness) for the bits are recorded as 37, 39, 42, 43, 49, 50, 54, 55, 59, and 63 hr. Assuming the failure times have a Weibull distribution, analyze the data using probability plotting. Estimate the shape and scale parameters.

PROBABILITY PLOTTING FOR THE NORMAL AND LOGNORMAL DISTRIBUTIONS

Normal Distribution

Let z be the standard normal variate; that is, $z = (x - \mu)/\sigma$, where (μ, σ^2) are the population mean and variance of a normal distribution. Recall that there is a one-to-one correspondence between z and the area under the normal PDF curve. For example, $z = 0$ corresponds to 50 percent, $z = -1.28$ to 10 percent, and so on. If the failure times of a population are normally distributed, then we can associate a z value with a given percent failure; that is, the percent failure will uniquely correspond to a z value. Following the notation in Chapter 5, designate this relation by the equation $z = \Phi^{-1}(F)$ with $\Phi^{-1}(F)$ denoting the inverse CDF function. Thus, we can write

$$\Phi^{-1}(F) = \frac{x - \mu}{\sigma}$$

or

$$x = \sigma\Phi^{-1}(F) + \mu = \sigma z + \mu$$

This last relation shows that if x is plotted on linear graph paper versus $\Phi^{-1}(F)$ for the normal distribution, one will get a straight line with slope equal to the standard deviation σ and intercept (at $z = 0$ or $F = 0.5$) equal to the mean μ. Thus, as we showed in Example 5.3, if the normal distribution holds, sample data of failure times plotted against the z value (determined from a normal table by treating the estimated CDF $F(x)$ as an area) should approximate a straight line. The slope of the line estimates σ; the intercept (50 percent point) estimates the mean μ. This method allows us to obtain an estimate of the mean and standard deviation from censored normal data.

Special-purpose normal probability plotting papers are available. However, using a normal table, one can generate the z values and plot time versus the z values on linear by linear paper. In either case, the slope can be obtained between any two convenient points whose corresponding areas represent a multiple of z. That is, we know that area for $z = -1.00$ is 0.1587, and for $z = 0$, the area is 0.5. So if we find on the time axis the times corresponding to 15.87 percent and 50 percent, the difference in these times is an estimate of the standard deviation σ. Similarly we could take the difference between the times at percent failures of 2.28 percent ($z = -2.00$) and 97.72 percent ($z = 2.00$) and divided the difference by 4 to get an estimate of the standard deviation.

If one has highly censored data and must use a low percent of failures to estimate the slope, the mean can be estimated by substituting the estimated slope σ into the equation $\mu = t - \sigma\Phi^{-1}(F)$ for any convenient time and percent failure pair. Thus, extrapolation to the 50 percent failure point to estimate the mean is unnecessary.

Exercise 6.9

One hundred units from a population of measurements believed to be normally distributed are randomly selected and measured using a method that destroys the units. In ranking the sample measurements from smallest to largest, the technician loses the largest 70 measurements because of a computer malfunction. The first 30 values plotted on probability paper provide a standard deviation estimate of 5.0. Estimate the mean of the population given that the best fitting line on the probability plot goes through the CDF = 10 percent point at the value 72.9.

Lognormal Distribution

As discussed in Chapter 5, if the natural logarithms of failure times are normally distributed with mean μ and variance σ^2, then the distribution of the times is lognormal with median $T_{50} = e^{\mu}$ and shape parameter σ. The median T_{50} and shape parameter σ are the standard parameters for expressing the lognormal distribution.

Similar to the normal, the rectifying equation is found by using the standard normal variate in the form

$$z = \frac{\ln t - \ln T_{50}}{\sigma}$$

or since $z = \Phi^{-1}(F)$,

$$\ln t = \sigma \Phi^{-1}(F) + \ln T_{50} = \sigma z + \ln T_{50}$$

Thus, if the lognormal distribution applies, a plot of the natural logarithms of failure times versus the inverse transformation of the estimated CDF percent failures (that is, the z value corresponding to a given area) should approximate a straight line which can be used to estimate the shape parameter σ from the slope, and the T_{50} from the fiftieth percentile point.

Again, we have many options. Using the term "percent failures" loosely to refer to the estimated CDF values, we can plot: (a) corresponding to the percent failures, the z values obtained from a normal table versus the log of times on linear-linear paper, (b) the percent failures versus the log of times on normal probability paper, (c) the percent failures versus the times on lognormal probability paper, or (d) the estimated cumulative hazard versus the times on lognormal hazard paper. All methods are basically equivalent, but computationally, the use of either lognormal probability or hazard paper is easiest.

The slope can be estimated from the equation

$$\sigma = \frac{\ln\left(\frac{t_2}{t_1}\right)}{\Phi^{-1}[F(t_2)] - \Phi^{-1}[F(t_1)]}$$

or, since $z = \Phi^{-1}(F) = 0$ at $F = 50$ percent and $z = -1.00$ at $F = 15.9$ percent,

$$\sigma = \ln\left(\frac{T_{50}}{T_{15.9}}\right)$$

As in the normal case, any convenient z multiple can also be used to estimate σ. Similarly, if T_{50} is outside of the range of the graph, the slope can be estimated and T_{50} solved from the equation

$$T_{50} = te^{-\sigma\Phi^{-1}[F(t)]} = te^{-\sigma z}$$

using any convenient time and percent failure combination.

Example 6.5   Lognormal CDF Plotting

Six hundred transistors are tested under stress for 1000 hr. Each failure is individually recorded in time. At the end of the experiment, 17 units have failed for a particular failure mode. The data is given in Table 6.6. It is assumed that the lognormal distribution applies. Check this assumption by plotting the data on lognormal probability paper. Estimate the parameters of the distribution. How long a test would be required to get 10 percent of the product to fail?

Solution

The estimated CDF at each point using median ranks is shown in Table 6.6. The data is graphed as Figure 6.10. The slope is estimated from the equation

$$\sigma = \frac{\ln(t_2/t_1)}{\Phi^{-1}[F(t_2)] - \Phi^{-1}[F(t_1)]}$$

by choosing (from the graph) the pairs $t = 10$ hr, $F(10) = 0.00165$, implying $\Phi^{-1}(F) = -2.95$ from the normal table, and $t = 1000$ hr, $F(1000) = 0.0275$, $\Phi^{-1}(F) = -1.92$. Then,

$$\sigma = \ln\frac{(1000/10)}{(-1.92 + 2.95)} \approx 4.47$$

The median is found from

TABLE 6.6 Lognormal Example (n = 600)

Failure Time t	Failure Count	Median Rank $F = (1 - 0.3)/(n + 0.4)$
3.7	1	0.0012
25.9	2	0.0028
58.6	3	0.0045
78.4	4	0.0062
146.7	5	0.0078
162.3	6	0.0095
224.1	7	0.0112
228.6	8	0.0128
275.9	9	0.0145
282.9	10	0.0162
481.4	11	0.0178
689.5	12	0.0195
720.0	13	0.0215
770.0	14	0.0228
851.2	15	0.0245
871.7	16	0.0261
999.3	17	0.0278

$$T_{50} = te^{-\sigma z} = 1000e^{-4.47 \times -1.92} \approx 5.3 \times 10^6$$

The time to reach 10 percent failures is

$$t = T_{50}e^{\sigma z_{10}} = 5.3 \times 10^6 e^{4.47 \times -1.28} \approx 17,350 \text{ hours}$$

Exercise 6.10

Simulate 15 random z values from a standard normal distribution. Using these z values, simulate 15 values of time to failure t from a lognormal distribution with parameters T_{50} and σ by applying the equation $t = T_{50}e^{\sigma z}$, where $\sigma = 1.0$ and $T_{50} = 500$ hours. Arrange the failure times from smallest to largest. Assume the stressing stops at 750 hr. Thus, only those times less than 750 hr can be counted as failures. Failure times above 750 hr are censored observations. Using median ranks, plot the observed failures on lognormal probability paper. Estimate T_{50} and σ. Compare to the original simulation values.

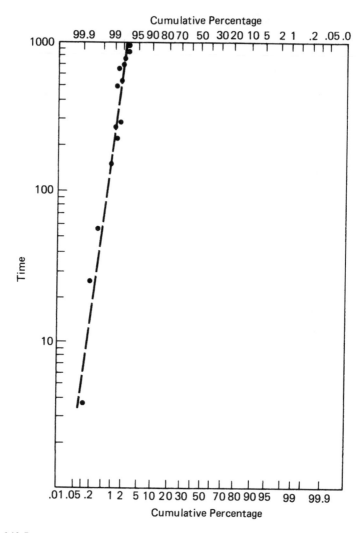

Figure 6.10 Lognormal Example

MULTICENSORED DATA

Kaplan-Meier Product Limit Estimation

The Kaplan-Meier (1958) product limit estimator is a long name for a very simple and useful procedure for calculating survival probability estimates at various
times. The power of this method lies in its ability to handle multicensored data.

Basically, the probability of a component surviving during an interval of time is estimated from the observed number of units at risk of failure during the interval. The estimates for each interval are multiplied together to provide a survival or reliability estimate from time zero. The procedure is best illustrated with an example.

Suppose eight objects are place on life stress and failures occur at 200, 300, 400, 500 and 600 hr. In addition, two good units are removed for destructive examination: one at 250 hr and another at 450 hr. Such nonfailed units effectively provide information only up to the time of removal, called a censoring time. Alternatively, we state that "losses" occurred at 250 and 450 hr. Another common terminology is to state that these units have "running times" of 250 and 450 hr.

Since we started with 8 units and one failed by 200 hr, the simple probability estimate for a unit surviving 200 hr is 7/8 = 0.875. At 250 hr, we had a loss. Thus, only the six remaining units were at risk of failure until the next failure time at 300 hr. The probability of surviving from 200 to 300 hr is estimated by 5/6. The loss at 250 hours contributes information only up to the previous failure time of 200 hr. The probability of surviving from time zero to 300 hr is estimated by

$$P(T > 300) = (7/8)(5/6) = 0.729$$

In words, the probability of surviving to 300 hr is given by the probability of surviving to 200 hr × the probability of surviving to 300 hr, given survival to 200 hr.

Similarly, the survival probabilities at the failure times are estimated below:

Time (Hr)	Survival Estimate (Reliability)
400	$(7/8)(5/6)(4/5) = 0.583$
500	$(7/8)(5/6)(4/5)(2/3) = 0.389$
600	$(7/8)(5/6)(4/5)(2/3)(1/2) = 0.194$
700	$(7/8)(5/6)(4/5)(2/3)(1/2)(0/1) = 0.000$

The cumulative distribution function can be estimated at each failure time by subtracting the survival values from 1.000.

We can write the general expression for the Kaplan-Meier product limit estimator. Denote the ordered failure and censoring times for each unit by t_i, where $i = 1, 2, \ldots, n$, and n is the total number of units observed. Let S represent the set of subscripts j of the times that t_i correspond to failures. The notation $j \varepsilon S$ indicates that j is an element of the set S. The general expression for the CDF estimate at any time t_i is

$$F(\hat{t_i}) = 1 - \prod_{\substack{j \varepsilon S \\ t_j \leq t_i}} \frac{n - j}{n - j + 1}$$

where the multiplication symbol

$$\prod_i X_i = X_1 X_2 \ldots X_n$$

This expression says that to find the CDF at any time t_i, multiply only those terms $(n-j)/(n-j+1)$ where j represents a failure time t_j less than or equal to t_i. Note the subscript j denotes the jth failure time only if there is no censoring; in general, j indicates the total number of failures and losses occurring up to time t_j.

For example, using the data above, we see that t_1, t_3, t_4, t_6, t_7 are failure times and t_2 and t_5 are censoring times. To estimate the CDF at, say, 300 hours, we use

$$F\overset{\wedge}{(t_3)} = F(\overset{\wedge}{300}) = 1 - \frac{(8-1)}{(8-1+1)} \frac{(8-3)}{(8-3+1)} = 1 - \frac{7}{8} \frac{5}{6} = 0.271$$

The KM estimator needs to be modified, however, because it reduces, under no censoring, to the simple i/n plotting positions which we previously stated were not best. Instead, we adopt a modification proposed by Michael and Schucany (1986). The following expression handles the situations with censoring, but it also yields the median rank estimators if we have only failure times and no censored run times less than or equal to t_i:

$$F\overset{\wedge}{(t_i)} = 1 - \frac{n+0.7}{n+0.4} \prod_{\substack{j \varepsilon S \\ t_j \le t_i}} \frac{n-j+0.7}{n-j+1.7}$$

For example, using the same data as above, we see that

$$F\overset{\wedge}{(t_3)} = F(\overset{\wedge}{300}) = 1 - \frac{(8+0.7)}{(8+0.4)} \frac{(8-1+0.7)}{(8-1+1.7)} \frac{(8-3+0.7)}{(8-3+1.7)}$$

$$= 1 - 0.780 = 0.22$$

We recommend this modified expression for probability plotting positions involving censoring.

Thus, from knowledge of the failure and censoring times we can construct an estimate of the CDF for multicensored data. Then, the estimated CDFs at each failure time can be plotted on any type probability paper to test the applicability of any distributional form, using the methods described earlier in this chapter.

Exercise 6.11

A computer manufacturer had three major shipments to customers during the past year. At customer A, 100 new computers were installed on February 1. At customer B, 83 days later, 200 new units were placed into operation. Finally, 200 days after the work at customer B, another 150 units started functioning at customer C's facility. All computers operate 24 hours per day, 7 days per week. By year end, customer A had reported failures on three computers, occurring at 512, 2417, and 7012 hr. Customer B had one failure at 3,250 hr and another at 5,997 hr. Customer C reported one 105 hr failure. Failing components were not replaced. Estimate the CDF at each failure time for all computers using the modified Kaplan-Meier Product Limit Estimator.

Cumulative Hazard Estimation

An alternative procedure for handling multicensored data using the cumulative hazard function is described by Nelson (1982). This technique involves estimating the cumulative hazard function at each failure time and then plotting the points on cumulative hazard paper. The method of estimation is detailed in Table 6.7.

TABLE 1.7   Cumulative Hazard Calculation

Time of Failure or Readout	Number on Test Just Prior to Failure(s)	Number of New Failures	Hazard Value	Cumulative Hazard Value
T_1	N_1	r_1	$h_1 = \dfrac{r_1}{N_1}$	$H_1 = 100 \times h_1$
T_2	N_2	r_2	$h_2 = \dfrac{r_2}{N_2}$	$H_2 = H_1 + 100 \times h_2$
T_3	N_3	r_3	$h_3 = \dfrac{r_3}{N_3}$	$H_3 = H_2 + 100 \times h_3$
.	.	.	.	.
.	.	.	.	.
.	.	.	.	.
T_k	N_k	r_k	$h_k = \dfrac{r_k}{N_k}$	$H_k = 100 \times h_k$

Example 6.6   Cumulative Hazard Plotting

Using the same data as in the Kaplan-Meier section, calculate the cumulative hazard estimates at each failure time.

Solution

Failure Times	Number on Test	Number of Failures	Hazard Value	Cumulative Hazard Value
200	8	1	1/8 = 0.125	12.5
300	6	1	1/6 = 0.167	29.2
400	5	1	1/5 = 0.200	49.2
500	3	1	1/3 = 0.333	82.5
600	2	1	1/2 = 0.500	132.5

The above cumulative hazard values could now be plotted on hazard paper to check against a specific distribution type.

SUMMARY

This chapter has reviewed graphical procedures for supplementing the analytical tools covered in previous chapter. We have discussed straight line concepts, least squares regression, and linear rectification. We have shown how various distribution models could be checked for applicability by the use of probability or hazard plotting. In addition, we have demonstrated how quick parameter estimates could be obtained from the graphs. Special commercial probability papers can be used or equivalent results may be obtained with linear or log papers coupled with simple transformations of CDF estimates. These methods are easily applied to censored data. Also, with Kaplan-Meier or cumulative hazard procedures, we can analyze multicensored data.

PROBLEMS

6.1   Plot the (readout) data up to the "greater than 550 hours" point as given for the light bulb data in Chapter 3, Table 3.3 using an exponential model. Do the points appear to follow a straight line?

6.2   Plot the 20 exponential data points generated in Chapter 3, Exercise 3.16. Do the simulated data appear to follow an exponential model?

6.3   Plot the 25 failure times from Chapter 4, Example 4.5, assuming a Weibull model. Repeat using the 10 readout intervals of Example 4.5. In each case, estimate the Weibull parameters from the plot.

6.4   Assume a proposed model for life data is

$$F(t) = \frac{1}{1 + at}, \qquad 0 \le t < \infty, a > 0$$

Describe how you would go through a "rectification" analysis allowing you to construct "probability" graph paper for this distribution. (Hint: The slope of lines on this paper will be $1/a$; the intercept is 0.) Plot the five data points given in Chapter 1, Example 1.7 on your constructed paper. Do they appear to follow a straight line with slope 1, as expected?

6.5 Plot the 30 failure times generated in Chapter 5, Exercise 5.15, assuming a lognormal model. Do the simulated data appear to follow a lognormal model? Estimate the lognormal parameters from the plot.

Chapter 7

Physical Acceleration Models

If we have enough test data, the methods described in the preceding chapters will allow us to fit our choice of a life distribution model and estimate the unknown parameters. However, with today's highly reliable components, we are often unable to obtain a reasonable amount of test data when stresses approximate normal use conditions. Instead, we force components to fail by testing at much higher than the intended application conditions. By testing this way, we get failure data that can be fitted to life distribution models, with relatively small test sample sizes and practical test times.

We pay a price to overcome our inability to estimate failure rates by testing directly at use conditions (with realistic sample sizes and test times), and it is paid in the form of required additional modeling. How can we predict from the failure rate at high stress what a future user of the product is likely to experience at much lower stresses?

The models used to bridge the stress gap are known as *acceleration models*. This chapter develops the general theory of these models and looks in detail at what acceleration testing under the exponential or Weibull or lognormal means. Several well known forms of acceleration models, such as the Arrhenius and the Eyring, are described. Practical use of these models, both for actual failure data and for degradation data, is discussed.

ACCELERATED TESTING THEORY

The basic concept of acceleration is simple. We hypothesize that a component, operating under the right levels of increased stress, will have exactly the same failure mechanisms as seen when used at normal stress. The only difference is "things happen faster." For example, if corrosion failures occur at typical use

fail mechanisms
what stress caused fail

temperatures and humidities, then the same type of corrosion happens much quicker in a humid laboratory oven at elevated temperature.

In other words, we can think of time as "being accelerated," just as if the process of failing were filmed and then played back at a faster speed. Every step in the sequence of chemical or physical events leading to the failure state occurs exactly as at lower stresses; only the time scale measuring event duration has been changed.

When we find a range of stress values over which this assumption holds, we say we have *true acceleration*. From the film replay analogy, it is clear that true acceleration is just a transformation of the time scale. Therefore, if we know the life distribution for units operating at a high laboratory stress, and we know the appropriate time scale transformation to a lower stress condition, we can mathematically derive the life distribution (and failure rate) at that lower stress. This is the approach we will use.

In theory, any well behaved (order preserving, continuous, etc.) transformation could be a model for true acceleration. However, in terms of practical applicability, we almost always restrict ourselves to simple constant multipliers of the time scale. *When every time of failure and every distribution percentile is multiplied by the same constant value to obtain the projected results at another operating stress, we have linear acceleration.*

Under a linear acceleration assumption, we have the relationship (time to fail at stress S_1) = $AF \times$ (time to fail at stress S_2), where AF is the acceleration constant relating times to fail at the two stresses. "AF" is called the acceleration factor between the stresses.

If we use subscripts to denote stress levels, with U being a typical use set of stresses and S (or $S_1, S_2, \ldots$) for higher laboratory stresses, then the key equations in Table 7.1 hold no matter what the underlying life distribution happens to be.

In Table 7.1, t_U represents a random time to fail at use conditions, while t_S is the time the same failure would have happened at a higher stress. Similarly, F_U, f_U and h_U are the CDF, PDF and failure rate at use conditions, while F_S, f_S and h_S are the corresponding functions at stress S.

Equation 1 of Table 7.1 is the linear acceleration assumption, and the other equations all follow directly from this and the definitions of the functions

TABLE 7.1   General Linear Acceleration Relationships

1. Time to fail	$t_U = AF \times t_S$
2. Failure probability	$F_U(t) = F_S(t/AF)$
3. Density function	$f_U(t) = (1/AF)f_S(t/AF)$
4. Failure rate	$h_U(t) = (1/AF)\,h_S(t/AF)$

involved. For example, $F_U(t)$ is the probability of failing by time t at use stress, which is equivalent to failing at time (t/AF) at stress S (by Eq. 1). This relationship is stated in Equation 2, and the next two equations follow by standard change of variable methods. (For example, see Mendenhall, et al., 1990).

Table 7.1 gives the mathematical rules for relating CDFs and failure rates from one stress to another. These rules are completely general and depend only on the assumption of true acceleration and linear acceleration factors. In the next three sections, we will see what happens when we apply these rules to exponential or Weibull or lognormal life distributions.

Exercise 7.1

Use the fact that $f(t) = F'(t)$ and $h(t) = f(t)/R(t)$ to verify Equations 3 and 4 in Table 7.1.

Exercise 7.2

Derive comparable equations to 2, 3, and 4 from Table 7.1 under the assumption of "quadratic" acceleration as given by $t_U = AFt_s^2$. What can happen under this kind of acceleration to stress failure times under 1 hr? Failure on stress before a "crossover" time would be expected to fail even earlier at use stress. Find the crossover time where the time to failure at use equals the time to failure at stress.

Exercise 7.3

Derive comparable equations to 2, 3, and 4 from Table 7.1 under the assumption of "exponential" acceleration as given by $t_U = AFe^t$. Note that at use conditions there is 0 probability of failure before time AF.

EXPONENTIAL DISTRIBUTION ACCELERATION

We add the assumption that $F_S(t) = 1 - e^{-\lambda_s t}$. In other words, times to fail at high laboratory stresses can be modeled by an exponential life distribution with failure rate parameter λ_s. Using Equation 2 to derive the CDF at use condition, we get

$$F_U(t) = F_S(t/AF) = 1 - e^{-\lambda_s t/AF} = 1 - e^{-(\lambda_s/AF)t}$$

By letting $\lambda_u = \lambda_s/AF$, we see that the CDF at use conditions remains exponential, with new parameter λ_s/AF.

This equation demonstrates that an exponential fit at any one stress condition implies an exponential fit at any other stress within the range where true linear acceleration holds. Moreover, when time is multiplied by an acceleration factor AF, the failure rate is reduced by dividing by AF.

Exercise 7.4

Use the time transformation given in Exercise 7.3 and the assumption the life distribution at stress S is exponential with parameter λ_S to derive $F_U(t)$. Is $F_U(t)$ still exponential? Next derive $h_U(t)$. Is $h_U(t)$ still constant?

The fact that the exponential failure rate varies inversely with the acceleration factor sometimes misleads engineers to assume this is always the case with linear acceleration. This belief is not correct. In general, the failure rate changes in a very nonlinear fashion under linear acceleration of the time scale. The simple results of this section apply only for the exponential distribution.

Exercise 7.5

Assume you have found that the normal distribution with parameters $\mu = 500$ hr and $\sigma = 60$ hr gives an excellent fit to high stress test failures for a certain component. In addition, you believe that there is a linear acceleration factor of 2000 between stress hours and use condition equivalent hours. Show that the distribution of failures at use will still follow a normal distribution but both μ_U and σ_U are 2000 times larger than the corresponding stress parameters.

$t_u = A.F[t_s]$

Example 7.1   Exponential Acceleration Factors

A component, tested at 125° C in a laboratory has an exponential distribution with *MTTF* 4500 hr. Typical use temperature for the component is 32° C. Assuming an acceleration factor of 35 between these two temperatures, what will the use failure rate be and what percent of these components will fail before the end of the expected useful life period of 40,000 hr?

$A.F[\lambda_U] = \lambda_S$

Solution

The *MTTF* is the reciprocal of the failure rate and varies directly with the acceleration factor. Therefore the *MTTF* at 32° C is 4,500 × 35 = 157,500 hr. The use failure rate is 1/157,500 = 0.635%/K. The cumulative percent of failures at 40,000 hr is given by $1 - e^{-0.00635 \times 40} = 22.4\%$.

$A.F\left[\frac{1}{MTTF_u}\right] =$

Exercise 7.6

In Example 7.1, suppose the requirement was that no more than 10 percent of the components fail by 40,000 hr at use conditions. In order to achieve this reliability, it is proposed to redesign the box environment so that the component operating temperature is significantly lower. What acceleration factor is needed from this lower operating temperature to the 125° C laboratory stress?

WEIBULL DISTRIBUTION ACCELERATION

This time we start with the assumption that $F_S(t)$ follows a Weibull distribution with characteristic life c_S and shape parameter m_S. The equation for the CDF is

$$F_s(t) = 1 - e^{-(t/c_s)^{m_s}}$$

and, transforming to use stress, we have

$$F_s(t) = F_U\left(\frac{t}{AF}\right) = 1 - e^{-[(t/AF)/c_s]^{m_s}}$$

$$= 1 - e^{-[t/(AF \times c_s)]^{m_s}} = 1 - e^{-(t/c_U)^{m_U}}$$

where $c_U = AF \times c_S$ and $m_U = m_S = m$.

This result shows that if the life distribution at one stress is Weibull, the life distribution at any other stress (assuming true linear acceleration) is also Weibull. The shape parameter remains the same while the characteristic life parameter is multiplied by the acceleration factor.

The equal shape result is highly significant. It is often mistakenly added as an assumption, in addition to assuming a linear acceleration model and a Weibull life distribution. As we have seen, however, it is a necessary mathematical consequence of the other two assumptions. If different stress cells yield data with very different shape parameters, then either we do not have true linear acceleration or the Weibull distribution is the wrong model for the data.

In Chapter 6, when we discussed Weibull probability plotting, we saw that the shape parameter turned out to be the slope of the line fitted to the data, when plotted on an appropriate type of graph paper. Therefore, when we have Weibull acceleration and we plot several stress cells of data on the same sheet of graph paper, we should end up with parallel lines. The lines will not be exactly parallel, of course, since we are dealing with sample data and estimates of the underlying distributions. However, lines that are very far from appearing parallel would indicate either model or data problems.

By calculating the Weibull failure rate at stress and use conditions, it is easy to see how $h(t)$ varies due to acceleration. For a stress failure rate, we have $h_s(t) = (m/c_s)(t/c_s)^{m-1}$. By expressing the characteristic life parameter for use as $c_U = AF \times c_S$, we obtain

$$h_U(t) = \frac{m}{AF \times c_s}\left(\frac{t}{AF \times c_s}\right)^{m-1} = \frac{1}{AF^m}\frac{m}{c_s}\left(\frac{t}{c_s}\right)^{m-1} = \frac{h_s(t)}{AF^m}$$

This is a linear change in the failure rate, but the multiple is $1/AF$ only when $m = 1$ and the distribution is exponential; otherwise the failure rate is multiplied by $1/(AF)^m$.

Exercise 7.7

An engineer is able to gain an additional 100× acceleration for a key component by reducing a box's operating temperature. He claims that will reduce the component's failure rate by at least 100×. Assuming the component has a Weibull life distribution, is his claim correct? If you disagree, can you give a counterexample?

Example 7.2   Weibull Multiple Stress Cells

Random samples of a manufacturer's capacitors were tested to determine how temperature accelerates failure times. Three temperatures were used; one cell at 85° C, one at 105° C, and one at 125° C. Each cell, or controlled oven, contained 40 capacitors operating at the oven temperature. At the following readout times new failures were determined: 24, 72, 168, 300, 500, 750, 1000, 1250, and 1500 hours. All testing was completed in about 10 weeks, with results as given in Table 7.2.

Assuming a Weibull distribution applies, plot all the three cells of test data on the same sheet of Weibull paper, estimate parameters, and check whether the equal slope (same m) consequence of true acceleration looks reasonable. If maximum likelihood data analysis programs are available, also use these to estimate parameters and test for equal shapes. Use the cell characteristic life estimates to compute acceleration factor estimates between 85 and 105° C, 85 and 125° C, and 105 and 125° C.

TABLE 7.2   Weibull Temperature Stress Failure Data

Readout Time	85° C New Failures	105° C New Failures	125° C New Failures
24	1	2	5
72	0	1	10
168	0	3	13
300	1	2	2
500	0	2	3
750	3	4	2
1000	0	5	2
1250	1	1	1
1500	2	4	0
	—	—	—
Total	8	24	38

Solution

Figure 7.1 shows the cumulative failures in each cell plotted at the corresponding readout time. The plotting position for each readout time's data is slightly different than the unadjusted cumulative percent failures because the plot was generated by the AGSS statistical program, and AGSS uses estimates of the CDF developed by Turnbill (1976).

The lines fitted through the points for each cell were obtained as described in Chapter 6, using least squares (regression) on the equation

$$\ln \left\{ -\ln \left[1 - \hat{F}(t) \right] \right\} = -m \ln c + m \ln t$$

When using a regression routine on this equation, the left hand side, evaluated at each readout time with $\hat{F}(t)$ estimated by the cumulative fraction failures to time t, is the dependent variable. The natural logarithm of each readout time is the independent variable. The resulting line should visually match the points placed at the plotting positions recommended in Chapter 6 (or the slightly modified plotting positions used by AGSS).

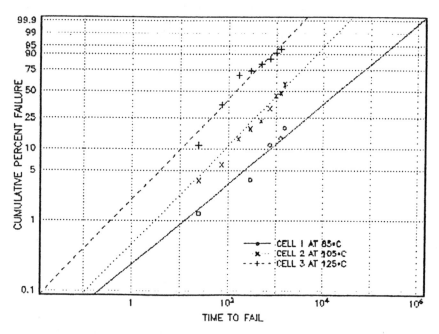

Figure 7.1   Weibull Plot of Table 7.2 Data

The graphical least square estimates are:

85° C cell	$\hat{m} = 0.57$	$\hat{c} = 40194$
105° C cell	$\hat{m} = 0.70$	$\hat{c} = 2208$
125° C cell	$\hat{m} = 0.71$	$\hat{c} = 242$

Maximum likelihood estimates for the same cells are:

85° C cell	$\hat{m} = 0.81$	$\hat{c} = 9775$	acceleration to 105° = 5.6
105° C cell	$\hat{m} = 0.82$	$\hat{c} = 1776$	acceleration to 125° = 7.9
125° C cell	$\hat{m} = 0.68$	$\hat{c} = 221$	acceleration from 85° = 44.2

Note that there are wide disparities between the estimates obtained using maximum likelihood versus those based on the graphical approach. The differences are likely to be largest when there is little data, as is the case in the 85° cell where there were only eight failures. As noted in earlier chapters, the maximum likelihood method will generally work best, although no method can assure accurate estimates when there are only a few failures.

Since the accuracy of the estimate of the shape parameter m plays a critical role in failure rate projections, and we know that under the assumption of *true acceleration* each cell has the same m, it makes sense to try to use the data from across all three cells to estimate a single "best" $\hat{m}$. Maximum likelihood estimation can easily accomplish this by a simple extension of the discussion on writing likelihood equations given in the section on Weibull parameter estimation in Chapter 4. Instead of maximizing three separate LIK equations, one from each cell, form one overall LIK equation by multiplying the three likelihoods together. This LIK equation has four unknown parameters; the different c values for each cell and the one overall m. Maximizing this LIK with respect to the four unknown parameters and the observed data from all the cells, gives a set of estimates based on the hypothesis that all the shapes are the same.

If MLE is done in this manner, constraining all cells to have the same shape parameter, the common $\hat{m}$ is 0.72, and the $\hat{c}$ estimates change to 12,270, 1849, and 229 hr for the 85, 105, and 125° C cells, respectively.

From the plot, or the more sophisticated MLE analysis, it can be seen that the equal slope "true acceleration" assumption is reasonable for these three data cells.

Exercise 7.8

Do you think it is reasonable that the MLE of the best single m in Example 7.2 turned out to be closer to the 0.68 MLE for m in the 125° C cell than the 0.81 and 0.82 MLEs for m in the 85° C and the 105° C cells? Give an explanation for your answer.

A very general quantitative method of testing hypotheses such as equality of shapes or fit of models is described in the next section.

LIKELIHOOD RATIO TESTS OF HYPOTHESES

In Chapter 4, in the section titled Weibull Parameter Estimation, a methodology for writing the likelihood function for any set of data was described. This procedure was generalized in the preceding section by multiplying together the likelihoods (or LIK functions) from different sets or cells of data to form one overall LIK function. If each cell has its own independent parameters, the maximum likelihood estimates will be no different from a situation where each cell is analyzed separately. If, however, one or more parameters are assumed to be the same across several cells, then the resulting MLEs are valid only if the hypothesis of equality is true, and the maximized value of LIK is a measure of "how likely that hypothesis is."

As an example, assume two vendors submit samples of the same kind of component and you have to decide whether they are both equivalent, as far as reliability is concerned. After running high stress life tests on both samples, how would you decide whether the results indicated no significant differences between the vendors or, at the other extreme, that one vendor was significantly better than the other? Clearly this kind of decision can have major financial impacts on the vendors as well as on the reliability of the products in which the components are used. An objective statistical test is needed rather than subjective decisions based on how good each set of data looks on its own.

Depending on the distribution assumptions and the kind of data obtained, there may be many different statistical procedures that will test the hypothesis that both vendors are equal versus that one vendor is significantly better than the other. One method, known as the likelihood ratio test, will generally apply. In particular, in the case of censored life test data, or the case of readout data, it is usually the only method that can be used.

The likelihood ratio test starts out by computing the likelihood value assuming the hypothesis is true (for example, in the case of the two vendors, both cells come from the same distribution). Call this likelihood LIK_1. It is computed by forcing both cells to have the same distribution parameters and maximizing the likelihood of the observed data. Next, compute the maximized likelihood of all the data with each cell allowed to have its own distribution parameters. Call this

LIK_2. It is the best possible likelihood value if the chosen distribution is correct. Clearly, the closer LIK_1 is to LIK_2, the more reasonable is the hypothesis of equality. On the other hand, if the likelihood is much larger with each cell allowed to have its own parameters, then the hypothesis of equality is much less likely.The likelihood ratio test gives us a way to decide *how large a difference is significant*. If we let $\lambda = (LIK_1/LIK_2)$, then, assuming the hypothesis is true, $-2 \times$ ln λ will have approximately a chi-square distribution. The degrees of freedom for the chi-square are equal to how many fewer parameters need to be estimated under the hypothesis underlying the LIK_1 calculation compared to the LIK_2 calculation. In the two vendor case, assuming a two parameter Weibull applies, the number of parameters under LIK_1 is 2, while the number of parameters under LIK_2 is 4. So $-2 \times$ ln λ has approximately a chi-square distribution with $4 - 2 = 2$ degrees of freedom. If the calculated value of $-2 \times$ ln λ exceeds a high percentile value (say 95th or 97.5th) of the chi-square distribution, then there is good evidence for rejecting the hypothesis of vendor equality. Otherwise, we could state that the assumption of equality is not contradicted by the data.

A more general statement of the likelihood ratio test procedure is as follows. Assume we have one or more cells or samples of data. First, we write the likelihood function of all the data, allowing distribution parameters to be unknown and different from sample to sample. We take the natural logarithm of this equation and maximize it with respect to all the unknown parameters and the sample data. Call this maximized log likelihood L_2. Next, we make whatever assumptions we want about the distributional parameters across the data cells. These assumptions may be that some parameters have known values, or that one or more parameters are unknown but equal across several cells, or even that there is an equation or model that determines many of the parameters across the cells (and this model may include stress terms and "acceleration model" parameters).

We then write the likelihood function of all the data under this set of assumptions and again maximize the log likelihood with respect to the new set of unknown parameters and the sample data. Call this second log likelihood L_1. Let $T = -2(L_1 - L_2)$. In addition, let d be the difference between the number of parameters assumed when calculating L_1 and the number of parameters used to find L_2. Then, if the second set of assumptions is correct, T will have a chi-square distribution with d degrees of freedom. If the calculated value of T is too large (as compared to $\chi^2_{d;95}$, $\chi^2_{d;97.5}$, or another preselected high percentile of the chi-square distribution with d degrees of freedom) then we have sample evidence that conflicts with our assumptions of equality.

We can use the likelihood ratio test (or L. R. test) to decide whether a set of Weibull data can be modelled reasonably well by an exponential distribution (just test whether $m = 1$), or whether several cells of Weibull data have the same slope on Weibull graph paper (just test whether all the cells have the same m). We will also use the L. R. test later in this chapter to test whether fitted acceleration models are consistent with the observed data. Of course, good computer software is nearly always necessary in reliability model applications whenever procedures based on maximum likelihood techniques are used.

Example 7.3   Weibull Likelihood Ratio Equal Shape Test

Use the L. R. test to determine whether the assumption of equal shapes for the three different temperature cells of data in Example 7.2 is reasonable.

Solution

The log likelihoods for the 85°, 105°, and 125° cells are -37.17, -79.43, and -80.19, respectively. That gives an L_2 of -196.79. Under the assumption of equal shapes, $L_1 = -197.16$. The chi-square statistic is $2 \times (197.16 - 196.79) = 0.74$. The degrees of freedom are $6 - 4 = 2$. This is around the 30th percentile of the chi-square distribution and, therefore, there is no reason to question the assumption of equal shapes.

Exercise 7.9

Assume the data set called DATA3 in Problem 1.5 at the end of Chapter 1 comes from a Weibull distribution. Use the method of maximum likelihood to estimate Weibull parameters with both m and c unknown. Then assume that $m = 1$ and the distribution is an exponential. Estimate the failure rate and use the L. R. test to see whether an exponential model is reasonable.

Exercise 7.10

Compare DATA3 and DATA4 from Problem 1.5 at the end of Chapter 1. Use the L. R. test to decide whether it is reasonable to assume they both can be modelled by Weibull distributions with the same shape parameter.

LOGNORMAL DISTRIBUTION ACCELERATION

Now we assume that at laboratory stress $F_S(t)$ can be adequately modeled by the lognormal distribution

$$F_s(t) = \Phi\left[\frac{\ln(t/T_{50s})}{\sigma_s}\right]$$

where Φ is the standard normal CDF defined in Chapter 5 and T_{50s} and σ_s are the lognormal parameters for the laboratory stress life distribution.

Making the acceleration transformation of the time scale given by $F_U(t) = F_S(t/AF)$, we find

$$F_U(t) = F_S\left(\frac{t}{AF}\right) = \Phi\left[\frac{\ln\left(\frac{t/AF}{T_{50S}}\right)}{\sigma_S}\right] = \Phi\left[\frac{\ln\left(\frac{t}{AF \times T_{50S}}\right)}{\sigma_S}\right] = \Phi\left[\frac{\ln\left(\frac{t}{T_{50U}}\right)}{\sigma_U}\right]$$

which is again a lognormal distribution with $\sigma_U = \sigma_S = \sigma$, and $T_{50U} = AF \times T_{50S}$, where σ_U and T_{50U} are the use stress lognormal parameters. This result is similar to that obtained for the Weibull: true linear acceleration does not change the type of distribution or the shape parameter. Only the scale parameter is multiplied by the acceleration factor between the two stresses.

Since shape or σ is equivalent to the slope on standard lognormal paper, again we expect different stress cells of data to give rise to nearly parallel lines when plotted on the same graph paper. The ratio of the times to reach any percentile, such as T_{50}, gives an estimate of the acceleration factor between the stresses.

Again, it is true that the parallel lines or equal sigmas result is a consequence of true linear acceleration and a lognormal life distribution model, and not an additional assumption. Also, the relationship between failure rates before and after acceleration is complicated and depends on the particular time point under evaluation. Failure rates must be calculated using the basic definition for $h(t)$ given in Chapter 5.

Example 7.4   Lognormal Multiple Stress Cells

A semiconductor module has been observed to fail due to metal ions migrating between conductor lines and eventually causing a short. Both temperature and voltage affect the time it takes for such failures to develop. It is decided to model the kinetics of failure by conducting a stress matrix experiment using six different combinations of temperature and voltage. The stress cells are: 125° C, 8 volts; 125° C, 12 volts; 105° C, 12 volts; 105° C, 16 volts; 85° C, 12 volts, 85° C, 16 volts. Normal use stress is 25° C and 4 volts. Sample size is 50 per cell. Table 7.3 gives the experimental design matrix for the experiment. No units were run at

TABLE 7.3   Experimental Design Matrix

	8 Volts	12 Volts	16 Volts
85° C		✔	✔
105° C		✔	✔
125° C	✔	✔	

85° C, 8 volts and 105° C, 8 volts because of limited resources and a concern that few failures would occur at the lower stress combinations.

Each cell is read out for new failures at 24, 100, 150, 250, 500, 750, and 1000 hours. Table 7.4 summarizes the failure data.

Assuming a lognormal failure distribution, plot the six cells of data and check whether the equal shape hypothesis of "true acceleration" is reasonable. Estimate stress cell parameters by graphical methods and the method of maximum likelihood.

Solution

Cumulative failure data points from the six stress cells are plotted on lognormal probability paper in Figure 7.2. The lines through the points were obtained by using least squares on the equation

$$\ln t = \ln T_{50} + \sigma \Phi^{-1} \left[F(t) \right]$$

where Φ^{-1} is the inverse of the standard normal distribution, and $F(t)$ is the cumulative fraction that failed up to time t. Here, the natural logarithm of time is the dependent variable and the normal inverse of the CDF estimate is the independent variable.

It can be seen from the graph that all the lines have very similar slopes; only cell 5, with significantly less data than the other cells, has a somewhat different looking σ. Estimates of the cell T_{50} values and sigmas are given in Table 7.5, using the least squares graphical method and maximum likelihood estimation (MLE). In addition, MLEs with sigma assumed to be equal across all cells are shown. These last estimates use the best statistical method possible under the assumption of linear acceleration and lognormality.

TABLE 7.4   Lognormal Stress Failure Data

Readout	Cell 1 125° 8 V	Cell 2 125° 12 V	Cell 3 105° 12 V	Cell 4 105° 16 V	Cell 5 85° 12 V	Cell 6 85° 16 V
24	0	0	0	0	0	0
100	1	5	0	0	0	0
150	6	3	1	1	0	0
250	6	23	3	10	1	1
500	23	13	19	21	1	5
750	9	5	15	10	3	15
1000	3	1	6	5	5	6
TOTAL	48	49	44	47	10	27

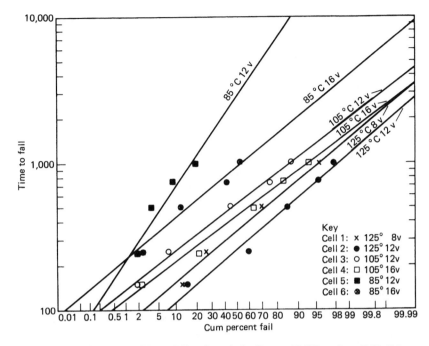

Figure 7.2   Plot of Lognormal Data Cells—Cumulative Percent Fall Data from Table 7.3

Exercise 7.11

Use the L. R. test to check whether the assumption of equal sigmas for the data in Table 7.5 is reasonable.

TABLE 7.5   Lognormal Stress Cell Parameter Estimates

	Graphical Estimates (least squares)		MLE Estimates (separate cells)		MLE Estimates (with one sigma)	
Cell	T_{50}	Sigma	T_{50}	Sigma	T_{50}	Sigma
1	340	0.62	346	0.64	346	0.60
2	254	0.63	244	0.61	244	0.60
3	516	0.57	523	0.55	524	0.60
4	422	0.54	410	0.56	410	0.60
5	2760	1.10	2121	0.88	1596	0.60
6	918	0.61	925	0.60	927	0.60

ACCELERATION MODELS

From the preceding sections we have seen that knowing the scale parameters (either c or T_{50}) for the life distributions at two stresses allows us to immediately calculate the acceleration factor between the stresses. Alternatively, if we already know the acceleration factor between a laboratory stress test and the field use condition, we can convert the results of our test data analysis to failure rate projections at use conditions. Indeed, this practice is often done as an ongoing process monitor for reliability on a lot by lot basis.

But what can be done if the acceleration factor to use conditions is not known, and data can be obtained in a reasonable amount of time only by testing at high stress? The answer is that we must use the high stress data to fit an appropriate model that allows us to extrapolate to lower stresses.

There are many models in the literature that have been used successfully to model acceleration for various components and failure mechanisms. These models are generally written in a deterministic form which says that time to fail is an exact function of the operating stresses and several material and process dependent constants.

Since all times to fail are random events that cannot be predicted exactly, and we have seen that acceleration is equivalent to multiplying a distribution scale parameter, we will interpret an acceleration model as an equation that calculates a distribution scale parameter, or a given percentile, as a function of the operating stress.

For example, if a failure mechanism depends on two stresses, and the associated failure times follow a lognormal model, an equation $T_{50} = G(S_1, S_2)$ that predicts the T_{50} based on the values of these two stresses, is an acceleration model. In the next two sections we shall see several common and useful forms for G.

Before proceeding to these models, however, one point should be stated. Just as, in general, different failure mechanisms follow different life distributions, they will also have different acceleration models. In Chapter 8, we will discuss how, using the competing risk model, we can study each failure mode and mechanism separately and derive the total component failure rate as a sum of the individual failure rates from each mechanism. This "bottoms-up" method is virtually the only way to do acceleration modeling successfully (although a top-down "black-box" approach is also discussed in Chapter 8).

For example, one failure mechanism might involve a chemical reaction and be accelerated by temperature. A component with this type of failure mode might also fail due to metal migration, which could be highly dependent on voltage and humidity in addition to temperature. At the same time, there could also be a mechanical wear-out failure mode dependent on the frequency of *on* and *off* cycles. Each of these modes of failure will follow completely different acceleration models and must be studied separately.

Therefore, when we discuss acceleration models and how to analyze stress cell failure data, we are presupposing the experiments have been carefully designed to produce data from only one failure mechanism, or any other types of failures have been "censored" out of the data analysis. The way we censor them out is by pretending they were removed from test without having failed. (The time of removal to use [i.e., the censoring time] is the time they actually failed due to the other failure mechanism.)

THE ARRHENIUS MODEL

When only thermal stresses are significant, an empirical model, known as the Arrhenius model, has been used with great success. This model takes the form

$$T_{50} = A e^{\frac{\Delta H}{kT}}$$

where

A and ΔH = unknown constants

k = Boltzmann's constant

T = temperature measured in degrees Kelvin at the location on the component where the failure process occurs

For example, in integrated circuits, it is common practice to calculate the internal (called the "junction") temperature of the device as the ambient (use or oven stress) temperature plus the incremental self-heating associated with the power dissipation of the device. (See Tummala and Rymaszewski, 1989, for a discussion of thermal considerations in microelectronics packaging.) Boltzmann's constant has the value 8.617×10^{-5} in eV/°K or 1.380×10^{-16} in ergs/°K. Temperature in degrees Kelvin is obtained by adding 273.16 to temperature in degrees Centigrade.

The parameter ΔH (pronounced "delta H") plays the key role in determining temperature acceleration. ΔH is sometimes referred to as the "activation energy," and its value depends on the failure mechanism and the materials involved. (See the section on The Eyring Model for a physical explanation of activation energy.) Alternatively, E_A is often used instead of ΔH in the reliability literature. Note that we can write the Arrhenius model in terms of T_{50}, or the c parameter (when working with a Weibull), or the $MTTF = 1/\lambda$ parameter (when working with an exponential), or any other percentile of the life distribution we desire. The value of the constant A will change but, as we are about to see, this constant will have no effect on acceleration factors. For convenience, we will use c or T_{50} values in this chapter, but the formulas work equally well with other percentiles (e.g., at 1 percent, 10 percent, etc.) used in place of the characteristic life or the median.

We solve for the acceleration factor between temperature T_1 and temperature T_2 by taking the ratio of the times it takes to reach any specified CDF percentile. In other words, the acceleration factor AF between stress 1 and stress 2 is defined to be the ratio of time it takes to reach $P\%$ failures at stress 1 divided by the time it takes to reach $P\%$ failures at stress 2. The assumption of true acceleration makes this factor the same for all percentiles. Using the Arrhenius model and the 50th percentile, we have

$$AF = \frac{T_{50_1}(at\ T_1)}{T_{50_2}(at\ T_2)} = \frac{Ae^{\Delta H/kT_1}}{Ae^{\Delta H/kT_2}}$$

from which

$$AF = e^{(\Delta H/k)\,[\,(1/T_1) - (1/T_2)\,]}$$

This result shows that knowing ΔH alone allows us to calculate the acceleration factor between any two temperatures. Conversely, if we know the acceleration factor, we can calculate ΔH as follows:

$$\Delta H = \frac{k\ln\left(\dfrac{T_{50_1}}{T_{50_2}}\right)}{\left(\dfrac{1}{T_1}\right) - \left(\dfrac{1}{T_2}\right)} = \frac{k\ \ln(AF)}{\left(\dfrac{1}{T_1}\right) - \left(\dfrac{1}{T_2}\right)}$$

This last equation shows us how to estimate ΔH from 2 cells of experimental test data consisting of times to fail of units tested at temperature T_1 and times to fail of units tested at temperature T_2. All we have to do is estimate a percentile, such as T_{50}, in each cell, then take the ratio of the corresponding times and use the preceding equation to estimate ΔH. This procedure is valid for any life distribution. It is useful to have a quick table look-up of acceleration factors as a function of temperatures and ΔH. Table 7.6 shows the acceleration factor from a low T_1 to a high T_2 for two values of ΔH; the acceleration factor if ΔH is 0.5 is shown in normal print, and the acceleration factor for $\Delta H = 1.0$ is in parentheses.

Table 7.7 and Figures 7.3 and 7.4 provide a two-step procedure for calculating acceleration factors for a wide range of ΔH. First the T_1 and T_2 values are used to find a TF value from Table 7.7. Then this TF value and the appropriate ΔH line are used in Figure 7.3 or 7.4 to find the acceleration factor. The acceleration factor can also be calculated from $AF = e^{\Delta H(TF)}$.

TABLE 7.6   Arrhenius Acceleration Factors for $\Delta H = 0.5$ and ($\Delta H = 1.0$)

Lower Temperature T_1 (°C)	Higher Temperature T_2 (°C)									
	65	75	85	95	105	115	125	135	145	155
25	10.0 (100)	16.4 (268)	26.1 (679)	40.5 (1637)	61.4 (3767)	91.1 (8305)	133 (17,597)	190 (35,938)	266 (70,933)	368 (135,628)
35	5.3 (28)	8.7 (76)	13.9 (192)	21.5 (463)	32.6 (1065)	48.5 (2348)	70.5 (4976)	101 (10,163)	142 (20,059)	196 (38,354)
45	2.9 (8.6)	4.8 (23.2)	7.7 (58.8)	11.9 (142)	18.1 (326)	26.8 (719)	39.0 (1524)	55.8 (3111)	78.4 (6141)	108 (11,743)
55	1.7 (2.8)	2.8 (7.6)	4.4 (19.3)	6.8 (46.6)	10.4 (107)	15.4 (237)	22.4 (501)	32.0 (1021)	44.9 (2020)	62.2 (3865)
65		1.6 (2.7)	2.6 (6.8)	4.0 (16.4)	6.1 (37.7)	9.1 (83.1)	13.3 (176)	19.0 (360)	26.6 (710)	36.8 (1358)
75			1.6 (2.5)	2.5 (6.1)	3.8 (14.1)	5.6 (31.0)	8.1 (65.7)	11.6 (134)	16.3 (265)	22.5 (507)
85				1.6 (2.4)	2.4 (5.5)	3.5 (12.2)	5.1 (25.9)	7.3 (52.9)	10.2 (104)	14.1 (200)
95					1.5 (2.3)	2.3 (5.1)	3.3 (10.8)	4.7 (22)	6.6 (43.3)	9.1 (82.9)
105						1.5 (2.2)	2.2 (4.7)	3.1 (9.5)	4.3 (18.8)	6.0 (36)
115							1.5 (2.1)	2.1 (4.3)	2.9 (8.5)	4.0 (16.3)
125								1.4 (2.0)	2.0 (4.0)	2.8 (7.7)
135									1.4 (2.0)	1.9 (3.8)
145										1.4 (1.9)

TABLE 7.7   *TF* Values for Calculating Acceleration (Use *TF* value and Figure 7.3 or 7.4 to calculate acceleration.)

Lower Temperature T_1 (°C)	Higher Temperature T_2 (°C)									
	65	75	85	95	105	115	125	135	145	155
25	4.6	5.6	6.5	7.4	8.2	9.0	9.8	10.5	11.2	11.8
35	3.3	4.3	5.3	6.1	7.0	7.8	8.5	9.2	9.9	10.6
45	2.2	3.1	4.1	5.0	5.8	6.6	7.3	8.0	8.7	9.4
55	1.0	2.0	3.0	3.8	4.7	5.5	6.2	6.9	7.6	8.3
65		0.99	1.9	2.8	3.6	4.4	5.2	5.9	6.6	7.2
75			0.93	1.8	2.6	3.4	4.2	4.9	5.6	6.2
85				0.98	1.7	2.5	3.3	4.0	4.6	5.3
95					0.88	1.6	2.4	3.1	3.8	4.4
105						0.79	1.5	2.3	2.9	3.6
115							0.75	1.5	2.1	2.8
125								0.71	1.4	2.0
135									0.68	1.3
145										0.65

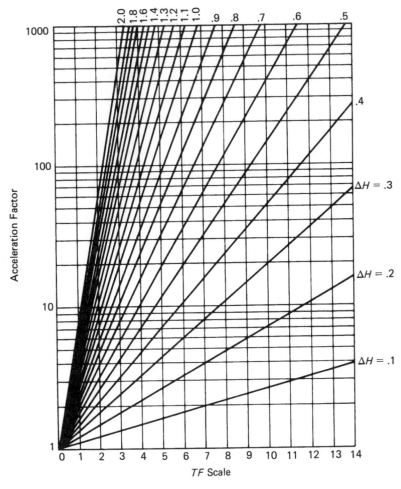

Figure 7.3   Acceleration Factor Graph. Choose *TF* value from Table 7.6 and find acceleration for a given Δ*H*.

Since the *TF* values in Table 7.7 are calculated from $TF = [(1/T_1) - (1/T_2)] \times 11{,}605$, where 11,605 is just the reciprocal of k, Boltzmann's constant, we can rewrite the equation for Δ*H* as $\Delta H = \ln(T_{50_1}/T_{50_2})/TF$. This is a convenient form for using Table 7.7 to calculate Δ*H* when two cells of data have been run.

Example 7.5   Arrhenius Acceleration

What is the acceleration from a use condition temperature of 35° C to a laboratory testing stress of 125° C if Δ*H* is 1.0? Use the two-step procedure involving

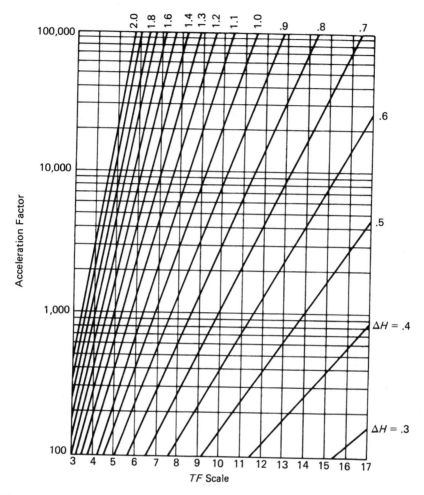

Figure 7.4   Acceleration Factor Graph. Choose TF value from Table 7.6 and find acceleration for a given ΔH.

Table 7.7 and Figure 7.3 or 7.4 to check the result obtained from Table 7.6. What would the acceleration be if ΔH is only 0.7?

Solution

Table 7.6 gives a direct look-up value for the acceleration factor when ΔH is 1.0 of 4976. Using Table 7.7, a TF value of 8.5 is obtained for the two temperatures. Going up from this value on the TF scale of Figure 7.4 to the line for $\Delta H = 1.0$, and then going over to the vertical acceleration scale, yields the more approximate estimate of 5000. For a ΔH of 0.7, the acceleration corresponding to the same TF value is about 390 (from Figure 7.3).

Example 7.6   Calculating ΔH with Two Temperature Cells

Use the 85° C and 125° C cells of Weibull data from Example 7.2 and Table 7.2 to calculate an estimate of ΔH.

Solution

Using the graphical estimates for c from Example 7.2, and the TF value of 3.3 from Table 7.7, we calculate ΔH to be

$$\Delta H = \frac{\ln\left(\dfrac{40194}{242}\right)}{3.3} = 1.55$$

The same calculation using the generally more accurate MLE values of 9775 and 221 for c estimates yields $\Delta H = 1.15$. Using the constrained equal shape MLE values of 12,270 and 229 gives a ΔH estimate of 1.21. Such differences between methods of estimation are not uncommon and will be seen again in later examples in this chapter.

Exercise 7.12

Use the MLE estimates of T_{50} (with 1 σ) from Table 7.5 to estimate ΔH assuming an Arrhenius model applies. First use the 85° and 125° 12 volt cells, then repeat with the 105° and 125° 12 volt cells and finally try using the 105° and 85° 16 volt cells.

Exercise 7.13

Calculate ΔH using the temperatures and acceleration factor given in Example 7.1.

Exercise 7.14

Use the ΔH calculated in Exercise 7.13 to find the new operating temperature needed for the components in Example 7.1 to achieve the goal of no more than 10 percent component failures by 40,000 hr (see Exercise 7.6).

Exercise 7.15

Determine the effect on the acceleration factor of doubling the activation energy from 0.65 eV to 1.3 eV when the stress temperature is 125° C and the use temperature is 45° C.

ESTIMATING ΔH WITH MORE THAN TWO TEMPERATURE CELLS

In the last example, we made use of only the 85° C and the 105° C cells from Example 7.2. If we had used the graphical c values from the 85° and the 105° cells, we would have calculated a ΔH estimate of [ln (40,194/2208)]/1.7 = 1.71. Or, if we had used graphical c values from the 105° and the 125° cells, the estimate would be [ln (2208/242)]/1.5 = 1.47. Which of these three estimates is the correct (or best) one to use?

The answer is none of them. We need a procedure that makes use of all the cells simultaneously to derive a ΔH estimate. This analysis can be done either graphically or using regression techniques or, even better, using the MLE method. The key equation is the Arrhenius model written in logarithmic form

$$\ln c = \ln A + \Delta H(1/k\mathrm{T})$$

This equation is linear in the (independent) variable $\{1/k\mathrm{T}\}$ and the (dependent) variable $\ln c$, with slope ΔH and intercept $\ln A$. So a plot of the data points, in the form $(\ln c, 1/k\mathrm{T})$ from each cell, should line up almost in a straight line. The slope of this line is an estimate of ΔH. Slope (as discussed in Chapter 6) is calculated by taking any two points on the line and calculating the ratio of the increase in height going from the left to the right, over the distance between the points along the x-axis (or the $1/k\mathrm{T}$ axis). If semilog paper is not used, units must be adjusted because of the logarithm calculation so that one decade on the y-axis is equivalent to 2.3 units on the x-axis.

Special graph paper can be constructed where the y-axis is logarithmic and the x-axis is calibrated in terms of reciprocal temperature multiplied by Boltzmann's constant, simplifying the plotting of cell data points. The point for a cell is located by looking up c (or T_{50} for the lognormal) on the y-axis, and the cell temperature on the x-axis. Various commercial brands of this paper are available, such as those made by Keuffel and Esser.

Once the points are plotted, the best straight line through the points represents the fitted Arrhenius model. This plot can be used as an instant calculator for finding the c corresponding to any temperature by going up from the temperature to the line, and then to the left to the c value on the y-axis. The line can be drawn subjectively by eye or, better yet, fitted using least squares on the log equation. By looking at how well the points follow a straight line, one also obtains a subjective evaluation of how appropriate the Arrhenius model is for the experimental data.

Example 7.7   Calculating ΔH with Three Temperature Cells

Use the three Weibull cells of example 7.2 to obtain a graphical fit of the Arrhenius model equation and an estimate of ΔH using all the cells.

Solution

The points are plotted in Figure 7.5, along with the least squares (regression) line. For the regression, the independent variable $1/kT$ values are 32.40, 30.69, and 29.15, corresponding to the temperatures of 85°, 105°, and 125°. The matching dependent variable values are ln $\hat{c}$ or 10.60, 7.70, and 5.49, corresponding to the graphical c estimates of 40,194, 2208, and 242. The ΔH estimate turns out to be 1.58. The actual points fit the line very closely.

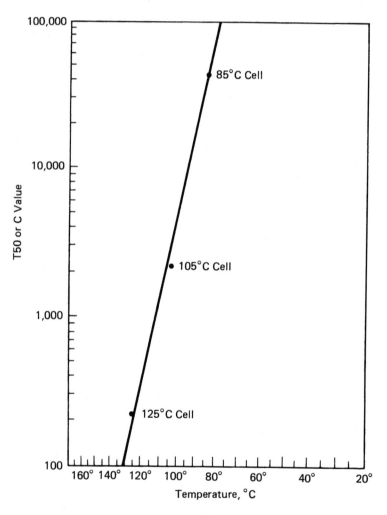

Figure 7.5   Arrhenius Plot

If a line had been drawn by eye through the plotted data points in Figure 7.5, the slope could be calculated by picking two convenient points at opposite ends of the line and using the formula for ΔH.

Since it is generally more accurate to use maximum likelihood estimates for $\hat{c}$, this would lead to an independent variable with values of 9.18, 7.47, and 7.7. The resulting ΔH estimate is 1.16. If the still better estimates from constrained maximum likelihood are available (where m is forced equal across the three cells), the estimate of ΔH is 1.22. Finally, if software is available to do likelihood estimation with the Arrhenius model assumption built into the likelihood equations, a direct MLE estimation of ΔH would yield the value 1.27.

Now that we have analyzed the data from Example 7.2 by several methods and fit an Arrhenius model to it, is how well did we actually do? In a real analysis, we may not know the answer to that question for years, if at all. In this case, however, all the data points were obtained by simulation from known Weibull distributions and a known Arrhenius equation. Thus we can display the results of our analysis and compare them to the "real" values. This comparison is shown in Table 7.8. The results show that all the methods based on maximum likelihood estimation did roughly as well and were considerably better than the completely graphical analysis. This result agrees with what we would expect from theory. We would also expect that, over the long run, using the fully MLE approach (the Full MLE column in Table 7.8) would consistently yield the best results.

The use of graph paper was illustrated for Weibull data and the c parameter in Figure 7.5, but the methodology is really independent of any particular distribution. This approach can be applied to the T_{50} or any other percentile. Since percentiles can be estimated, as described in Chapter 1, without knowing what the distribution is, it is not necessary to assume a form for the failure mechanism life

TABLE 7.8   Weibull Data Analysis Summary

	Estimation Method			
Parameter	Graphical Method	MLE Plus Regression	Equal Shape MLE Plus Regression	Full MLE
$\Delta H = 1.20$	1.58	1.16	1.22	1.27
$\ln A = -29.52$	-40.50	-28.35	-30.06	-31.48
$m = 0.70$	0.66 (3 cell avg.)	0.77 (3 cell avg.)	0.72	0.72
c at 85° (11,424)	37,597	10,706	13,708	14,659
c at 125° (1,461)	2,542	1,466	1,616	1,670
c at 125° (230)	224	244	246	237

distribution in order to fit an Arrhenius model. An analysis without assuming any form of distribution is called a non-parametric analysis. What we gain from assuming a suitable distribution model is accuracy (provided the assumption is approximately valid) and the ability to project results to percentiles that cannot be observed using the small sample sizes reliability analysts typically work with. Hence, parametric models are the general rule.

Example 7.8   Likelihood Ratio Test of an Arrhenius Model

Use the L. R. test to check on whether the Arrhenius model assumption is consistent with the data in the three temperature cells of Table 7.2. In other words, can we accept that the c parameters for the three cells vary with temperature according to the Arrhenius equation.

Solution

A calculation of the log likelihood under the equal shapes and Arrhenius model assumption yields -197.3. Assuming there is true acceleration (and therefore the shapes are equal) but some other model applies, the log likelihood is 197.16 (as previously given in Example 7.3). Therefore, the test statistic T is $2(197.30 - 197.16) = 0.28$. Since we replaced three "c" parameters (one for each stress cell) by the two Arrhenius parameters A and ΔH, the degree of freedom for T is 1. A chi-square table lookup shows that 0.28 is around the 40th percentile, and therefore the data is consistent with the Arrhenius model.

Exercise 7.16

Analyze the data in Table 7.2 and fit an Arrhenius model, this time assuming a lognormal distribution applies. Compare the new estimate of ΔH to those given in Table 7.8. Use the results of both the Weibull analysis and the lognormal analysis to project the cum population failures at 100,000 hr for a use stress of 25° C. Note that the lognormal cum fail projection is several orders of magnitude smaller than the Weibull projection. This shows how important it is to chose the correct life distribution model. Based on the plots of the data under the two distribution assumptions, would you have preferred a Weibull model? Compare the log likelihoods of the two distribution assumptions. There is no likelihood ratio test, since the two models have the same number of unknown parameters and hence there are no degrees of freedom for the test. However, a larger log likelihood is qualitative evidence to help support a choice of a model.

For additional information on the Arrhenius model, including a useful table of typical activation energies for a variety of microelectronic failure mechanisms, see Pecht, Lall and Hakim (1992).

THE EYRING MODEL

The Arrhenius model is an empirical equation that justifies its use by the fact that it "works" in many cases. It lacks, however, a theoretical derivation and the ability to model acceleration when stresses other than temperature are involved. Example 7.4, with both temperature and voltage playing key roles, could not be handled by the Arrhenius model.

The Eyring model offers a general solution to the problem of additional stresses. It also has the added strength of having a theoretical derivation based on chemical reaction rate theory and quantum mechanics. In this derivation, based on work by Eyring, the parameter ΔH has a physical meaning. It represents the amount of energy needed to move an electron to the state where the processes of chemical reaction or diffusion or migration can take place. For further details see Eyring, Glasstones, and Laidler (1941).

The Eyring model equation, written for temperature and a second stress, takes the form

$$T_{50} = \left[AT^\alpha e^{\frac{\Delta H}{kT}} \right] \left[e^{\left(B + \frac{C}{T} \right) S} \right]$$

The term in the first bracket is the temperature term, while the second bracket contains the general form for adding any other type of stress. In other words, if a second nonthermal stress was needed in the model, a third bracket multiplier exactly the same as the second, except for replacing B and C by additional constants D and E, would be added to the equation. The resulting Eyring model for temperature and two other stresses would then be:

$$T_{50} = AT^\alpha e^{\frac{\Delta H}{kT}} e^{\left(B + \frac{C}{T} \right) S_1} e^{\left(D + \frac{E}{T} \right) S_2}$$

As before, T_{50} can be replaced by c or any other percentile, or even the mean time to fail.

It is interesting to look at how the first term, which models the effect of temperature, compares to the Arrhenius model. Except for the T^α factor, this term is the same as the Arrhenius. If α is close to zero, or the range over which the model is applied is small, the term T^α has little impact and can be absorbed into the A constant without changing the practical value of the expression. Consequently, the Arrhenius model is successful because it is a useful simplification of the theoretically derived Eyring model.

When we try to apply the Eyring model to life test data with several critical stresses, however, we often run into several difficulties. The first of these is the general complexity of the model. The temperature term alone has three parameters to estimate: A, α, and ΔH. Each additional stress term adds two more unknown constants, making the model difficult to work with.

As a minimum, we need at least as many separate experimental stress cells as there are unknown constants in the model. Preferably, we have several more beyond this minimal number, so that the adequacy of the model fit can be examined. Therefore, a three stress model should have about nine or ten cells. Obviously, designing and conducting an experiment of this size is not simple.

Another difficulty is finding the proper functional form, or units, with which to express the nonthermal stresses. Temperature is in degrees Kelvin. But how should voltage or humidity be input, for example? The theoretical model derivation doesn't specify, so the experimenter must either work it out by trial and error or derive an applicable model using arguments from physics and statistics.

As an example, consider one form of the temperature and voltage acceleration model given by

$$T_{50} = A e^{\frac{\Delta H}{kT}} V^{-B}$$

This expression may not look much like a two stress Eyring model but, in fact, it is. Make the substitutions $\alpha = 0$, $C = 0$ and $S_1 = \ln V$ (where V is volts), and the two-stress Eyring model reduces to this simple acceleration equation with three unknown parameters to estimate.

Example 7.9   Two-Stress Acceleration Model

Use the results of the analysis of the six stress cells of lognormal data from Example 7.4 to find the three unknown parameters in the acceleration model for temperature and voltage given above. Estimate the average failure rate at a typical use condition of 25° C and 4 volts.

Solution

There are several ways to proceed, depending on the statistical tools (usually computer programs) available to the analyst. The simplest approach is to look at the three 12 volt cells, with temperatures of 125°, 105°, and 85°. Since voltage is constant, and the temperature portion of the model is Arrhenius, we can analyze just these three cells with the methods described in Example 7.7. This procedure gives us estimates of A and ΔH. To estimate B, we note that there are pairs of cells at each temperature which only differ in the voltage used. Taking the ratio of the T_{50} values for any pair cancels out the Arrhenius portion of the model and yields a simple equation for B.

For example, using the graphical estimates for T_{50} from the 125° cells yields $(254/340) = (8/12)^B$. Taking logarithms of both sides and solving for B gives an estimate of 0.72. The same method with the two 105° cells yields 0.70. The 85° cells give the much higher B estimate of 3.83. An average of these three numbers gives 1.75.

The numbers obtained in this calculation should, however, be disquieting to the analyst. The 125° and 105° cells yielded consistent estimates in the 0.7 range. These cells had most of the failures (a total of 188). The 85° cells, with only 37 failures between them, gave a very different estimate. A practical, ad hoc procedure to correct for this would be to weight the estimates by the numbers of failures and then divide the sum by the total number of failures. This weighted method would result in a B of 0.92.

A better way to fit the acceleration model would be to use (multiple) regression analysis on the all the cell T_{50} values simultaneously (see Draper and Smith, 1981). First we write the model equation in logarithmic form as

$$\ln T_{50} = \ln A + \Delta H/k\mathrm{T} + B(-\ln V)$$

which can be rewritten in the standard multiple regression format as

$$Y = a + \Delta H X_1 + BX_2$$

after making the obvious substitutions. Then we input the Y, X_1, and X_2 vectors to the regression program to get the model estimates.

The regression procedure can be improved by using weighted regression, with the number of failures in a cell used as the weight for that cell's $\ln T_{50}$.

Once we have estimated the model parameters, straight substitution gives the T_{50} for the use condition of 25° and 4 volts. Combined with the best overall sigma estimate, this estimate can be used to calculate the lognormal CDF at 40,000 hr and then the average failure rate over the first 40,000 hr.

As in Example 7.7, all the answers are known for this data because it was simulated from known model parameters and lognormal distributions. Table 7.9 shows the results we would have obtained using regression and weighted regression on graphical or MLE cell T_{50} estimates. The weighted sigma estimates were obtained by weighting the cell sigma estimates shown in Table 7.5 by the cell number of failures. The column labeled Full MLE used a software package (AGSS) that can embed a wide range of acceleration models within the likelihood equations and then use the method of maximum likelihood estimation to directly estimate all the parameters of interest, including ΔH and B.

Several very important lessons can be learned from the results in Table 7.9. First, note how poorly the straight unweighted graphical analysis did. This result was caused by the high T_{50} estimate in the lowest stress cell (85° C, 12 V), based

TABLE 7.9   Lognormal Data Analysis Summary

	Estimation Method			
Parameter	Graphical + Regression	MLE + Regression	Full MLE	True Value
ΔH	0.69 0.57 wtd.	0.63 0.56 wtd.	0.55	0.55
B	1.57 1.07 wtd.	1.40 1.08 wtd.	1.07	1.00
ln A	−10.76	−9.57	−7.83	−8.09
σ	0.68 avg.	0.64 avg.	0.60	0.60
Cell 1 T_{50}	340	346	360	348
Cell 2 T_{50}	254	244	233	232
Cell 3 T_{50}	516	523	542	542
Cell 4 T_{50}	422	410	399	407
Cell 5 T_{50}	2760	2121	1384	1391
Cell 6 T_{50}	918	925	1019	1043
Use T_{50} (25° C, 4 V)	968,719 199,460 wtd.	489,323 177,187 wtd.	157,940	150,673
Use AFR @ 40K hr in %/K	0.0000035 0.012 wtd.	0.00011 0.018 wtd.	0.028	0.034

Weighted (wtd.) estimates are those obtained using a weighted regression procedure to fit the acceleration model.

on only 10 failures (nearly twice the true value). Consequently, we obtained high ΔH and B estimates and a projected use T_{50} more than six times too high. As a result, the use condition average failure rate estimate was optimistic by an incredible four orders of magnitude! This example underlines the dangers inherent in fitting models and using them to extrapolate way beyond the range of the data. We extrapolate because we have no choice, but we must appreciate the risks involved. If we can be this far off, using simulated data that really follows the assumed models and has no "maverick" points, what can happen with real data?

On the other hand, improving our methods of estimation and model fitting paid handsome dividends. Merely using the cell numbers of failures as weights for a weighted regression program improved the graphical estimation to a bottom line average failure rate estimate less than three times better than the true value. Weighted regression, starting with MLE values, gave a result within two times, while the full MLE method was only off by about 20 percent. Moreover, the better methods were right on the money for ΔH and σ, and very close for B.

This improvement, from four orders of magnitude off to only 20 percent error, shows the value of using the best techniques of estimation available. Programs may have to be developed or purchased, and the analysis becomes more complicated and less intuitive, but the increase in precision is worth the effort. At the very least, using a weighted regression approach with graphical estimates can provide significant improvements. A weighted regression will not let cells with few failures have as much influence as cells with better data.

It should be recalled, however, that estimating distribution and model parameters from random data carries no absolute guarantees that apply on a case-by-case basis. If we use larger sample sizes, we will generally have more accurate results. The same is true with using "better" statistical techniques. The only assurance we have is that, in the long run, over many sets of data, the better technique will prove itself. As an example of how in any given analysis the methodology might not matter, the reader might want to try redoing the Weibull/Arrhenius analysis summarized in Table 7.8 using weighted regression on the graphical estimates. The ΔH estimate improves only slightly, going from 1.58 to 1.53.

Exercise 7.17

Use an MLE program that will estimate one σ across all the lognormal stress cells in Table 7.5 to estimate cell T_{50} values. Fit the two-stress acceleration model used in Example 7.9 using weighted regression with these T_{50} estimates. Good weights to use are the reciprocals of the asymptotic variances of the T_{50} estimates (usually given by any good MLE program). Show that with these weights, the estimate of the AFR at 40,000 hr is only off by 3 percent. (This degree of accuracy is a chance result for this set of data).

Exercise 7.18

Use the L. R. test to check whether the equal slope assumption across all cells (required for "true" acceleration) is reasonable. Then use the L. R. test to check whether the fit of the chosen acceleration model can also be accepted.

Exercise 7.19

Redo the analysis of the data in Table 7.5, this time assuming a Weibull life distribution applies. How do the estimates of ΔH and B compare to those shown in Table 7.9? How does the Weibull $AFR(40,000)$ projection compare to the lognormal estimate? Use comparisons of the log likelihoods and the cell data plots to decide which model is a better choice.

OTHER ACCELERATION MODELS

There are many other models, most of which are simplified forms of the Eyring, that have been successful. A model known as the Power Rule model has been used for paper impregnated capacitors. It has only voltage dependency, and takes the form AV^{-B} for the mean time to fail (or the T_{50} or the c parameter). This equation is similar to the model discussed in Example 7.7, without the temperature term. It is also referred to as the Inverse Power Rule, since lifetime varies inversely with voltage raised to the power B.

Another way to model voltage is to have a term such as Ae^{-BV}. This kind of term is easy to work with after taking logarithms.

Humidity plays a key role for many failure mechanisms, such as those related to corrosion or ionic metal migration. The most successful models including humidity have terms such as $A(RH)^{-B}$ or $Ae^{-B(RH)}$, where RH is relative humidity.

A useful model for electromigration failures uses current density as a key stress parameter:

$$T_{50} = AJ^{-n}e^{\frac{\Delta H}{kT}}$$

with J representing current density. This mechanism produces open short failures in metal thin film conductors due to the movement of ions toward the anode at high temperature and current densities. A typical ΔH value is between 0.5 and 1.2 electron volts, while $n = 2$ is common. The lognormal life distribution adequately models failure times, with σ normally in the 0.5 to 1.5 range.

Models for mechanical failure due to cracks and material fatigue or deformation often have terms relating to cycles of stress or changes in temperature or frequency of use. The (modified) Coffin-Manson model used for solder cracking under the stress of repeated temperature cycling (as an electronic component is powered and unpowered) is an example of such a model. As published by Landzberg and Norris (1969), this model takes the form

$$N_f = Af^{-\alpha}\Delta T^{-\beta}G(T_{max})$$

where

N_f = the number of cycles to a given percent failures

f = the cycling frequency

ΔT = the temperature range

$G(T_{max})$ = an Arrhenius factor term evaluated at the maximum temperature reached in each cycle

Typical values for α and β are 2 and 1/3, respectively, while the ΔH activation energy term in $G(T_{max})$ is around 1.25.

A trial-and-error approach to model building, using something like a full Eyring and working down, based on fitting to data, should only be tried as a last resort. An experiment based on this approach is likely to be costly and unsuccessful. It is much better to have a model in mind before designing the experiment. This model would either come from a theoretical study of the mechanism or a search of the literature. The simplest model that can be found or derived should be used for as long as it matches experimental data and makes predictions that have not been contradicted by experience.

Exercise 7.20

Consider the two forms of models described in this chapter for voltage acceleration. Express the acceleration factors for each of these models in terms of the unknown parameters.

DEGRADATION MODELS

In the preceding examples, many test failures in several different stress cells were needed to estimate the life distribution and model parameters. In certain cases, however, it is possible to do an analysis even without actual failures. This occurs when there is a measurable product parameter that is degrading over time towards a level that is defined to be a "failure" level.

For example, a component may start the test with an acceptable resistance value reading. Over time the resistance reading "drifts." Eventually, it reaches a certain unacceptable value or undergoes an unacceptable percent change, and the part is considered to have failed. At every test readout, the resistance reading can be measured and recorded, even though failure has not yet occurred. If we call each measurement a data point, a test cell of n components yields n data points at every readout time, even though few or none of the components may fail during the test.

We need one key assumption to make use of this data. There has to be some function of the measurable parameter that is changing linearly with time. (In some cases, it is more convenient to have the change be linear with log time or time to an exponent, but we will restrict ourselves to the simple case in this treatment.) For example, let Q_t be a threshold voltage level which degrades over time according to the formula $Q_t = Q_0 e^{-Rt}$. Then, the function $G(Q_t) = \ln Q_t = \ln Q_0 - Rt$ will change linearly with time.

If we call Q_t the measurable product parameter at time t, and the appropriate function $G(Q_t)$, then the linear change assumption implies

$$G(Q_t) = G(Q_0) + R(S)t \text{ or } G(Q_t) = I + R(S)t$$

where $R(S)$ is the degradation slope at stress S, and I is the initial value of G at time zero [or $G(Q_0)$]. Of course, there will also be error introduced in every observation due to experimental error and component to component differences. These types of errors will be discussed as we look at ways of estimating $R(S)$ from data and using these estimates to fit acceleration models. We will look at two ways we can use degradation data to fit acceleration models.

Method 1

This method is similar to other previously described graphical techniques in that it is informal but easy to understand and carry out without sophisticated statistical tools.

On a sheet of graph paper, let the y-axis represent $G(Q_t)$ and the x-axis represent time. Draw a horizontal line across the paper at the $G(Q_t)$ value that is defined to be a failure. Next plot the data points corresponding to the $G(Q_t)$ readings for a test unit at each readout. If the right $G(Q_t)$ has been chosen, these points should line up approximately on a straight line. They will not be exactly on a line because of experimental measurement errors, which we assume are independent and identically distributed. Fit a line through the points, either by eye or by means of a regression program. Then extend the line until it crosses the failure line and note the corresponding time. This value is the derived time of fail for the first unit.

Repeat this procedure for every unit in every stress cell. The lines will have different slopes because of component to component variation in degradation slope — this variation is in addition to measurement error and is the reason component lifetimes vary in a population. When this part of the analysis is complete, the derived failure times make up a data set similar in appearance to a life test experiment where every unit was tested until failure. These derived failure times can be used to estimate life distribution parameters. Then we can fit an acceleration model, using the methods we have already discussed.

One problem with this procedure is that the many readout measurement errors introduce additional variability into the data, which might inflate the measurement of the life distribution shape parameter. For this reason, it is good to have a σ or m measurement based on actual failures in one high stress cell, if possible.

Method 2

First, we place the additional restriction that $G(Q_0)$ be zero. That is the same as saying that the initial value I is zero, and only the amount of degradation over time is important — not a unit's starting point. Typically, this assumption is valid, since very common functions for $G(Q_t)$ are percent change or absolute change from the initial time zero value. Call D the failure level value for $G(Q_t)$ that signals us that a unit no longer meets functional requirements. D is the "distance" to go until failure is reached. The model now is of the form $D = R \times t_f$, or, *distance to failure equals rate of degradation multiplied by time to failure.* Solving for time to failure, $t_f = D/R(S)$. Consequently, time to failure is proportional to the reciprocal of the stress dependent slope.

The lognormal life distribution is a common model for degradation failures (see the section in Chapter 5 on "Lognormal Distribution Areas Of Application"). But the reciprocal of a lognormal distribution is also lognormal (with the same σ). This property means that the random degradation slopes of units operating at the same temperature T will have a lognormal distribution. In other words, each unit can be considered to have its own degradation slope chosen from a population of slopes following a lognormal model. The shape parameter for $R(T)$ is the same as for t_f, and the median $R_{50}(T)$ is related to T_{50} by $R_{50}(T) = D/T_{50}$.

If we now assume T_{50} varies with temperature according to an Arrhenius relationship, we can write

$$R_{50} = Ae^{\frac{-\Delta H}{kT}}$$

where A is a the constant term, replacing D.

To use the Arrhenius equation to estimate ΔH, we need estimates of $R_{50}(T)$ for each temperature cell. In the cell with temperature T_1, we calculate the set of slopes $[G(Q_{tmax})/t_{max}]$, one for each component, and treat this set as if it were a sample of lognormal observations. Here, t_{max} is the last readout time in the cell and $G(Q_{tmax})$ will generally be the largest amount of degradation observed for each component (although there may be exceptions to this rule in low-stress cells). By dividing the largest degradation amount by the largest readout time, the experimental or observational error should be significantly reduced, leaving a good slope observation for that component. Earlier readout $G(Q_t)$ measurements may have errors that are much more significant when compared to the amount of degradation, so we are better off not using them.

To get $\hat{R}_{50}(T_1)$ for a temperature cell with n units on test, we analyze the sample of n slopes using standard lognormal methods. Since the sample is complete, the simplest procedure would be to compute the sample mean and standard deviation of the natural logarithms of the slopes. The sample mean is $\ln \hat{R}_{50}(T_1)$ and the standard deviation s is an estimate of the life distribution σ (which is equal to the degradation slope σ).

If there are j temperature cells, and we let $y_1 = \ln \hat{R}_{50}(T_1)$, $y_2 = \ln_{50}(T_2)$, and so on with $y_j = \ln \hat{R}_{50}(T_j)$, then

$$y_j = a + bx_j + \varepsilon_j$$

where $a = \ln A$, $b = -\Delta H$, $x_j = 1/(kT_j)$ and ε_j represents random, independent errors that we will also assume are identically distributed (which should be approximately true if all the stress cells have the same number of test components).

Standard regression programs can be used to estimate a and b, (with confidence bounds, if we further assume the ε_j terms are normally distributed). A use $\hat{T}_{50}$ can then be projected by dividing D (the "distance" to fail) by (T_{use}). As was the case with Method 1, a use shape parameter should be estimated from actual failure data. Weighted regression, using the square root of the cell sample sizes, should be used if the cells have different numbers of components. Note from the previous equation that the estimate $-\hat{b}$ is also the estimate of ΔH, and therefore it can be used to calculate the acceleration factor between any two temperatures using the formulas described earlier in this chapter. If only two temperature cells are run, and each cell has the same number of test units and readouts, then the estimate of ΔH is

$$\Delta \hat{H} = k\,(y_1 - y_2)\left(\frac{1}{T_2} - \frac{1}{T_1}\right)^{-1}$$

where y_1 is the average of all the $\ln R(T_1)$ estimates from the cell at temperature T_1, and y_2 is the average of all the $\ln R(T_2)$ estimates from the cell at temperature T_2.

More complicated acceleration models like an Eyring for several stresses are handled the same way. The median cell slope estimates (measuring the rate of degradation) are set equal to the reciprocal of the acceleration model equation. This expression for $R(t)$ can usually be transformed into a linear form by taking logarithms and changing variables. A program for multiple regression can then be used to solve the resulting equations.

Using degradation data gives us many data points, even from small numbers of units on test. We can also include stress cells that are close to use conditions as long as the amount of parameter drift we are measuring stands out from the instrument measurement error. We are protected from misjudging the proper sample sizes and stress levels needed to obtain adequate failures. These are several very desirable properties of degradation or drift modeling.

The disadvantage of using this data to model acceleration is that it takes us one step farther away from reality when we deal with parameter drift instead of actual failures. What do we do about units that do not appear to drift at all? What about those that degrade and then seem to improve or recover? All these situations present both mathematical and conceptual difficulties that can cast doubt on the analysis validity.

Our recommendation is to use drift or degradation analysis only when the drift mechanism is understood and relates directly to actual failures. Even then, plan to have at least one cell with high enough stress to produce actual failures. This cell can be used to test the validity of the T_{50} projections obtained from degradation modelling, and to estimate the distribution σ.

Example 7.10   Degradation Data Analysis

A resistor used in a power supply is known to drift over time, degrading eventually to failure. Increasing the operating temperature speeds up resistor degradation. Once a change of 30 percent is reached, the power supply can no longer function. Preliminary studies indicate the Arrhenius model applies. Two stress cells, each containing 10 resistors, are run at 105° C and 125° C. Percent change in resistance from time zero is measured for each unit at 24, 96, and 168 hr readouts. The results are given in Table 7.10.

Use both method 1 and method 2 to estimate ΔH and project an average failure rate over 100,000 hr of field life at a use temperature of 30° C.

Following method 1, least squares was used to fit straight lines through the both the origin and the three degradation readings for each resistor on test. Next, the time where each line crossed the 30 percent failure point was calculated. The lines and the projected failure times for each resistor in the 105° C are shown in Figure 7.6. The same information for the units in the 125° C cell is shown in Figure 7.7.

The T_{50} estimates, using the projected failure times given in Figures 7.6 and 7.7, are 1224 hr and 309 hr for the 105° C and 125° C cells, respectively. A pooled overall σ estimate is 0.92, obtained using standard methods for normal data (working with the ln t_f from each cell).

TABLE 7.10   Degradation Data

	105° Cell % Degradation			125° C Cell % Degradation		
Component	24 Hr	96 Hr	168 Hr	24 Hr	96 Hr	168 Hr
1	1.0	6.0	11.1	7.1	27.1	46.8
2	0.0	0.4	1.3	0.6	3.6	5.3
3	1.1	4.5	6.4	3.0	12.9	21.7
4	0.0	1.7	2.7	2.5	11.6	18.5
5	1.5	5.4	9.4	0.3	3.1	7.5
6	0.8	4.3	6.4	0.9	3.1	5.4
7	0.9	2.6	3.9	2.8	11.2	21.2
8	0.0	3.0	5.5	1.8	10.8	18.2
9	2.0	6.0	10.1	5.3	22.0	38.4
10	0.6	0.7	0.1	3.5	13.4	22.4

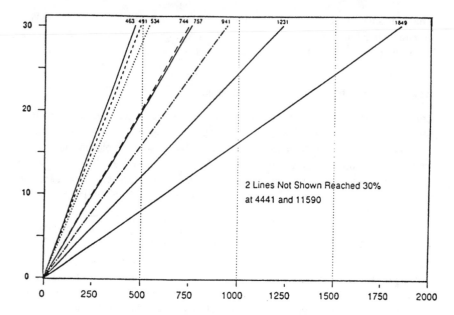

Figure 7.6   Projected Degradation Failure Times, 105° Cell

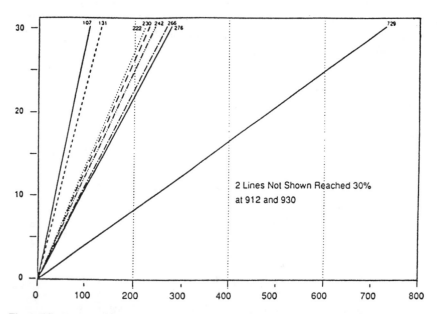

Figure 7.7   Projected Degradation Failure times, 125° Cell

The estimate of ΔH is calculated by

$$\Delta H = k \left[\ln \left(\frac{T_{50_1}}{T_{50_2}} \right) \right] \left(\frac{1}{T_1} - \frac{1}{T_2} \right)^{-1} = 0.89$$

and the acceleration factor to 30° C is 3,391. Finally, $AFR(100K)$ is estimated to be 53 FITs.

For a method 2 slope analysis, we start by dividing all the 168 hr readings by 168. This gives us 10 slope estimates from each temperature cell. For the 105° C cell, the data are 0.06607, 0.00774, 0.0381, 0.01607, 0.05595, 0.0381, 0.02321, 0.03274, 0.06012, 0.0006. For the 125° C, we have 0.2786, 0.03155, 0.1292, 0.1101, 0.04464, 0.03214, 0.1262, 0.1083, 0.2286, 0.1333.

Analyzing these two sets of data as lognormal complete samples gives an $\hat{R}_{50}$ for 105° C of 0.0212 and an $\hat{R}_{50}$ for 125° C of 0.0971. The pooled σ estimate is 1.14. The ΔH estimate is obtained by letting $y_1 = \ln 0.0212$ and $y_2 = \ln 0.0971$ and using the equation for ΔH given in the explanation of method 2. The result is $\Delta \hat{H} = 0.99$ and the acceleration from 125° C to 30° C is 8,452. The $\hat{T}_{50}$ estimate for the 125° C cell is just $D/\hat{R}_{50}$ or 30/0.0971 = 309 (the same as with method 1). The use AFR at 100,000 hr is estimated to be 21 FITs.

After studying the two proposed methods and Example 7.10, some obvious questions arise. Which method is better? And if method 2 is a good way to analyze degradation data, why bother to take any intermediate degradation readouts?

If no intermediate readouts are made, both methods will give the same answers. However, the earlier readouts allow one to visually examine the key assumption of linearity. Therefore, these readouts should be included in at least one cell, even if the simpler method 2 is chosen for the final analysis. As to which method is "better," there is no clear answer. Both methods are "quick and dirty" in the sense that they tend to ignore the experimental error in the observations made at each readout. And both methods, as shown in Example 7.10, will give nearly the same results. The important thing to focus on is that either method allows us to calculate an estimate for ΔH and project use failure rates from an experiment where one cell had no actual failures and the other cell had only two failures.

STEP STRESS DATA ANALYSIS

Another technique used to ensure enough failures when conducting stress tests is to periodically increase the stress within a cell until almost all of the units have failed. The primary drawback to changing stresses by steps, while an experiment is running, is the difficulty of analyzing the resulting data and constructing models. This section will briefly describe a way of conducting an Arrhenius model analysis, using only one cell of units and periodically increasing that cell's operating temperature.

The method relies on repeated use of the concept that changing stress is equivalent to a linear change in the time scale. If we knew the value of ΔH, we could calculate acceleration factors that relate time intervals at several temperatures to equivalent time intervals at one fixed reference temperature. The step stress experiment would then be reduced to a single stress experiment, with artificial, calculated time intervals for readouts.

To show how this works, assume the actual readout times are $R_1, R_2, \ldots, R_k$, and the test ends at R_k. At each readout time, new failures are recorded and the temperature of the cell is increased. Let the temperature during the interval 0 to R_1 be T_1. The temperature between R_1 and R_2 is T_2, and so on, with the final (highest) temperature T_k occurring during the kth interval between times R_{k-1} and R_k. At readout time R_i, the number of new failures observed is $F_i \geq 0$. N units start test, and the survivors at the end of test are $S = N - \Sigma\, F_i$. The real-time events of this experiment are diagrammed in Figure 7.8.

Now assume we know ΔH. Then we can normalize all the time intervals to equivalent times at the first temperature T_1. First we calculate all the acceleration factors between later temperatures T_j and the starting temperature T_1. These factors, denoted by A_j, are given by

$$A_j = e^{(\Delta H/k)\,[\,(1/T_1) - (1/T_j)\,]}$$

The transformed length of the ith time interval becomes $A_i \times (R_i - R_{i-1})$ and the equivalent readout times at the single stress T_1 are

$$R_1 = R_1$$

$$R_i = R_1 + \Sigma\, A_j(R_j - R_{j-1})$$

$$= R_{i-1} + A_i(R_i - R_{i-1})$$

Using these transformed readout times and the numbers of observed failures, lognormal or Weibull or exponential estimation can be carried out.

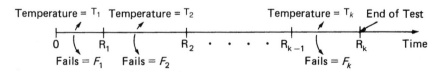

Figure 7.8   Arrhenius Step Stress Data Schematic

Of course, the problem with all this is that we do not know ΔH in advance. If we did, we would not have to run a step stress experiment. So the trick is to keep assuming different values for ΔH and fitting life distribution parameters to the resulting transformed data. The "best" ΔH is the one where the transformed data most closely fits the assumed life distribution. If maximum likelihood estimation is used, the MLE for ΔH is the assumed value that yields the highest likelihood in the analysis. If graphical techniques are used, the best ΔH gives points that line up closest to a straight line on the appropriate life distribution graph paper.

It is obvious from the above discussion that analyzing data from a multiple stress cell where several different stresses are increasing would be very difficult. A procedure would have to be specially worked out for the assumed acceleration model and the step stress levels.

Exercise 7.21

A stress test is performed on N units for 500 hr at a temperature of $55°$ C. The next 500 hr of test is conducted on the surviving units at T = $75°$ C. Finally, another 500 hr is done at $100°$ C. What are the equivalent readout times for a test that remains at $55°$ C, assuming $\Delta H = 0.7$ eV?

Example 7.11   Step Stress Model

A sample of 250 components from a population with a temperature dependent lognormal life distribution was tested to determine a T_{50} and σ at any one temperature and an estimate of the Arrhenius ΔH parameter. Since only one test chamber was available, a step stress method was chosen. The initial temperature was $85°$ C.

Readouts were made at 500, 1000, 1500, and 2000 hr. There were no failures at 500 hr, one failure at 1000 hr, seven failures at 1500 hr, and five failures at 2000 hr. At 2000 hr, the temperature was increased to $100°$. There were two failures at a readout at 2100 hr. Then the temperature was increased to $115°$. The next readout, at 2200 hr, showed four failures. At this point, a final temperature increase to $125°$ was made. The last three readouts, at 2300, 2400, and 2500 hr, showed 36, 38, and 29 failures, respectively. At 2500 hr the test ended, with 128 unfailed units. What are the lognormal parameter and ΔH estimates?

Solution

The iterative MLE method described in this section was used. The results were: at $85°$ C, $T_{50} = 8506$ and $\sigma = 0.80$ (σ is the same for any temperature). The estimate of ΔH is 0.86.

Since the data were simulated, we know the "true" values to compare to these estimates. These were: $T_{50} = 8000$; $\sigma = 0.9$, and $\Delta H = 0.85$. The estimation, especially for ΔH, was fairly close.

Most readers will not have a program capable of this kind of analysis, so this example just shows what can be done and provides check values for anyone who writes his own program to use.

Table 7.11 and Figure 7.9 show how ΔH could be estimated by an iterative graphical procedure. Table 7.11 gives the actual readout times and cumulative percent failures, as well as equivalent readout times (referenced to the 85° temperature) assuming values of 0.5 and 0.86 and 1.0 for ΔH. For example, at 100° C, the acceleration factors are 1.92, 2.07, and 3.68, respectively. So 100 hr at 100° C is equivalent to 192, 307, and 368 hr at 85° C, depending on the value of ΔH. Figure 7.9 shows a plot on lognormal probability paper of the cumulative percent failures versus the equivalent readout times for each choice of ΔH. The MLE of $\Delta H = 0.86$ gives the best line, with the high and low ΔH value lines fanning out to the right and the left of the "correct" line. Fitting successive least squares graphical lines to a range of ΔH values and picking the ΔH that gives the minimum residual line would be the graphical estimation procedure.

One last comment on the design of the experiment described in this example. The most accurate results are obtained by staying at a low-level stress as long as needed to get some data, then going quickly to the highest possible stress—as was done in this case.

Exercise 7.22

Use the "true" values for T_{50}, σ, and ΔH in Example 7.12 to estimate the expected cumulative percent failures at each readout in Table 7.11, and compare to results.

TABLE 7.11   Step Stress Example Data

		Equivalent Readout Times			
Stress (°C)	Readout Times	$\Delta H = 0.5$	$\Delta H = 0.86$	$\Delta H = 1.0$	Cumulative % Failures
85	500	500	500	500	0.0
85	1000	1000	1000	1000	0.4
85	1500	1500	1500	1500	3.2
85	2000	2000	2000	2000	5.2
100	2100	2192	2307	2368	6.0
115	2200	2542	3168	3591	7.6
125	2300	3051	4812	6184	22.0
125	2400	3560	6455	8776	37.2
125	2500	4069	8099	11368	48.8

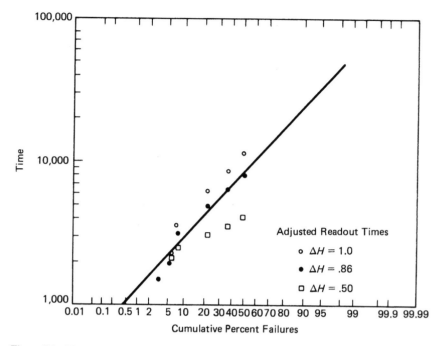

Figure 7.9 Plot Step Stress Data for ΔH = 0.5, 0.86, and 1.0

ACCELERATION AND BURN-IN

Up until now, we have considered stress testing and acceleration modeling as a means to obtain failure data and project use failure rates. There are two other important reasons why we might run components for a number of hours at high stress. The first would be to monitor incoming lot reliability, accepting or rejecting lots based on the number of failures obtained from a sample put on stress test. In Chapter 9, in the section on "Relating an OC Curve to Lot Failure Rates," we will show how to use lot acceptance sampling techniques to reject lots that might have an unacceptably high field failure rate. This use of a high stress reliability monitor assumes we already know the acceleration factor between use conditions and stress conditions, and the life distribution is known, with only the scale parameter possibly changing from lot to lot.

The second reason for running components at high stress prior to shipment or use in an assembly is to produce and remove failures that might otherwise occur early in the field life of the product. This practice is known as "burn-in," and its goal is to reduce the component failure rate throughout the useful lifetime of component operation. As opposed to using a sample to monitor reliability, burn-in is generally applied to all components to guarantee a desired failure rate below that obtainable from non-burned-in components.

In Chapter 8, we will evaluate the benefits of burn-in using a sophisticated general algorithm for component reliability projection. Here, we look at some simple consequences of acceleration and burn-in.

We assume a Weibull life distribution model applies, and we know the acceleration factor between use and stress is "a". Let $h_{bi}(t)$ be the use failure rate curve for the component population after a burn-in of t_{bi} hours and let $h(t)$ be the use failure rate curve for the population without any burn-in. Then $h_{bi}(0) = h(at_{bi})$ and the failure rate at time t is given by $h_{bi}(t) = h(t + at_{bi})$. Note that the failure rate curve after burn-in is just the continuation of the old pre-burn-in failure rate curve starting at at_{bi} hours. This relation is true because the failure rate is already a "conditional" failure rate curve giving the rate of failure of the survivors at time t. The equations for $f_{bi}(t)$ and $F_{bi}(t)$ would be more complicated, requiring conditional probabilities to account for having survived the burn-in period:

$$F_{bi}(t) = \frac{F(t + at_{bi}) - F(at_{bi})}{1 - F(at_{bi})}$$

$$f_{bi}(t) = \frac{f(t + at_{bi})}{1 - F(at_{bi})}$$

What does all this mean when $F(t)$ is a Weibull distribution with shape parameter m? If $m < 1$, then $h(t)$ is a decreasing function and $h_{bi}(t)$ is always less than the pre-burn-in failure rate. With the proper choice of burn-in stress and time, the post burn-in failure rate can be reduced to meet any given requirements. If $m = 1$, then $h(t) = \lambda$ and $h_{bi}(t) = \lambda$. That means there will be no gain from burn-in at all. Even worse, if $m > 1$, then $h_{bi}(t) > h(t)$, and the failure rate is higher after burn-in. In Chapter 8 we shall see, however, the useful result that burn-in can be effective even for increasing $h(t)$ if there is a defective subpopulation.

Example 7.12   Burn-In Failure Rate Improvement

A component population follows a Weibull life distribution with $m = 0.8$ and $c = 2.582 \times 10^8$. The *AFR* for the first 3 months of field use (estimated to be $t = 720$ power-on hours) is projected to be

$$\frac{H(t)}{t} = \frac{(t/c)^m}{t} = \frac{(720/c)^{0.8}}{720} = 50 \text{ FIT}$$

However, the customer requirement is 20 FITs for the first 90 days. A burn-in is necessary, and the highest practical burn-in temperature for this component is 105° C (with use temperature 25° C, and $\Delta H = 1.0$). How long a burn-in period is needed?

Solution

The acceleration factor from 25° C to 105° C for $\Delta H = 1.0$ is 3,767 from Table 7.6. The after burn-in average failure rate (in FITs) is given by

$$\frac{H(at_{bi} + t) - H(at_{bi})}{t} = 10^9 \times \frac{\left(\dfrac{3767 t_{bi} + 720}{2.582 \times 10^8}\right)^{0.8} - \left(\dfrac{3767 t_{bi}}{2.582 \times 10^8}\right)^{0.8}}{720}$$

By trial and error, we find that a burn-in time of $t_{bi} = 6$ hr gives us an *AFR* of 20.01 FITs, which is equivalent to the desired after burn-in failure rate for all practical purposes.

Exercise 7.23

For the component population described in Example 7.12, how long a burn-in at 105° C is needed to reduce the first 90-day *AFR* to about 15 FITs (to the nearest half hour)? What would the first 90-day *AFR* be if a 1 hr burn-in is chosen?

One important and somewhat puzzling consequence of burn-in should be noted. The failure rate of the original unburned-in population in Example 7.12 would show a rapidly decreasing failure rate when measured by field data. A typical field monitoring program might calculate the empirical cum hazard function $H(t)$ and plot it versus time on log versus log paper (see the sections in Chapter 6 on "Probability Plotting For The Weibull Distribution" and on "Cumulative Hazard Estimation"). The slope of a line fit through the $H(t)$ points would estimate m and, with sufficient amounts of field data, we would expect this slope to be close to 0.8. People familiar with Weibull cum hazard plotting would take comfort in the knowledge that a slope less than 1 implies a decreasing failure rate, so field data will improve with time.

But what does the cum hazard plot look like for field data from the burned-in population? True, the failure rate starts out much better than was the case for the original population. However, the slope of the cum hazard plot will now appear to be 1, implying the failure rate is no longer getting better. It seems as if the burn-in has converted the population into following an exponential life distribution model.

The population life distribution has not actually changed from a Weibull with shape 0.8 to an exponential, however. It only appears that way because we are plotting the cumulative values of $h(t)$ starting at at_{bi} versus a time axis starting at time 0. The function

$$H_{bi}(t) = H(t + at_{bi}) - H(at_{bi})$$

is no longer a Weibull cum hazard when plotted on log versus log paper. It has a changing slope that starts out to be 1 and eventually asymptotes to 0.8. In general, for any value of m and equivalent burn-in time of at_{bi}, the slope of the cum hazard plot will start at 1 and gradually approach m. In practice, the slope is likely to remain close to 1 throughout the field lifetime of the components. The period of rapid failure rate decrease takes place during the burn-in, and little visible improvement will be noticed during the field lifetime.

Exercise 7.24

Compute the cum hazard function for the burned-in components described in Example 7.12 at 200, 400, 600, and 720 hr of field use. Use least squares to fit a line through these points when plotted on log versus log paper and estimate the slope of the line (i.e., fit the model $H_{bi}(t) \approx A + m \ln t$ using regression to obtain m). You should obtain a slope greater than 0.998. Estimate the slope again after 100,000 field hours using the cum hazard at 100,200, 100,400, 100,600 and 100,720 hr points. Show it is still nearly 0.88. Finally, show that after 5,000,000 hr of time, the slope falls to under 0.81.

CONFIDENCE BOUNDS AND EXPERIMENTAL DESIGN

So far in this chapter, only point estimates of acceleration model parameters and use condition failure rates were mentioned. To understand the precision of the data and method, and do proper risk assessment, it is important to have at least approximate confidence intervals for these estimates. In Chapter 3, when we discussed the exponential distribution, it was easy to calculate upper and lower bounds. When we deal with more complicated distributions, and several stress cells of interval or multicensored data, confidence level calculations become much more difficult.

In particular, the quick and easy graphical methods do not allow any confidence interval estimation. Good regression programs will give confidence bounds on acceleration model parameter estimates and, if weighted regression is used, these bounds appear to be accurate (although theory to support their accuracy is lacking). The asymptotic theory of maximum likelihood estimation offers the best general approach for calculating confidence bounds. If programs for this kind of analysis are used, they will usually give bounds.

Another important topic, having a direct relationship to the precision of the final estimates, is the proper design of the experiment. By design, we mean choice of stress levels and sample sizes. This concept is complicated for life testing data, since the number of failures in a cell is not known in advance and precision depends more on the number of failures than on the sample sizes put on test.

Basic design theory, in the non-life test case, says best results are obtained by choosing stress levels as far apart as possible. In the life test case, this guideline must be tempered by two considerations: stresses too high may introduce new failure modes and violate true acceleration; stresses too low may not yield any failures. Consequently, stress level and sample size determination often becomes more of an art, based on past experience and "feel," than an exact science.

Some analytic approaches are possible, however. Two of these are:

1. Assume values for the unknown parameters and, using these values as if they are correct, pick sample sizes and stresses that will produce an adequate number of failures in each cell. Typically, at least 10 — and preferably over 20 — failures in each cell is adequate. If the data is readout data, failures must be spread out over three or more intervals. Assumed values generally come from typical literature ΔHs and shape parameters. Scale constants such as use T_{50} are calculated by assuming the population just meets its use condition failure rate objective.
2. Make the same unknown parameter value assumptions as in 1. Then simulate cells of failure data for a given experimental design and analyze the data. Repeat this many times to get a feel for the precision of the results. Try again with a different design and see what results are obtained. By iterating on this procedure, a good design will be determined.

Both these methods rely on a good pre-guess of the true model parameters. Kielpinski and Nelson (1975) have carried this approach even farther for the lognormal, and Meeker and Nelson have studied the Weibull (1975). They have derived optimal life test schemes assuming exact time of failure data and pre-guessed acceleration model and distribution shape parameters. Guidelines for planning experiments are given by Meeker and Hahn (1985).

An approach for evaluating and planning acceleration model experiments is described by Nelson (1990, p. 349–361). A computer program routine to carry out method 2 above is incorporated within the IBM AGSS program package, and a proposed version of AGSS will include the capability for designing locally optimal life test schemes. Example 7.13 shows how a method 1 analysis might be carried out.

Example 7.13   Planning Acceleration Experiments

We are testing a new component to verify that it meets a use average failure rate objective of 0.001%/K over 100,000 hr. The failure mode is temperature dependent and is believed to follow an Arrhenius model. Use temperature is 25° C. We would like to test at 65° C, 85° C, and 125° C for 2000 hr. How many units should be put in each cell, and is our choice of temperatures and test time reasonable?

Solution

Assume that for the failure mode, a lognormal distribution is appropriate. Our pre-guess for σ is 1, and for ΔH it is 0.7. The average failure rate requirement implies that $F(100,000)$ is approximately 0.001, allowing us to calculate a use T_{50} from

$$T_{50} = 100,00e^{-\sigma\Phi^{-1}(0.001)} = 100,000e^{-\Phi^{-1}(0.001)}$$

where Φ^{-1} is the inverse of the standard normal distribution and can be looked up in normal tables. The result is a T_{50} of 2,198,840 hr.

Now we use the ΔH of 0.7 to calculate acceleration factors between 25° and 65°, or 85°, or 125°. These factors turn out to be 25, 96, and 937. That leads to T_{50} values of 87,600, 22,900, and 2,347 for these temperature cells. With such a high T_{50} in the 65° cell, it is very unlikely that we would see any failures at this stress unless we put many thousands of units on test. In the 85° cell, if we go out to 5000 hr, we would expect 6 percent of the units to fail. So we modify our design to start at 85° and place 200 units on test in this cell for 5000 hr.

Since we have decided against the 65° cell, we could add a cell at 100°. The acceleration from 25° to 100° is 239, so the T_{50} is 2,198,842/239 = 9200. This cell will have 27 percent failures expected at 5000 hr, so a sample size of 100 is adequate.

Finally, we calculate that the 125° cell will have an expected 44 percent failures in 2000 hr. One hundred units on test in this cell, for only 2000 hr, will be sufficient.

Note how, if the pre-guessed values are close to the real parameter values, the original experimental design would have yielded insufficient data. An easy calculation ahead of time corrected the situation.

Exercise 7.25

Failures have been observed for a component family due to dendritic migration of metal between conductors, leading to shorts. Increasing temperature and voltage will accelerate the failure time according to the Arrhenius and the inverse power law models. Use condition is 2 volts and 25° C. The highest practical test conditions are 12 volts and 125° C. If you have 150 components to test, and three test ovens, design an experiment lasting 2000 hr to determine whether the components will meet a requirement of no more than 0.01 percent cumulative failures by 100,000 hr at use conditions. Assume the voltage acceleration parameter is 1.5, ΔH is 1.0, and the life distribution is Weibull with a shape parameter of 1.3. Show that the experiment you design is expected to produce enough failures in each cell for a successful analysis, assuming your pre-guesses of the unknown parameters are approximately correct and the components exactly meet use condition failure rate requirements.

SUMMARY

Many important concepts and techniques were introduced in this chapter. Some were surveyed briefly, while others were developed in detail by means of examples. These ideas should be studied carefully, for they are typical of situations often encountered in the analysis of life test data.

The theory of acceleration followed from equating "true" acceleration with a linear change in the time scale. From this it followed that lognormal or Weibull shapes remained constant from stress condition to stress condition. Acceleration factors could be calculated by taking the ratios of T_{50} values, or any other percentile, or even the mean time to fail. Acceleration models describe how the time scale changes as a function of stress. These models can be set equal to sample estimates of some convenient percentile to solve for the unknown model parameters and project to a low stress application.

The following seven steps describe how an acceleration model study might be carried out.

1. Choose as simple an acceleration model for the failure mechanism under investigation as seems appropriate from past experience, or a literature study, or a theoretical derivation. Many models have an Arrhenius temperature term and are simplified Eyring equations.
2. Design an experiment consisting of enough different stress cells to estimate the model parameters. Make sure sample sizes and stresses are such that each cell has an adequate number of failures. More than 10 failures spread out over at least three readout intervals is a rule of thumb for adequate data from a cell. An assessment of proper sample sizes and stresses can often be made by working backward from the use failure rate objective and guessing at reasonable model and shape parameters.
3. Choose an appropriate life distribution for the failure mechanism.
4. Analyze the failure data in each cell with as accurate a technique as is available. In any case, do a graphical analysis as well to see how well the data fits the life distribution and follows the equal slope consequence of true acceleration.

5. Fit the acceleration model parameters, again using the best technique available. If a program is available that uses maximum likelihood techniques to estimate both shape and acceleration model parameters, that will generally be the best possible method.

6. If using regression, be sure to use weights, especially if the numbers of failures differ widely from cell to cell. The numbers of failures associated with each T_{50} or c estimate make good weights, except when constrained maximum likelihood estimation is used. In that case, use the reciprocal of the asymptotic variance estimates.

7. Substitute the estimated model parameters into the model equation, along with use stresses, to project a T_{50} or c. The use shape parameter is the best single value fitting all the stress cells. Obtain the shape estimate as a weighted average of cell shape estimates, using cell failure numbers as weights, or by using constrained maximum likelihood estimation.

Some additional methods that offer the potential of much useful data, with a minimum of testing, are degradation modeling and step stress testing. These procedures require additional assumptions, however, and the resulting data may present analysis difficulties.

When acceleration factors between use and stress conditions are known, and the failure rate function decreases over time, it may be possible to improve the field average failure rate to a desired number by burning-in the component population. The required burn-in time can be determined in advance to achieve the required amount of improvement.

PROBLEMS

7.1   Assume that the life distribution for a population of components follows a model with a cumulative distribution function given by

$$F(t) = 1 - \frac{k}{k+t}$$

If there is true (linear) acceleration between the testing stress and the use stress, with acceleration factor a, write the CDF for the life distribution at use conditions.

7.2   Make up your own expanded version of Table 7.6, with acceleration factors for every 5 degrees temperature increase from 25° C to 125° C, and for ΔHs from 0.5 to 1.5 in steps of 0.1. At what value of ΔH does the acceleration roughly double for every 10° increase in temperature? At what value for ΔH does the acceleration roughly double for every 5° increase?

7.3   A component, tested at 150° C in a laboratory, has an exponential distribution with a *MTTF* of 3000 hr. Typical use temperature for the component is 45° C. Assuming an acceleration factor of 150 between these two temperatures, what is the expected use failure rate and what component failure percentage is expected during the first 40,000 hr? What is ΔH?

7.4   A component, tested at 150° C in a laboratory, has a Weibull distribution with characteristic life of 1000 hr and a shape factor of 0.5. Typical use temperature for the component is 55° C. Assuming an acceleration factor of 250, what is the expected average failure rate over the first 40,000 hr? What fraction of the components should typically survive 40,000 hr? Calculate ΔH for these components.

7.5   A component, running at 125° C in the laboratory, has a lognormal life distribution with a T_{50} of 15 hr and a σ of 1. Assuming an Arrhenius model with an activation energy of 1.3 eV, find the acceleration factor to a use condition of 45° C. What fraction of the components is expected to survive to 20,000 hr?

7.6   Using the electromigration model described in the section on "Other Acceleration Models," calculate an acceleration factor between a stress test run at 150° with a current density of 500,000 amps/cm^2, and a field condition of 35° C with 100,000 amps/cm^2. Assume an activation energy of 0.5 eV and a current density exponent of 2. If the failure distribution is lognormal with a sigma of 2.5, and the T_{50} at stress is 2500 hr, estimate the fraction of units expected to fail by 20,000 hr at field conditions.

7.7   Show how, by a proper choice of stress functions and constant values, the general Eyring equation for two stresses can be converted into the electromigration model.

7.8   Four cells of components were put on test at temperatures of 110° C, 120° C, 130° C, and 140° C to model a failure mechanism believed to follow an Arrhenius acceleration model. The lowest temperature cell had 100 units on test; all other cells contained 50 units. Readouts were taken at 24, 48, 96, 150, 500, and 1000 hr. The test ended at the last readout and the results are summarized in the table below.

Readout Time	110° Cell New Fails	120° Cell New Fails	130° Cell New Fails	140° Cell New Fails
24	0	0	1	2
48	0	0	1	1
96	0	2	3	3
150	0	0	1	1
500	3	3	3	10
1000	1	3	8	10
Unfailed	**96**	**42**	**33**	**23**

Use a Weibull life distribution model assumption and estimate m and ΔH and the use CDF at 40,000 hr for a use temperature of 60° C. Repeat the analysis using a lognormal assumption. Which model appears to fit the data better?

7.9 To model a failure mechanism that is accelerated by both temperature and voltage, six cells, each containing 50 components, were run at various temperature and voltage combinations. Readouts were taken at 24, 48, 168, 500, 1000, and 2000 hr, and the numbers of failures were recorded. The test ended at 2000 hr. The data is summarized in the following table.

Readout	8 V, 125° Fails	10 V, 115° Fails	10 V, 125° Fails	12 V, 95° Fails	12 V, 115° Fails	12 V, 125° Fails
24	2	0	1	0	0	0
48	0	0	0	0	0	0
168	4	0	2	0	3	5
500	6	3	10	0	6	15
1000	9	9	16	2	9	20
2000	9	10	13	3	18	8
Unfailed	**20**	**28**	**8**	**45**	**14**	**2**

Assume the Arrhenius, Inverse Power model (AI model) used in Example 7.7 applies and that the life distribution is lognormal. Estimate ΔH and B, and project a use CDF at 100,000 hr for use conditions of 2V, 25° C. Repeat the analysis assuming a Weibull life distribution. Which model appears to fit the data better?

7.10 Five sample components are measured for initial threshold voltages. Then they are placed on test at a stress with a known acceleration factor of 55, with respect to use conditions. Threshold voltages are readout at 500 and 1000 hr. At 1000 hr, the test had to be terminated. The data is given in the table below.

| | Threshold Voltage | | |
Component	Initial	500 Hr	1000 Hr
1	0.99	0.83	0.68
2	0.82	0.72	0.65
3	0.96	0.88	0.80
4	0.87	0.72	0.60
5	0.91	0.85	0.78

A failure value is considered to be a threshold voltage below 0.5 volts. Assuming a linear degradation model holds, estimate failure times for each component. Next, assume the life distribution is exponential and estimate the stress *MTTF* and the use failure rate.

7.11 Write the general expression for the field cum hazard function of a Weibull population that reaches the field after a burn-in equivalent to t_{bi} field hours. Show that a plot of the cum hazard function on log versus log paper has a slope that approaches 1 as field time approaches 0. Show also that the slope approaches the original Weibull shape parameter m, as time approaches infinity.

7.12 Components made by certain manufacturer have a failure mechanism described by a lognormal distribution. Typical use condition is 35° C. An accelerated test was run using these components with three temperature cells: 175° C, 150° C, and 125° C. Analysis of the data in each cell confirmed a lognormal distribution with a common sigma of 3.9. The T_{50} values estimated for the three cells were, respectively, 75, 150, and 500 hr. The values of $(1/kT)$ are 25.89, 27.42, and 29.15 for these cells. Plot $\ln T_{50}$ versus $1/kT$. Determine the slope and intercept to get the activation energy and proportionality constant. Project a field usage T_{50}, and estimate the fraction failing after 30,000 hr.

7.13 A family of components has a failure mechanism described by an exponential distribution. Typical use temperature is 45° C. The failure mechanism is accelerated by increasing radiation dosage, and a sample of 100 units was tested while undergoing a radiation dosage increased by a factor of 10 over normal use exposure.

A model for time to failure is believed to be

$$T = AD^{-\alpha}$$

where A is a proportionality constant, D is the radiation dosage, and α is an empirically determined constant equal to 1.5. If the stress failure rate was found to be 1.75 percent per thousand hours, what is the estimate of the *MTTF* under normal use conditions?

7.14 A single cell step stress experiment was run for the component family described in Exercise 7.23. The 150 test units underwent voltage and temperature stresses as described the table below, and the failures during each readout interval were recorded. Use the data in the following table to estimate ΔH, the voltage exponent B, and the Weibull shape parameter m. Finally, project a use CDF at 100,000 hr.

Redout Start	Readout End	Voltage (volts)	Temperature (°C)	Fails During Interval
0	24	6	125	0
24	72	6	125	1
72	168	6	125	10
168	500	6	125	21
500	1000	6	125	38
1000	1200	6	125	15
1200	1400	9	125	16
1400	1600	9	125	18
1600	1800	12	90	1
1800	2000	12	90	1
2000	2100	12	110	5
2100	2200	12	110	0
2200	2300	12	125	9
2300	2400	12	125	4
2400	2500	12	125	4

Remaining unfailed at end of test: 7 components

Chapter 8

System Models and Reliability Algorithms

Earlier chapters have described how to estimate reliability distribution parameters and probabilities for components operating at typical use conditions. How can these probabilities be used to predict total system performance? How does the design of the system affect reliability? What are the benefits of redundant design? How can complex systems be broken down for analysis into single subassemblies?

This chapter will answer questions like these. We will also look at how we can put together component models to form reliability algorithms that provide targets against which actual performance can be measured. In addition, some useful models for defect discovery and early so-called "weak unit" failures will be described.

SERIES SYSTEM MODELS

The most commonly used model for system reliability assumes that the system is made up of n independent components which all must operate for the system to function properly. The system fails when the first component fails. This model is called a *series*, *first-fail*, or *chain model* system. Even though either the independence assumption or the first-fail assumption may not be strictly valid for an actual system, this model is often a reasonable and convenient approximation of reality.

In the section on "Some Important Probabilities" in Chapter 2, we derived the formula for the reliability of a series system composed of n identical elements using the multiplication rule for probabilities. Now, we generalize that case to a series system of n, possibly all different, components.

Let the ith component have the reliability function $R_i(t)$. Then the probability the system survives to time t, or the system reliability function $R_s(t)$, is the probability that all the components simultaneously survive to time t. Under the independence assumption, this probability is the product of the individual probabilities of survival (multiplication rule). These probabilities are just the $R_i(t)$. The expression is

$$R_s(t) = \prod_{i=1}^{n} R_i(t) = R_1(t) \times R_2(t) \times \dots \times R_n(t)$$

or, in terms of the CDF functions,

$$F_s(t) = 1 - \prod_{i=1}^{n} [1 - F_i(t)]$$

For system failure rates, the relationship is even simpler:

$$h_s(t) = \sum_{i=1}^{n} h_i(t)$$

$$AFR_s(T_1, T_2) = \sum_{i=1}^{n} AFR_i(T_1, T_2)$$

These equations show that, for a series system, the failure rate can be calculated by summing up the failure rates of all the individual components. There are no restrictions on the types of distributions involved, and the result is exact rather than an approximation. The only requirements are the independence and first-fail assumptions. The proof of this convenient formula is very easy and is outlined in the following example.

Example 8.1   Series Systems

Derive the additivity relationship for series system failure rates by using the fact that the failure rate function can be defined as the negative derivative of the natural logarithm of the reliability function (see the section on "The Cumulative Hazard Function" in Chapter 2).

Solution

We have

$$-\ln R_s(t) \;=\; -\ln \prod_{i=1}^{n} R_i(t) \;=\; \sum_{i=1}^{n} -\ln R_i(t)$$

Therefore, the failure rate is

$$\frac{d\,[-\ln R_s(t)\,]}{dt} \;=\; \sum_{i=1}^{n} \frac{d\,[-\ln R_i(t)\,]}{dt} \;=\; \sum_{i=1}^{n} h_i(t)$$

from which the additivity of failure rates immediately follows.

THE COMPETING RISK MODEL (INDEPENDENT CASE)

The series model formulas apply in another important case. A single component with several independent failure modes is analogous to a system with several independent components. The failure mechanisms are competing with each other in the sense that the first to reach a failure state causes the component to fail. The series system probability arguments again apply, and the reliability of the component is the product of the reliability functions for all the failure modes. Failure rates are additive, mechanism by mechanism, to get the failure rate of the component.

The more general competing risk model, where the failure processes for each mechanism are not independent, becomes a separate research study, depending on how the mechanism random times of failure are correlated. This general model will not be treated in this text.

Example 8.2   Bottoms-Up Calculations

A home computer has most of its electronics on one board. This board has 16 memory modules, 12 assorted discrete components, and a microprocessor. The memory modules are specified to have an exponential failure rate with $\lambda = 0.01$ %/K. The discrete components each have a Weibull CDF with $m = 0.85$ and $c = 3{,}250{,}000$ hr. The microprocessor is thought to have two significant failure mechanisms. Each mechanism was modeled on the basis of accelerated testing designed to cause that type of failure. The results, adjusted to normal use conditions, yielded a lognormal with $\sigma = 1.4$ and $T_{50} = 300{,}000$ for one mode, and an exponential with $\lambda = 0.08\%/K$ for the other mode. Assuming all components and mechanisms operate independently (at least until the first fail), what is the board failure rate at 5000 hr? What is the chance a board has no failures in 40,000 hr?

Solution

First, we derive the microprocessor failure rate by adding the lognormal and exponential competing failure mode failure rates, evaluated at 5000 hr. This sum is 1600 PPM/K. The reliability at 40,000 hr is the product of the exponential and lognormal reliability functions. The result is 0.896. For this part of the example, the competing risk model was used.

Using the series model, we add the failure rates of the 16 memory modules together and obtain 1600 PPM/K. The product of their exponential reliability functions at 40,000 hr is 0.938.

The 12 discretes each have a Weibull failure rate of 700 PPM/K at 5000 hr, adding 8400 PPM/K to the board total. The product of the 12 reliability functions, evaluated at 40,000 hr, is 0.7515.

The sum of all the failure rates gives 11,600 PPM/K, or 1.16%/K, for the board total failure rate at 5000 hr. The probability the board lasts 40,000 hr without a failure is the board reliability, or the product of the component reliabilities: $0.896 \times 0.938 \times 0.7515 = 0.632$.

This last example showed how, starting with individual failure mode models and using the competing risk and series model as building blocks, a bottoms-up calculation of subassembly or system failure rates is performed. Since testing on the system level is usually limited due to time and cost constraints, this bottoms-up approach is of great practical value.

Exercise 8.1

Assume the shape parameter for the 12 discrete components in Example 8.2 is 1.0 instead of 0.85. What is the board failure rate at 5000 hr now? What is the reliability at 40,000 hr?

PARALLEL OR REDUNDANT SYSTEM MODELS

A system that operates until the last of its components fails is called a *parallel* or *redundant* system. This system model is the other extreme from the series model, where all components must work. Parallel systems offer great advantages in reliability, especially early in life. In applications where good reliability and low front-end failure rates have higher priority than component cost, designing in redundancy, at least for key parts of the system, is an increasingly employed option. The computer systems on the space shuttle represent an example of this: every system is replicated with several backup copies—even including independent versions of key software.

As before, let the ith system component have CDF $F_i(t)$. The probability that the system fails by time t is the probability that all the components have failed by time t. This probability is the product of the CDFs, or

$$F_s(t) = \prod_{i=1}^{n} F_i(t)$$

$$R_s(t) = 1 - \prod_{i=1}^{n} [1 - R_i(t)] = 1 - \prod_{i=1}^{n} F_i(t)$$

Failure rates are no longer additive (in fact, the system failure rate is smaller than the smallest component failure rate) but must be calculated using basic definitions.

Example 8.3   Redundancy Improvement

A component has CDF $F(t)$ and failure rate $h(t)$. The impact of this failure rate makes a significant adder to a system currently under design. To improve reliability, it is proposed to use two of these components in a parallel (redundant) configuration. Show that the improvement can be expressed as a factor k given by

$$k = \frac{1 + F(t)}{2F(t)}$$

where the old failure rate is k times the new failure rate. How much improvement results when $F(t) = 0.01$, as compared to later in life when $F(t) = 0.1$ or 0.5?

Solution

The CDF of the two components in parallel is $F^2(t)$, and the PDF, by differentiating, is $2F(t)f(t)$. The failure rate of the pair (leaving out the time variable t for simplicity) is

$$h_s = \frac{2Ff}{1 - F^2} = \frac{2Ff}{(1 + F)(1 - F)} = \frac{2F}{(1 + F)} h$$

This result shows that $h = k \times h_s$, with k as given above.

When $F = 0.01$, $k = 50.5$, or about a $50\times$ improvement. When F is 0.1, k is only 5.5. For $F = 0.5$, the failure rate improvement drops to 1.5 times. Thus, redundancy makes a large difference early in life, when F is small, and much less of a difference later on. The rule of thumb is that one gains by a factor of about $(1/2F)$.

This example can easily be generalized as follows: if a single component with CDF F is replaced by n components in parallel, then the failure rate is improved by the factor

$$k = \frac{1 + F + F^2 + \ldots + F^{n-1}}{nF^{n-1}}$$

There is about $(1/nF^{n-1})$ times improvement in early life.

Exercise 8.2

The microprocessor in Example 8.1 contributed more to the board failure rate than any other single component. What would the new failure rate be at 5000 hr, and the new board reliability at 40,000 hr, if a second (redundant) microprocessor were added to the board?

Note that the parallel model assumes the redundant components are operating all the time, even when there have been no failures. An alternative setup would be to have the redundant components in a backup or standby mode, only being called on to operate when needed. This model will be discussed next.

STANDBY MODELS AND THE GAMMA DISTRIBUTION

We treat here only the simple case where one or more identical units are on hand, to be used only as necessary to replace failed units. No allowance will be made for the failure rate of a switching device although, in practice, this element would have to be added into the overall system calculation.

The lifetime until a system failure is the sum of all the lifetimes of the original and standby components as they each operate sequentially until failure. For n components in the original plus standby group, the system lifetime is

$$T_n = t_1 + t_2 + \ldots + t_n$$

with the t_i independent random variables, each having the single component CDF $F(t)$.

For $n = 2$, the CDF for T_n can be derived using the convolution formula for the distribution of the sum of two independent random variables. For this application, the convolution is

$$F_2(t) = \int_0^t F(u) f(t-u) du$$

If we now add a third component lifetime and do another convolution, we derive F_3 and so on until, for F_n, we have

$$F_n(t) = \int_0^t F_{n-1}(u) f(t-u) du$$

For complicated life distribution, such as the Weibull or the lognormal, the convolution integrals would have to be evaluated numerically. In the exponential case, however, the calculations are much simpler.

Example 8.4   Standby Model

A subassembly has a high exponential failure rate of $\lambda = 2\%/K$. As an insurance backup, a second subassembly is kept in a standby mode. How much does this reduce the failure rate when the subassembly CDF is 0.01? What about when the CDF is 0.1 or 0.5?

Solution

Substituting the exponential CDF and PDF into the convolution formula gives

$$F_s(t) = \int_0^t \left(1 - e^{-\lambda u}\right) e^{-\lambda(t-u)} du = 1 - \lambda t e^{\lambda t} - e^{-\lambda t}$$

The standby model PDF is the derivative of this, or

$$f_s(t) = \lambda^2 t\, e^{-\lambda t}$$

Using 0.00002 for λ, the time when $F(t)$ is 0.01 corresponds to 502.5 hr. $F(t)$ is 0.1 and 0.5 at times 5268 and 34,657, respectively. Using these times and λ value, we find the improvement over 2%/K [by calculating h_s using $f_s/(1 - F_s)$] turns out to be about 100 times when the CDF is 0.01, and 10 times when the CDF is 0.1. The improvement factor is only about 2.4 times when the CDF is 0.5.

At the early times, the failure rate improvement given by the standby model with two exponential components was twice that given by the parallel model (see Example 8.3).

We can generalize Example 8.4 to the case of an r level standby system of exponential components. The system lifetime PDF can be derived, by repeated convolutions, to be

$$f_s(t) = \frac{\lambda^r t^{r-1} e^{-\lambda t}}{(r-1)!}$$

This expression is known as the *gamma distribution*. The parameters are λ and r. The PDF has values only for non-negative t, and λ and r must be positive numbers. The *MTTF* is r/λ, and the variance is r/λ^2.

In our derivation, r can have only integer values (the number of identical exponential components in the standby system). The general gamma distribution allows r to take on noninteger values and uses the gamma function (described in Chapter 4) to replace factorials. In this form of the PDF,

$$f(t) = \frac{\lambda^r t^{r-1} e^{-\lambda t}}{\Gamma(r)}$$

This is a very flexible distribution form, and it is often used empirically as a suitable life distribution model, apart from its derivation as the distribution of a sum of exponential lifetimes.

If we have an r standby exponential model, leading to the above gamma distribution, it can be shown, using approximations described in Gnedenko et al. (1969), that the improvement factor is approximately $r!$ times greater in early life than the improvement obtained from the parallel system model. For $r = 2$, this gives a two times improvement, as seen in Example 8.4.

Exercise 8.3

For the subassembly in Example 8.4, compute the approximate failure rate when the CDF is 0.01 for the case where two standby subassemblies are added to backup the original subassembly. (Hint: calculate the same failure rate if the two additional are in a parallel redundant mode and reduce this number by the improvement factor for standby versus redundant models.)

Another special case of the gamma is of interest. When $\lambda = 0.5$ and r is an integer, by substituting $r = d/2$, the PDF becomes

$$f(t) = \frac{(1/2)^{d/2} t^{(d-2)/2} e^{-t/2}}{\Gamma(d/2)}$$

This is the chi-square distribution with d degrees of freedom, used to obtain exponential confidence bounds in Chapter 3.

COMPLEX SYSTEMS

Models for systems that continue to operate as long as certain combinations of components are operating can be developed with great generality (see Barlow and Proschan, 1975). Here we discuss two types of complex systems: those that operate as long as at least r components (any r) out of n identical components are working, and those that can be diagrammed as combinations of series and parallel (not necessarily identical) components.

The formula for the reliability function when at least r out of n components must work is obtained by summing the probabilities "exactly r," "exactly $r + 1$," "exactly $r + 2$," and so on, all the way up to "exactly n" work. These exact cases are all disjoint events, and the sum is the probability that at least r out of n are working. Each of these probabilities can be evaluated using the binomial formula (see Chapter 9). The result is

$$R_s(t) = \sum_{i=r}^{n} \binom{n}{i} R^i(t) \left[1 - R(t)\right]^{n-i}$$

where

$$\binom{n}{i} = \frac{n!}{n!(n-i)!}$$

and

$$n! = n \times (n-1) \times (n-2) \times \ldots \times 1$$

Exercise 8.4

For a certain type of airplane to fly, at least two out of its three engines must function. The engine reliability at time T hours is 0.995. Find the probability that the airplane flies successfully for T hours, assuming identical and independent engines.

Exercise 8.5

Repeat Exercise 8.4 for engine reliability at T hours of 0.99, and again for engine reliability of 0.95. Note how the probability of airplane failure increases by orders of magnitude.

The CDF and failure rate for this model are derived from the reliability function using basic definitions.

Many systems can be broken down into combinations of components or subassemblies that are in parallel configurations, and combinations that are in series. These systems can be diagrammed like an electric circuit, with blocks logically "in parallel" and blocks logically "in series." The system "working" means there is a path for "electricity" to flow from one end of the diagram to the other. The system may or may not actually have electronic parts—or it may have a combination of electronic and mechanical components. The electric circuit diagram is used only as a convenient device that helps us reduce the system, by successive steps, to simpler systems with equivalent failure rates.

The three steps in this method are:

1. Diagram the system as if it were an electric circuit with parallel and series components and groups of components. Display each component as a circle and write R_i within the circles of all the components that have that reliability function.
2. Successively reduce combinations of components by replacing, for example, a group of components that are in series by one equivalent component. This step will produce a large circle that has the product of the Rs from each circle it replaced as its reliability. For components in parallel, the equivalent component has an R calculated using the parallel model formula (one minus the product of the individual circle CDFs).
3. Continue in this fashion until the entire system is reduced to one equivalent single component whose reliability function is the same as that of the entire original system.

This procedure sounds complicated and arbitrary. Actually, it turns out to be fairly automatic, after a little practice. A few examples will illustrate how it works.

Example 8.5   Complex System Reduction

A system has five different parts. Three of them must work for the system to function. If at least one of the remaining two components is working, along with the first three, the system will function. What is the system reliability function?

Solution

The analogous circuit is drawn below.

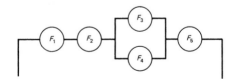

The simplest way to proceed is to replace R_3 and R_4 with one component of equivalent reliability.

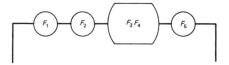

Now all the components are in series and a straight forward application of the series model provides the answer.

$$F_5 = 1-(1-F_1)(1-F_2)(1-F_3F_4)(1-F_5)$$

Example 8.6   Complex System Reduction

A system of six components can be broken up logically into three subassemblies. The first has three components, two of which are the same, and as long as any one of the three works, the subassembly will work. The second subassembly has two different identical components, either of which must work for this part of the system to function. The last (logical) subassembly consists of one critical part. The system works only as long as each subassembly functions. What is the system reliability?

Solution

Shown below are the diagrams for the system (left) and the successive reduction steps solving for the system reliability (right).

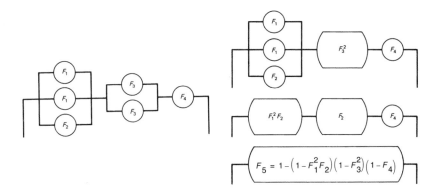

The hardest part of this procedure turns out to be making the initial diagram of the system. Not all systems can be broken down this way, even as an approximation. A simple generalization, left to the reader, would be to allow blocks of "r out of n" configured components. The single equivalent component replacing such a block would have a reliability function given by the binomial formula at the start of this section.

Exercise 8.6

A system has 12 components. Components 1 through 8 are different and have CDFs $F_1, F_2, \ldots, F_8$. Components 9 through 12 are the same, with CDF F. Components 4, 5, and 6 are critical, and each must operate for the system to function. However, only one of components 1, 2, and 3 has to be working, and the same for 7 and 8. At least two of the four identical components must work, as well. Diagram the system and write the probability the system survives.

Sometimes the diagram methods given in this section can be used in combination with simple probability arguments to solve complex system models. For example, the following diagram shows a simple parallel or redundant situation where the components A and B are backed up by C and D. Say A is a logic processor for a computer, B is a memory area with critical data, and C and D are redundant backups for A and B, respectively.

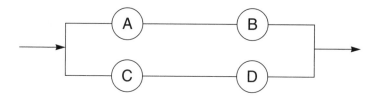

Obviously, this configuration is much more reliable than just having A and B. However, it is also flawed as a design, since if B and C fail we still have a processor and the critical data, but they are not linked together. What we need is a switch, shown in the schematic as component E.

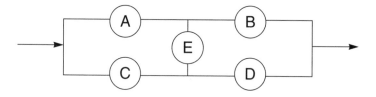

The probability that the system survives until time t can be written as follows:

P (system survives to t) $= P$ (system survives to $t|$E survives) P (E survives)

$+ P$ (system survives to $t|$E fails) P (E fails)

(See the law of Total Probabilities in Chapter 2.) The system surviving given the switch works is diagrammed by

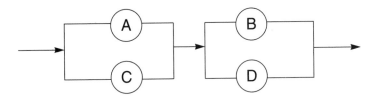

or two parallel blocks in series with each other. The system surviving without the switch is given by the original design. Using the diagram reduction rules we easily obtain

$$R_s = (1 - F_A F_C)(1 - F_B F_D) R_E + [1 - (1 - R_A R_B)(1 - R_C R_D)] F_E$$

Exercise 8.7

Find the system reliability if a third backup system processor F and memory G are added, and the switch E is able to couple any working processor to any working memory.

DEFECT MODELS AND DISCOVERY DISTRIBUTIONS

This section describes a very useful model that will complete the kit of tools we need to construct very general reliability algorithms that can match the failure rate characteristics of most components and systems encountered in practice.

Suppose a small proportion of a population of components have the property of being highly susceptible to a certain failure mechanism. Consider them to be manufacturing defects in a reliability sense—they work originally and cannot be detected as damaged by the standard tests and inspections; yet, after a short period of use, they fail. The rest of the components in the population are either not at all susceptible to this failure mechanism or they fail much later in time. A typical example might be a failure mechanism that only occurs when there are traces of a certain kind of contaminant left on critical spots within the component.

Let $F_d(t)$ be the CDF that applies to the small proportion of susceptible components. The CDF for the rest of the components will be denoted by $F_N(t)$. This CDF is usually derived by applying the competing risk model to the life distributions for all the failure mechanisms that apply to normal (non-defective) components.

The total CDF, $F_T(t)$, for the entire population is constructed by taking a weighted average of the CDFs for the defective and normal subpopulations. The weights are the proportions of each type of subpopulation. This situation is called a *mixture*. For example, if α proportion are reliability defects, and $(1 - \alpha)$ are normal, then

$$F_T(t) = \alpha F_d(t) + (1 - \alpha) F_N(t)$$

Typically, $F_d(t)$ will be a Weibull or a lognormal with a high probability of failing early, while $F_N(t)$ has a low early failure rate that either stays constant or increases very late in life.

The failure rate of the mixture CDF is

$$h_T(t) = \frac{\alpha f_d(t) + (1 - \alpha) f_N(t)}{1 - [\alpha F_d(t) + (1 - \alpha) F_N(t)]}$$

and this has a "bathtub" shape for the distributions mentioned above.

This model, where a (hopefully small!) proportion of the population follows an early life failure model, appears to have wide applicability (see, for example, Trindade, 1991). Some ways of spotting it from test or use data will be described later in this chapter (in the section on data analysis). Some authors describe these early failures as "infant mortality," a term that goes back to the actuarial origins of the hazard rate.

The defect mixture model can be extended to cover the almost universal problem of the field discovery of real manufacturing defects that escape in-house detection. No matter how carefully we test components or systems before shipment, some small fraction of defects may escape, to be discovered first by the customer. If this fraction of escapes is α, the mixture model and failure rate equation apply with only a change in the interpretation of $F_d(t)$. This distribution is no longer a life distribution for the failure times of reliability defects. Instead, it is a distribution that empirically models the rate at which customers discover the time-zero defects that escape the manufacturing plant. We denote this CDF by $F_e(t)$ and interpret it as a discovery distribution for manufacturing escapes.

A good model for this discovery rate is a Weibull distribution with shape parameter between 0.5 and 0.8, and a characteristic life parameter chosen so that 95 or 99 percent of the defects are discovered by 3000 to 5000 hr. After fine tuning to match the front-end curve of actual data, this model can be a standard part of most reliability algorithms.

Exercise 8.8

Approximately 5 percent of a certain product is known to contain defects that will cause the product to fail. Two hundred randomly chosen units are placed on stress. Failures are observed at the following times: 2, 3, 5, 7, 8, 10, 15, 30, 52, 70, 96, and 150 hr. The stress test continues to 1,000 hr without any further failures. Estimate the observed (or empirical) CDF $F_T(t)$ at each failure time. Estimate the fraction defective α. Estimate the defective subpopulation CDF $F_d(t)$ at each failure time. Assuming an exponential distribution for $F_d(t)$, estimate the MTTF. Write the estimated defect model for $F_T(t)$.

GENERAL RELIABILITY ALGORITHMS

By starting with a general and flexible form for component failure rates and using the previously discussed system models to build up to higher assemblies, we can reduce reliability evaluation to an algorithm that handles most applications. The field use CDF would include a term for discovery of escapes and terms for as many reliability defect subpopulations as are needed. To this would be added the typical population competing failure modes. These competing distributions could have any of the shapes discussed in the chapters on the Weibull and the lognormal distributions (i.e., an early life, a constant, or a wear-out shape). If we call the CDF for the normal population competing failure modes F_N, the complete algorithm is

$$F_T = \alpha F_e + \beta F_d + (1 - \alpha - \beta) F_N$$

where F_e is the discovery CDF for the α proportion of defective escapes, F_d is the early life distribution for the β proportion of reliability process defects (if there are more than one type, add as many more such terms as are needed), and F_N is derived from the n normal product competing failure modes as $F_N = 1 - R_1 \times R_2 \times \ldots \times R_n$. This model is most effective if acceleration models for the parameters of F_d and F_N are known. Then the algorithm can be written to give the use CDF as a function of use conditions, making the expression very useful for a product that has varied applications.

Note that the general algorithm is a mixture of the CDFs F_e, F_d, and F_N, with each of these given weights that add up to one. This mixture model corresponds to the physical assumption that all the defective subpopulations are separate from one another, with a defective component only having one type of defect and always eventually failing because of that defect. It would not be difficult to write a general algorithm where the defective subpopulation CDFs compete with one another and the normal failure CDFs via the competing risk model. This would lead to much more complicated equations, however, and little difference in failure rates for the typical cases where the weights for defective subpopulations are small. In many cases, two CDFs are sufficient to model the normal population competing failure modes. One CDF is an exponential and contributes a constant failure rate. The reliability function for this CDF will be denoted by R_I. The second CDF contributes an increasing or wear-out type failure rate. This could be a suitable lognormal or Weibull distribution. The reliability function for this CDF will be denoted by R_w. Then F_N is (by the series model) $F_N = 1 - R_I R_w$. The expression for F_T becomes

$$F_T = \alpha F_e + \beta F_d + (1 - \alpha - \beta)(1 - R_I R_w)$$

Example 8.7   General Reliability Algorithm

A dense integrated circuit module consists of an encapsulated microchip connected to a ceramic substrate. After months of life testing and modeling, distributions for the significant failure mechanisms at use conditions are known. There is a constant failure rate of 100 PPM/K and a lognormal wear-out distribution with $\sigma = 0.8$ and $T_{50} = 975,000$. In addition, it is estimated that about 0.2% of the modules have entrapped contaminants that lead to corrosion failure according to a lognormal distribution with $\sigma = 0.8$ and $T_{50} = 2700$.

If the test coverage and efficiency allows 0.1% defective modules to be shipped, and the assumed discovery model is a Weibull with $m = 0.5$ and $c = 400$, give the failure rate curve for the module.

Solution

All the parameters and distributions for the general reliability algorithm have been specified. The value of α is 0.001, and $\beta = 0.002$. The use CDF and PDF are calculated from

$$F_T = 0.001 \, F_e + 0.002 \, F_d + 0.997 \, (1 - R_I R_w)$$

$$f_T = 0.001 \, f_e + 0.002 \, f_d + 0.997 \, (f_I R_w + f_w R_I)$$

where F_e and f_e are the discovery Weibull CDF and PDF, F_d and f_d are the lognormal CDF and PDF for the reliability defects, R_I and f_I are the reliability function and PDF for the exponential (constant failure rate) failure mode, and R_w and f_w are the reliability function and PDF for the wear-out lognormal.

A graph of the failure rate $h_T(t) = f_T(t)/[1 - F_T(t)]$ is shown in Figure 8.1. Note the bathtub shape with a steep front end due to the discovery of escapes and reliability defect failures. The failure rate settles down to a little over the constant adder of 100 PPM/K after about 15,000 hr. Wear-out starts being noticeable by

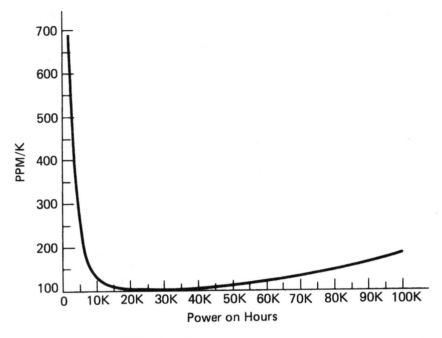

Figure 8.1   Example 8.7 Failure Rate Curve

50,000 or 60,000 hr, causing the failure rate to rise to nearly 180 PPM/K at 100,000 hr.

Exercise 8.9

Plot a graph of the failure rate curve (0 to 100K hr) under the following conditions: R_I is exponential with $\lambda = 50$ FITs, R_W is Weibull with $m = 2$ and $c = 500,000$ hr, F_e is exponential with $MTTF = 1,000$ hr, $\alpha = 0.5\%$, $\beta = 1\%$, and F_d is lognormal with $\sigma = 0.4$ and $T_{50} = 700$ hr.

BURN-IN MODELS

If the front end of a failure rate curve, such as the one shown in Example 8.7, is too high to be acceptable, and it is not possible to make significant improvements in the manufacturing process, there are two options available to improve the situation. By increasing test coverage and efficiency, the escape discovery portion of the front end can be reduced. Also, by stressing or "burning in" the product prior to shipment, the reliability defect portion of the front end failure rate can be virtually eliminated. This section discusses how to use our general reliability algorithm to do a mathematical analysis of the possible benefits of a burn-in.

Suppose the burn-in is for T hours at much higher than normal use stress. We need acceleration factors, to use conditions, for every failure mode (apart from defect discovery) in the general algorithm. These factors are obtained using methods such as described in Chapter 7. Assume the factor for the reliability defect mode is A_d and the factors for the constant and wear-out modes are A_I and A_w, respectively. If there are multiple mechanisms (and CDFs) for any of these types, more factors are necessary.

Now, we have to decide how efficient the burn-in testing is at catching all the failures produced by the burn-in stress. Are all these failures detected and removed prior to shipment? Or do some escape to be discovered by the end user along with the other manufacturing escapes? Are any previously undetected manufacturing escapes caught by the burn-in testing?

For simplicity, we assume none of the α manufacturing escapes are found and removed at the burn-in. Let the burn-in test efficiency be $(1 - B_e)$. This means that burn-in testing catches $(1 - B_e)$ of all the early life failures that happen during the burn-in period. B_e is the fraction of these failures that escape and merge with the other α manufacturing defect escapes. The fallout and escapes from burn-in are

$$\text{Fallout} = (1 - B_e)\, [\beta F_d(A_d T) + (1 - \alpha - \beta)]\, [1 - R_I(A_I T)\, R_w(A_w T)]$$

$$\text{New escapes} = \gamma$$

$$= B_e \{\beta F_d(A_d T) + (1 - \alpha - \beta)\, [1 - R_I(A_I T)\, R_w(A_w T)]\}$$

In addition, the new proportion of shipped product that is defective [and will be discovered according to $F_e(t)$] is

$$\text{Total escapes to field} = (\alpha + \gamma)/(1 - \text{fallout})$$

The $(1 - \text{fallout})$ term appears in the denominator because the population actually shipped has been reduced, and the defects are now a higher proportion. Similar correction terms appear in the expression for the field CDF after burn-in. Let

$$\alpha' = \frac{(\alpha + \gamma)}{1 - \text{fallout}}$$

$$\beta' = \beta \frac{1 - F_d(A_d T)}{1 - \text{fallout}}$$

Then the CDF is

$$F_{BI}(t) = \alpha' F_e(t) + \frac{\beta' [F_d(t + A_d T) - F_d(A_d T)]}{[1 - F_d(A_d T)]}$$

$$+ \frac{(1 - \alpha' - \beta') [R_i(A_i T) R_w(A_w T) - R_I(t + A_I T) R_w(t + A_w T)]}{R_I(A_I T) R_w(A_w T)}$$

where the correction terms that appear in the denominators can also be viewed as making the probabilities of failure or survival conditional upon surviving the burn-in.

Failure rates are obtained by taking the derivative of the after burn-in CDF $F_{BI}(t)$, and then calculating $h(t)$.

$$f_{BI}(t) = \alpha' f_e(t) + \frac{\beta' f_d(t + A_d T)}{R_d(A_d T)} + \frac{(1 - \alpha' - \beta')}{R_I(A_I T) R_w(A_w T)}$$

$$\times [f_w(A_w T + t) R_I(A_I T + t) + f_I(A_I T + t) R_w(A_w T + t)]$$

$$h_{BI}(t) = \frac{f_{BI}(t)}{1 - F_{BI}(t)}$$

Example 8.8   Burn-In Model

The reliability engineers responsible for the component described in Example 8.7 decide to try a 9 hr burn-in to improve the front end of the failure rate curve shown in Figure 8.1. The three failure modes described by the CDFs (F_d, F_I, and F_w) are accelerated by temperature according to an Arrhenius model, with ΔHs of 1.15, 0.5, and 0.95, respectively. The field use temperature is 65° C, and the proposed burn-in temperature is 145° C. Assuming perfect efficiency at catching failures generated by the burn-in, what will the expected burn-in fallout be? How will the new failure rate curve for burned-in product compare to the old curve? What are the old and new $AFR(10,000)$ values?

Solution

First we use the methods described in Chapter 7 to calculate the Arrhenius acceleration factors. Table 7.6 gives a TF value of 6.6. From Figures 7.3 and 7.4 we read off the approximate acceleration values of $A_d = 2,000$, $A_I = 27$, and $A_w = 500$. The fallout at burn-in is

Fallout $= 0.002 F_d (9 \times 2000) + 0.997 [1 - R_I (9 \times 27) R_w (9 \times 500)]$

$= 0.0020065$

by substituting the proper CDFs from Example 8.7.
    The after burn-in CDF and PDF are

$$F_{BI}(t) = 0.001002 F_e(t) + \frac{(0.00001776 [F_d(t + 18000) - 0.9911])}{0.008860}$$

$$+ 0.9990 [0.9999757 - R_I(t + 243) R_w(t + 4500)]$$

$$f_{BI}(t) = 0.001002 f_e(t) + 0.002004 f_d(t + 18000)$$

$$+ 0.9990 [f_w(t + 4500) R_I(t + 243) + f_I(t + 243) R_w(t + 4500)]$$

Figure 8.2 shows how $h_{BI}(t)$ compares to the old $h(t)$. The burn-in has made a significant improvement in the front-end failure rate, with little effect on the val-

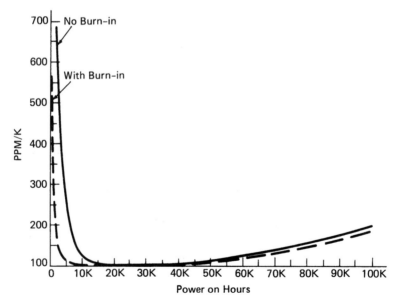

Figure 8.2   Burn-In Improvement Example

ues after about 18,000 hr. The new $AFR(10,000)$ is 201 PPM/K or FIT, as compared to the non-burn-in value of 390 PPM/K. This value was calculated from

$$AFR(T) = 10^9 \times \frac{(-\ln R(T))}{T}$$

In the last example, the theoretical analysis showed that a short burn-in would be beneficial. However, burn-in may not always be so effective, since burn-in affects every failure mode differently. Constant failure rate modes are not improved at all, while wear-out modes are made worse. Only the early life failure modes, or those due to a reliability defective subpopulation, are affected; for those to be influenced significantly by a short burn-in, there must be a high acceleration factor.

Exercise 8.10

Find the effect on the AFR from 0 to 1000 and 10,000 hours of a 24 hr burn-in on the component described in Exercise 8.9. Assume the same acceleration models and burn-in and field conditions as in Example 8.8. Plot both new and old failure rate curves.

For reasons of economy and effectiveness, it is best to improve early life performance by improving the manufacturing process. With a dedicated return and analysis program of field failures, the major contributors to early life failure can be traced back to mistakes or improvable procedures at specific stages of manufacture. Fixing these improves both yield and reliability.

The next most efficient way to improve early life performance is by more effective tests and screens. Finally, comes the burn-in, or last-minute stress screening option. This fix can be thought of as a process band-aid—sometimes necessary, no matter how much we would like to avoid it.

The burn-in model of this section can be a useful tool, helping us to choose the best stress level and time for burn-in or, in some cases, warning us that burn-in is not likely to produce the results desired. Careful data collection, along with analysis of both burn-in fallout and later life performance, should be part of any burn-in implementation plan. This is especially true when using stresses of the "shake, rattle, and roll" type, for which no known acceleration model may exist.

Some useful references on burn-in are Peck (1980) and Jensen and Petersen (1982). They call the defect subpopulation failures "freaks" and the escapes "infant mortalities." They also point out that it is common to introduce infant mortality defects during the manufacturing operations that occur after a component-level burn-in. These added defects may make the burn-in appear less effective.

DATA ANALYSIS

The competing risk model and the defect mixture model have been treated theoretically in the proceeding sections. Here we concern ourselves with how to analyze life test data when one of these models applies.

Assume we have run a life test of a component and are in the middle of plotting the failure data and running analysis programs when we are told by the failure analysis engineers that they can divide the failures into three very distinct modes. Moreover, based on the nature of the failure mechanisms, you might want to use different life distributions to model these modes. How can you statistically separate the different failure distributions and estimate their parameters?

This subject was mentioned briefly in the section titled "Failure Mode Separation" in Chapter 2. Assuming the independence of failure modes (i.e., they are not "looking over their shoulder at each other" and changing their kinetics depending on how the other is progressing) and a first-fail model, we can treat the data mode by mode as multicensored data. When we are analyzing mode 1, all mode 2 and 3 failures are "censored units," taken off test at the failure time. Multicensored data can be plotted using either hazard plotting procedures or the Kaplan-Meier product probability method (see Chapter 6). Good maximum likelihood estimation programs will be able to analytically handle multicensored data and estimate distribution parameters. The key point is to go through the analysis one mode at a time, treating all other modes as units taken off test.

The ability to separate data points by failure mode is critical to the analysis. This separation should be supported with physical analysis; trying statistically to separate failure modes is difficult and dangerous, especially if the distributions have considerable overlap.

Occasionally, a component with several known failure mechanisms has zero failures on life test. Even if we are willing to assume that each mode has an exponential life distribution, is there anything we can do with zero failures? Say there are three failure modes with unknown failure rates λ_1, λ_2, and λ_3. Then the component has the constant failure rate of $\lambda = \lambda_1 + \lambda_2 + \lambda_3$, and we can use the zero failures formulas of Chapter 3 to put an upper bound on λ. Unfortunately, this same upper bound applies individually to each of the three mechanism failure rates, so we have the paradoxical-sounding result that the upper bound on the total is 1/3 the sum of the upper bounds of the parts. If we have known acceleration factors A_1, A_2, and A_3 for the three modes, the best we can do is use the minimum of these to reduce the component test upper bound to a use-condition upper bound.

The final aspect of data analysis that we will discuss in relation to the models in this Chapter is how to detect and analyze data with failures from a reliability defective subpopulation. For example, if we test 100 units for 1000 hr and have 30 failures by 500 hr, and no more by the end of test, are we dealing with two populations or just censored data? If we continue the test, will we see only a few more failures because we have "used up" the bad ones? Or will the other 70 fail according to the same life distribution?

The easiest way to determine that data contains a defect subpopulation is by graphical analysis. Assume, for example, that the failure mode is one typically modeled by a lognormal distribution. If we plot the failures on lognormal graph paper, and instead of following a straight line they seem to curve away from the cumulative percent axis, this is a signal that a defect subpopulation might be present. If we run the test long enough, we would expect the plot to bend over asymptotic to a cumulative percent line that represents the proportion of defectives in the sample.

So the clues are a plot with points that bend and a slowing down or complete stop of failures, even though many units are still on test. In addition, a physical reason to expect reliability defects is highly desirable.

Returning to our 30 failures out of 100 on test example, look what happens if we replot the data, this time assuming only 30 units were on test. If these were truly the complete reliability defective subpopulation of units on test, and the lognormal model is appropriate for their failure mode, then the new plot should no longer bend. The T_{50} and sigma estimates obtained from the best fitting line (or a MLE program) would be the proper parameter estimates for the $F_d(t)$ of the defective subpopulation. The estimate of β, the proportion defective, would be 0.3. (See Trindade, 1991, for examples of this type of analysis.)

We would run into trouble, however, if there were a large number of defective units left on test that hadn't had enough time to fail. The original plot, with unadjusted sample size, would have less curvature, and it might not be clear where the asymptote is. If we worked by trial and error, adjusting the number on test from the full sample size down to the actual number of failures, we might be able to pick one plot where the points line up best. This fitting can be done by computer, using an iterative least squares fit and choosing the starting sample size that yields the smallest least squares error.

It is possible to work out more sophisticated techniques (such as maximum likelihood estimation) to estimate the proportion of defectives and the distribution parameters. The maximum likelihood approach will be described after a simple example using the graphical least squares approach is given. One clear advantage of the maximum likelihood approach is that it can be used to statistically test the hypothesis that there actually is a defective subpopulation.

Example 8.9   Defect Model

A certain type of semiconductor module has a metal migration failure mechanism that is greatly enhanced by the presence of moisture. For that reason, modules are hermetically sealed. It is known that a small fraction will have moisture trapped within the seal. These units will fail early, and it is desired to fit a suitable life distribution model to these reliability defects.

Test parts are made in such a way as to greatly increase the chance of enclosing moisture in a manner typical of the normal manufacturing process defects. One hundred of these parts were put on life test for 2000 hr. There were 15 failures. The failure times were: 597, 623, 776, 871, 914, 917, 1021, 1117, 1170, 1182, 1396, 1430, 1565, 1633, 1664.

Estimate the fraction defective in the sample and the life distribution parameters for this subpopulation (assuming a lognormal model applies). Use an iterative least squares method.

Solution

Figure 8.3 shows a lognormal plot of the fail times using $100(i - 0.3)/100.4$ as the plotting position for the ith fail. The points do appear to have curvature, although exactly where they will bend over to is not readily apparent. By iterative least squares trials, changing n from 100 down to 15, a "best" fit occurs at $n = 18$. This plot is shown in Figure 8.4. Here the points line up well. The graphical T_{50} and sigma estimates are 1208 and 0.43. (These results are quite good, since the data

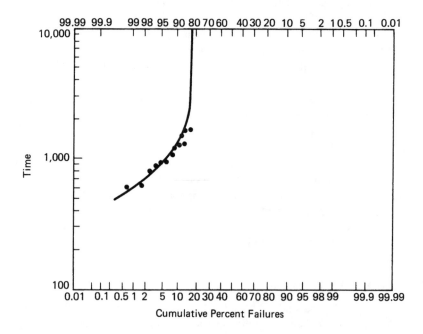

Figure 8.3   Data from Example 8.9 Plotted (unadjusted)

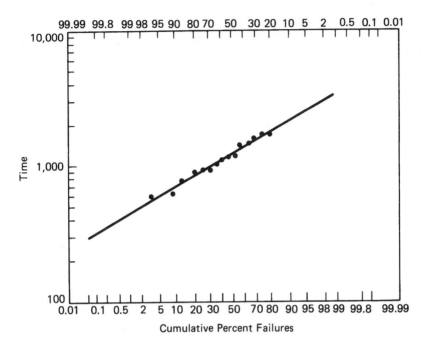

Figure 8.4   Data from Example 8.9 Plotted (adjusted)

were simulated from 20 defects with a T_{50} of 1200 and a sigma of 0.45.) The best fit fraction defective estimate of 18/100 is also close to the "true" value of 0.20.

Exercise 8.11

Check the fit of the model to the data in Example 8.9 by plotting the CDF F_T estimated according to the model against the empirical CDF F_T. (Use linear scales for both axes.)

Solving for the parameters in the defect model using the method of maximum likelihood turns out to be a simple extension to the maximum likelihood theory described in Chapter 4. In that Chapter, maximum likelihood equations were described for various types of censored data. The basic building blocks of the equations were the PDF f and the CDF F. If, however, only a fraction, p, of the population is susceptible to the failure mechanism modeled by $F(t)$, then $pF(t)$ is the probability that a randomly chosen component fails by time t. Similarly, the "likelihood" of a randomly chosen component failing at the exact instant t becomes $pf(t)$. The rule for writing likelihood equations for the defect model is to substitute pf and pF wherever f and F appear in the standard likelihood equation.

The standard likelihood equation for Type I censored data [n on test, r fails at exact times $t_1, t_2, \ldots, t_r$, and $(n - r)$ units unfailed at time $T =$ the end of test] is given by

$$LIK = \left[\prod_{i=1}^{r} f(t_i) \right] [1 - F(T)]^{n-r}$$

If we suspect that only a fraction p of the population is susceptible to failure, the (defect model) maximum likelihood equation becomes

$$LIK = p^r \left[\prod_{i=1}^{r} f(t_i) \right] [1 - pF(T)]^{n-r}$$

Maximum likelihood estimates are defined as the values of p and the population parameters that maximize LIK. These are the same values that maximize $L = \log_e LIK$, an easier equation to work with. As described in Chapter 4, the standard calculus technique of taking partial derivatives of L with respect to the unknown parameters and setting these derivatives equal to 0 gives a set of equations whose solutions are the desired estimates. Unfortunately, for reliability data, these equations are usually nonlinear and have to be solved using a numerical technique such as the Newton (or Newton-Raphson) method described in most textbooks on numerical analysis, (see, for example, Gerald and Wheatley, 1984).

If maximum likelihood estimates have been calculated for a suspected defect model, it only takes a little more work to test the hypothesis $p = 1$, versus the alternative that only a fraction of the population will actually fail. Let L_1 be the maximum log likelihood for the standard (non-defect) model and let L_2 be the maximum log likelihood for the defect model. The test statistic $\lambda = -2(L_1 - L_2)$ is known as the *likelihood ratio test statistic*. If the hypothesis $p = 1$ is true, λ will have approximately a chi-square distribution with 1 degree of freedom. If λ is larger than, say, $\chi_{1;.95}$, then we would reject the standard model and accept the defect model at the 95 percent confidence level.

Exercise 8.12

Derive the defect model likelihood equation for interval (or readout) data where the readout times are $T_1, T_2, \ldots, T_k =$ the end of test. Assume r_1 fails are discovered at time T_1, r_2 at T_2, and so on until the end of test. There are r total fails observed and $n - r$ units removed at the end of test.

Example 8.10   Maximum Likelihood Estimate for the Defect Model

Calculate the maximum likelihood estimates for data in Example 8.9, assuming a defect model applies. Use the likelihood ratio test to see whether the data supports the existence of a susceptible subpopulation.

Solution

A sophisticated computer program is needed to solve the complicated nonlinear likelihood maximization equations. Several commercially available programs exist for handling lognormal or Weibull censored data, but none to date has the capability for the defect model analysis described in this chapter. The general-purpose statistical package AGSS (A Graphical Statistical System), announced by IBM in March of 1992, will operate at a PC or workstation level as well as on a large computer, and there are plans to include a defect model analysis in a future release. Until such programs are available, the reader who has the knowledge will have to program custom solutions using the likelihood equations described above.

For the data in Example 8.9, the maximum likelihood answers are as follows: the T_{50} estimate is 1112, and the σ estimate is 0.35. The fraction defective estimate is 0.16. The maximum value of the log likelihood for the defect model has the value -150.5, while the maximum log likelihood assuming $p = 1$ is -153.6. A likelihood ratio test of the defect model hypothesis yields a chi-square statistic of 6.106, which is very unlikely to occur if there were no defective subpopulation (if $p = 1$, a chi-square statistic this large would occur only about 1.4 percent of the time).

A final word of caution: although the defect model is very useful and applies in many applications, a slight appearance of curvature on a probability plot does not automatically indicate its presence. Random data can often give the appearance of curvature, especially in the tails, even when all the samples belong to the same population. A long period without further failures is a more reliable clue, and a physical explanation should always be sought. The iterative least squares procedure makes use, in effect, of another parameter in fitting the points—so it will often find a better fit as if there were a defect population, even if none is present. Even a more analytic approach using a likelihood ratio test to confirm the presence of a defect subpopulation should be backed up by a reasonable physical explanation for the difference between susceptible and nonsusceptible units.

Exercise 8.13

Use the MLE method to estimate the parameters of the defect model for the data in Example 8.9, assuming the test ended at 1665 hr instead of 2000 hr. What does the likelihood ratio test say in this case?

THE "BLACK BOX" APPROACH: AN ALTERNATIVE TO "BOTTOMS-UP" METHODS

The approach described so far has been to model each failure mechanism within each component and use the competing risk model and the series and the parallel models (or the more complex models and algorithms described in the preceding sections) to build up to system level failure rates and reliabilities. This approach is highly recommended for several technical reasons: acceleration models make physical sense only when applied at the individual failure mechanism level, and this situation is also the natural place to model a homogeneous population of failure times using one of the lifetime distribution models discussed in this text. However, there are often compelling practical reasons to eschew a "bottoms-up" approach. Frequently, a company purchases a complete assembly of components and uses this assembly within a box or system it manufactures and sells. It may be difficult or impossible to obtain failure-mode-level, or even component-level, reliability information from the supplier of this assembly. It is a common practice to call this assembly a "black box" and test it as if it were a single component within the system. If the test is done at high laboratory stresses a single (best guess or "average") acceleration factor is used to translate the results to typical use conditions. A flexible life distribution such as the Weibull or the lognormal is then used to calculate failure rates, and the results are added to the other system components to obtain a total system failure rate projection.

This procedure saves the time and expense of either testing many unfamiliar components or doing extensive literature searches for reliability data on these components (with no guarantee of finding appropriate estimates for the actual system use conditions). For these practical reasons, "black box" testing will continue to be a common practice regardless of theoretical difficulties. It is important, however, to be aware of situations that can lead to inaccurate or misleading results. The following examples will show some of these situations and allow us to formulate guidelines to minimize the dangers inherent in "black box" testing.

Example 8.11   "BLACK BOX" TESTING I

A subassembly contains three components. Assume that the true Arrhenius ΔH for component 1 is 0.6, for component 2 it is 0.9, and for component 3 it is 1.2. Also assume that the true (but unknown) use failure rate for all three components is a constant 0.01%/K hr. It is proposed to test the assembly as if it were a single component with a ΔH of 0.9 (the actual "average" ΔH). Testing will be done at 125° C. What is the actual population use and test failure rate for this assembly? If the test results yield a "perfect" stress failure rate estimate (exactly equal to the theoretical test failure rate), what will the test engineers predict the use condition failure rate to be?

Solution

The true assembly failure rate is the sum of the failure rates of the three components, or 0.03%/K hr. The true acceleration factors based on the actual Arrhenius ΔH constants for the three components are 30, 167, and 1624 for components 1, 2, and 3, respectively (see Chapter 7 for details on acceleration factor calculations with the Arrhenius model). That means the (theoretical) test failure rate is 0.30%/K hr + 1.67%/K hr + 16.24%/K hr = 18.21%/K hr. If the test failure rate estimate actually turns out to be 18.21%/K hr, however, use of a single ΔH of 0.9 would yield a use failure rate estimate of 18.21/167 = 0.11%/K hr. This would be an overestimate of the use failure rate by a factor of almost 4×, even though the test results were as accurate as possible.

Example 8.12   "BLACK BOX" TESTING II

Assume the same situation as in the previous example except for the actual use failure rates of the three components. Instead of each having a 0.01%/K hr failure rate, component 1 now has a 0.1%/k hr failure rate, while components 2 and 3 both have 0.001%/K hr failure rates. What are the true use and stress failure rates, and what will the test engineers estimate the use failure rate to be, again assuming they obtain a perfect estimate of the stress failure rate?

Solution

The use failure rate is 0.1 + 0.001 + 0.001 = 0.102%/K hr. The stress failure rate is 3 + 0.167 + 1.624 = 4.791%/K hr. This time the use failure rate would be estimated by 4.791/167 = 0.03%/K hr. This understates the true use failure rate by a factor of more than 3×.

The two "black box" testing examples just given used the correct average ΔH, and there was no sampling error in the estimate of the stress failure rate of the assembly. Yet, in the first case we overestimated the true use failure rate by 4×, and in the second case we underestimated the true use failure rate by more than 3×. Obviously, a linear average of ΔHs leads to problems, since the corresponding acceleration factors are not linear. Similar examples could easily be constructed using any other kind of "average" ΔH or "average" acceleration factor.

On the other hand, if the failures occurring in the first example were analyzed down to type of component and failure mode, it would become obvious that most of the failures were due to component 3 (about 16 out of 18, or almost 90 percent of the failures, typically would be due to component 3). The appropriate acceleration factor for those failures is 1624—much higher than the 167 used in the calculation. No wonder the failure rate was overestimated!

In the second example, more than 62 percent of the failures would be from component 1. These are only accelerated by a factor of 30 and are treated optimistically by the use of an acceleration factor of 167. These results explain why the use failure rate obtained in the example was an underestimate.

So when are we likely to have trouble using the "black box" approach? Clearly, when the average acceleration factor used in the analysis is inappropriate for a large number of the failures that occur during the stress test, we stand a good chance of running into difficulty. We might also come to incorrect conclusions if a mechanism that contributes a significant failure rate at use conditions has a very low ΔH. The sample size at stress might not be high enough to see even one of these failures—yet they might be the dominant cause of failure at use conditions.

We can put together these observations into a set of general guidelines for "black box" testing:

1. Assess in advance the failure mechanisms likely to cause use condition failures. They should have approximately equal ΔHs, or the approach is likely to be invalid. Use this ΔH for planning the test assuming an exponential model (see Chapter 3).

2. After the test is over, verify by failure analysis that the failure mechanisms were as expected. If other mechanisms are observed that might have widely varying acceleration factors, follow-up experiments using a bottoms-up approach are called for.

3. If the possibility of components in the assembly having failure mechanisms with very low acceleration factors cannot be ruled out, extreme overstressing of a few assemblies is recommended. Many failures will occur, some of which are extraneous results of the overstress, and others of which are relevant. Analyze these failures to gain confidence that nothing has been overlooked. (This kind of testing is sometimes referred to as "elephant testing." The derivation of this term comes from picturing the assemblies placed one by one in the path of a walking elephant. It may be impossible to project a use failure rate from the results of this kind of testing, but the weak points of the assembly are certainly exposed!)

SUMMARY

This chapter introduced the series and competing risk models, the escape discovery model, and the defect model as building blocks to system reliability. In conjunction with a general algorithm for component failure rates, we have a flexible set of tools to model and project reliability. If we have stress dependency incorporated into our general algorithm, we can also make trade-off analysis calculations concerning burn-in and test efficiency changes.

Methods were described to graphically spot the possible presence of a reliability defective subpopulation. Maximum likelihood methods, requiring sophisticated software, were used to improve upon the graphical results.

We also saw how system redundancy provides enormous improvement in early life failure rates. As component costs go down and high reliability becomes more essential, this design option will apply more and more.

While a "bottoms-up" approach to reliability analysis and failure rate projection is generally recommended, we also presented guidelines for "black box" testing.

PROBLEMS

8.1 A VCR has most of its components on one board. The board has 12 type "A" units, 15 "B" type, and 6 "C" type. All components operate independently. The first failure of any component causes the VCR to fail. Type "A" modules have an exponential failure rate with the *MTTF* equal to 300,000 hr. The "B" modules follow a lognormal distribution with T_{50} = 30,000 hr and sigma = 1.3. The "C" modules have three significant failure mechanisms. All are described by Weibull distributions with shape parameters equal to 0.8 and characteristic lives of 500,000 hr, 800,000 hr, and 1,100,000 hr, respectively. What is the average failure rate of the VCR during the first 1,000 hr? What is the AFR during the next 2,000 hr? What is the chance that the VCR has no failures during the first 500 hours?

8.2 Certain radar systems have reliability described by the exponential distribution with *MTTF* equal to 600 hours. To ensure better reliability, three identical systems are configured such that one system is operating while the other two are in standby mode. The failure of the first system causes an immediate switch to one of the remaining two. If the second one fails, the third in standby mode is also immediately switched on. Assuming a perfect switching device that never fails, find the PDF for the system lifetime. What is the system *MTTF*? What fraction of systems survive 2,000 hr? Now assume the switch has a probability of 0.001 of not working any given time it is used. What are the system PDF, *MTTF,* and survival probability at 2,000 hr now? Hint: with probability 0.999^2, the system lifetime is the sum of three exponential lifetimes; with probability 0.001 × 0.999, the system lifetime is the sum of two exponential lifetimes and, with probability 0.001, the system lifetime ends after the first failure. Answer the questions for each case (or possible branch), and use these probabilities as weights to get the complete answer.

8.3 A system consists of seven units: A, B, C, D, E, G, and H. For the system to function, unit A *and* either unit B or C *and* either D and E together *or* G and H together must be working. Draw the diagram for this setup and write the equation for the CDF of the system in terms of the individual component CDFs.

8.4   One hundred components are put on stress test. Readouts occur at 25, 50, 100, 250, and 500 hours. The numbers of new failures discovered at the respective readouts are 15, 9, 12, 5, and 1. Plot the data on lognormal probability paper. Does the fit appear reasonable? Repeat the plotting, assuming a defect model applies and trying different values for the fraction susceptible to failure. Does the fit improve? Use an iterated least squares graphical procedure to estimate T_{50} and sigma for the defect subpopulation and the fraction defective. If maximum likelihood defect model estimation programs are available, estimate the parameters using the method of maximum likelihood and test whether the assumption of a defect subpopulation is warranted.

8.5   Fifty components are put on stress test. Readouts are taken at 10, 25, 50, 100, 200, 500, and 1000 hours. The numbers of new failures discovered at the respective readouts are 2, 2, 4, 5, 4, 3, and 0. Repeat the plotting and estimation procedures described in problem 8.4, this time assuming a Weibull model for the defect subpopulation.

8.6   Write the maximum likelihood equation for Type I censored data, assuming that a defect model applies and the exponential distribution adequately models the defective subpopulation. Derive the equations that must be solved to obtain maximum likelihood estimates (by taking partial derivatives of the log likelihood).

8.7   You have been asked to review the field failure rate projection made for a complex electronic board your company purchases from a vendor. You are told that 50 boards were tested at high stress for 2000 hr with 6 failures observed (failed boards were immediately repaired and put back on test). Projections were made assuming an acceleration factor of 1000, which you are told is believed to be a conservative overall factor for the many different types of components on the board. The analysis was straightforward: 6 failures from 100,000,000 use-equivalent board hours gives a failure rate estimate of 60 FITs, with a 90 percent upper bound of 106 FITs.

What further questions would you ask to validate this projection? What answers would lead you to accept the projection? What answers would lead you to recommend additional testing? What risks would you point out are inherent with this kind of testing and analysis? Can you think of other recommendations you might make that would give early warning if the test had failed to disclose a possibly serious field reliability exposure?

Chapter 9

Quality Control in Reliability: Applications of the Binomial Distribution

Manufacturers often conduct various tests on samples from individual lots in order to infer the expected level of reliability of product in customer applications. In addition, stressing of consecutive groups of samples from production may be employed to monitor the reliability of a process. Such results may be plotted on statistical control charts to verify that the manufacturing process is "under control." However, since such studies can be costly and time consuming—and often destructive to units—it is important that efficient sampling designs be selected to provide the necessary information while using the minimum quantities of product. This chapter covers the implementation of various types of sampling plans for attribute data, the associated risks, the operating characteristic curves, and the choice of minimum sample sizes. We discuss the applications of various discrete distributions, especially the binomial. Also, the calculation of confidence limits is treated. Finally, we look briefly at the application of statistical process control charting for reliability.

SAMPLING PLAN DISTRIBUTIONS

There are several important considerations an engineer must keep in mind when choosing appropriate sampling plans. First, one must establish the scope of the inference: are the results to be used to draw conclusions about an individual lot or about an ongoing process? If the former, then the size of the sample relative to the lot size must be considered. Obviously, if the sample size is a significant portion of the lot size, say over 10 percent, then there is more information about the quality of the remainder of the lot than there would be in sampling from a process capable of producing an infinite number of units. Indeed, different probability distributions are required to treat each case: the hypergeometric distribution applies when lot size must be considered; the binomial distribution holds for inference to a process. The binomial distribution has so many important applications it will be described in detail. However, to understand the simple derivation of the binomial distribution, we must digress for a moment and discuss permutations and combinations.

PERMUTATIONS AND COMBINATIONS

A *permutation* is an arrangement of objects in which order is important. For example, suppose we have three seating positions and three people: call them A, B, and C. We have three choices for placement in the first position: A, B, or C. Given occupancy of position one, there remain two possible choices for position two. That is, if B is in position one, only A and C are left to sit in position two. When both the first and the second positions are occupied, only one choice remains for the third seat. A quick way of determining the number of potential arrangements uses the following basic principle of counting: if an operation has x possible outcomes and each of these outcomes can be followed by an operation consisting of y possible outcomes, then the total number of outcomes for the successive operations is xy. Extension to three or more operations is obvious. See Ross (1994) for further treatment on this topic.

So by this product rule, for the seating arrangements there are a total of $3 \times 2 \times 1 = 6$ possible orders. We use $n!$, called "n factorial," to denote the descending product of numbers. Thus, $3 \times 2 \times 1 = 3!$. By convention, $0! = 1$. We list the possible arrangements of the three objects below:

Possible Arrangements of Three Objects
ABC
ACB
BAC
BCA
CAB
CBA

Each order listed is called a *permutation*. Thus, the number of permutations of n objects is $n!$.

Suppose we have eight objects, and we wish to determine the number of ordered arrangements we can form using four objects at a time. Consider four available positions. The first position may be occupied by any of eight objects, the second by any of the remaining seven objects, the third by any of the six, and the fourth by any of the five left. Thus, the total number of permutations is $8 \times 7 \times 6 \times 5 = 1{,}680$. We won't list these, but the general formula for the number of permutations of n objects taken r at a time is

$$_nP_r = n(n-1)(n-2) \dots (n-r+1)$$

Note there are r separate terms in the product above.

Suppose we have eight objects and we want to select four at a time, but we don't care about the order of the four selected. Such would be the case, for example, if we were forming teams consisting of four players each from eight possible choices. Such selections in which order is immaterial are called *combinations*. To determine the number of combinations, one simply divides the number of permutations $_8P_4$ by $4!$, the number of permutations of four objects at a time. Thus, the number of combinations of eight objects taken four at a time is

$$_8C_4 = \frac{8 \times 7 \times 6 \times 5}{4 \times 3 \times 2 \times 1} = 70$$

In general, the expression for the number of combinations can be multiplied by $1 = [(n-r)!/(n-r)!]$ to give a simpler appearing expression:

$$_nC_r = \frac{_nP_r}{r!}$$

$$= \frac{n(n-1)(n-2)\dots(n-r+1)}{r!}$$

$$= \frac{n(n-1)(n-2)\dots(n-r+1)(n-r)(n-r-1)\dots1}{[(n-r)(n-r-1)\dots1]\,r!}$$

$$= \frac{n!}{(n-r)!\,r!}$$

$$= \binom{n}{r}$$

where: $\binom{n}{r}$ is the special symbol used to denote the number of combinations of n objects taken r at a time.

Exercise 9.1

How many different eleven-letter words can be formed from the letters in the word MISSISSIPPI?

Exercise 9.2

How many ways can we seat r individuals in any order on $n \geq r$ chairs. (Note: this problem is analogous to the forming of n letter words from r Os (where O designates an occupied chair) and $(n - r)$ Es (where E designates an empty chair.)

Exercise 9.3

How many different ways can r failures occur among n units on stress? (Note that time to failure is not the issue here. We are allocating the designation "failure" in any order among n stressed units.) Hint: see previous exercise and let O indicate a failure and E a surviving unit.

THE BINOMIAL DISTRIBUTION

The exponential, Weibull, normal, and lognormal distributions are examples of continuous distributions. However, data do not always occur in a continuous form. For example, the following questions involve discrete concepts:

1. If the probability of failure of a component at time t is 0.1, what is the probability of at least one failure among ten similar units placed on stress and run for time t?
2. What is the probability of exactly no failures, or one, two, ..., or ten failures by time t?

Such questions usually can be handled via the binomial distribution. In fact, four conditions are necessary for the binomial distribution to apply, as follows:

1. Only two outcomes are possible (e.g., success, failure).
2. There is a fixed number (n) of trials.
3. There exists a fixed probability, p, of success from trial to trial.
4. The outcomes are independent from trial to trial.

In general, for reliability work, the first two conditions are met; the third is assumed to hold, at least approximately; and the fourth is usually applicable. For example, an engineer stresses a fixed number of units (condition 2). Each unit will either survive or fail (condition 1) the test, and the failure of one unit does not affect the probability of failure of the others (condition 4). For similar reasons, we assume all units are equally likely to fail (condition 3) at any given point in time.

The general expression for the binomial distribution, which gives the probability of exactly x failures in n trials with probability of success p per trial, is

$$P(X = x) = \binom{n}{x} p^x (1-p)^{n-x}$$

Although we have not derived previous distributions, it is very illustrative of the application of the probability rules and combination concepts covered earlier in this text to develop the binomial distribution directly. Consider a component with "known probability of failure" by time t given by p. That is, $p = F(t)$, where $F(t)$ is the known CDF value. (Note that p can represent success or failure probabilities since only two outcomes are possible; one with probability p and the other with probability $q = 1 - p$.) Suppose we have n such units and we run them on stress to t hours. One possible result is that at the end of the experiment there are no failures. The probability of a single unit surviving is $1 - p$. The probability of all n independent units surviving is given by the product rule as

$$P(X = 0) = (1-p)(1-p)\dots(1-p) \qquad (n \text{ terms})$$

$$= (1-p)^n$$

Consider the case of one failure occurring. A possible sequence (where F = failure and S = survival) might be

$$FSSS\dots SSS$$

That is, the unit in position one fails and the remaining $n - 1$ units survive. The probability of this specific sequence is

$$p(1-p)(1-p)\dots(1-p) = p(1-p)^{n-1}$$

However, we normally do not specify the order in which a unit failed—only that some unit did not survive. Thus, we are interested in how many ways can any one part fail out of n items, where order doesn't matter. As Exercise 9.3 shows, the answer is given by the combination formula $_nC_1 = n$. Since all sequences with one failure and $n - 1$ survivors have the same probability, and there are n such mutually exclusive sequences, then, by the union rule for probabilities, the probability for exactly one fail is

$$P(X = 1) = np(1 - p)^{n-1}$$

Similarly, the probability of a given sequence in which two failures occur and $n - 2$ survive is given by

$$p^2(1 - p)^{n-2}$$

Since there are $_nC_2 = n(n-1)/2$ different ways of two failures occurring among n units where order is immaterial, the probability of exactly two failures in n items is

$$P(X = 2) = \frac{n(n-1)p^2}{2}(1 - p)^{n-x}$$

Continuing this way, we see that, in general the probability of getting exactly x failures from n items on stress is

$$P(X = x) = \binom{n}{x} p^x (1 - p)^{n-x}$$

For the binomial distribution, the mean or expected number of failures is np; the variance is $np(1 - p)$.

Example 9.1   Binomial Calculations

One hundred light bulbs will be stressed for 1000 hr. The probability of a bulb failing by 1000 hr has been determined to be 0.01 or 1 percent from previous experimental work. Assuming the bulbs on stress are from the same population,

    a. What is the probability that all bulbs survive 1000 hr?
    b. What is the probability of exactly one bulb failing?
    c. What is the probability of at least one bulb failing?

Solution

    a. $(1 - p)^n = (1.00 - 0.01)^{100} = 0.99^{100} = 0.366$

       Therefore, approximately 1 chance out of 3 exists that no bulbs will fail by 1000 hr.

    b. $np(1 - p)^{n-1} = (100)(0.01)(0.99)^{99} = 0.370$

c. We could calculate the individual probabilities of 1, 2, 3, and so on, failures and add these together to get the answer. However, a much simpler procedure recognizes the fact that the probability of at least 1 failure equals 1 minus the probability of no failures. Thus,

$$P(X > 0) = P(X \geq 1) = 1 - P(X = 0) = 1 - 0.366 = 0.634$$

In roughly two out of three such experiments, we would expect at least one failure, but in only about one-third of the experiments will there be exactly one failure.

Exercise 9.4

Fifty devices are placed on stress for 168 hr. The probability of a device failing by 168 hr is 0.05 or 5 percent.
  a. What is the probability that all devices survive 168 hr?
  b. What is the expected number of failures?
  c. What is the probability of at least 1 failure?

Example 9.2   Binomial PDF

Figure 9.1 shows a plot of the binomial probability density function for the case of 20 units sampled from a population with individual probability of surviving past 100 hr given by 0.2. Recall the binomial distribution has an expected number of successes given by np and variance given by $np(1 - p)$. Here, the expected number is $20 \times 0.2 = 4$, and the variance is $20 \times 0.2 \times 0.8 = 3.2$. Note that np corresponds to the peak in the probability distribution function.

Exercise 9.5

Sketch out the binomial CDF for the situation where $n = 10$ and $p = 0.05$. Repeat for $n = 10$ and $p = 0.5$. Estimate the mean and variance for each case. Comment on the appearance of the PDFs.

NONPARAMETRIC ESTIMATES USED WITH BINOMIAL DISTRIBUTION

Suppose we have insufficient data to determine the underlying distribution for a component, but we know from previous studies that about 95 percent of units survive 100 hr. Such an estimate is called *nonparametric* because no specific, continuous distribution of the failure times is assumed. We are planning a 100 hr mission requiring 20 such units to be operational. What is the probability of success for the mission, defined as no failures?

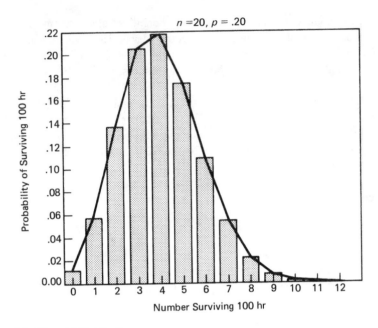

Figure 9.1   Binomial Distribution

The nonparametric studies have provided the CDF estimate which we equate to the probability of failure $p = 1 - 0.95 = 0.05$. Thus, the probability of no failures among 20 units is the same as the probability that all survive; that is,

$$(1-p)^n = (0.95)^{20} = 0.358$$

We realize that there is only about one chance in three, roughly, of having all components operational during the mission. We must go back to the drawing board to improve our reliability!

Exercise 9.6

In terms of survival probability for mission duration, how reliable must the 20 components be to achieve a mission success level of at least 0.99?

CONFIDENCE LIMITS FOR THE BINOMIAL DISTRIBUTION

Suppose we have a population of 50,000 integrated circuits, and we wish to sample 100 units and stress for a particular mode of failure. At the end of the experiment, we wish to make a statement of inference about the probable range of percent defective in the population for this failure mode. That is, we'll make an interval statement with a certain degree of confidence about the population; e.g., we're 90 percent confident that the true percent defective is between 4.8 and 9.3 percent.

The simplest way to do this based on sample results is to refer to the classic charts for confidence limits for p in binomial sampling by Clopper and Pearson (1934), shown as Figures 9.2–9.5. Given the sample fraction defective x/n in an experiment, the confidence limits are obtained directly from the charts by the intersections of the curved n lines with the abscissa (x values) and then reading the ordinate (y values). For example, in our stress, if we find 20 defective units out of 100, then $x/n = 20/100 = 0.20$. Referring to the chart for the 95 percent confidence level, we read the upper confidence limit from the upper curved $n = 100$ line intersecting with the abscissa $x/n = 0.20$ to be about 0.29. Similarly, the lower confidence limit appears to be about 0.13. Thus, we state with 95 percent confidence that the true population value lies somewhere within 0.13 to 0.29.

In using confidence limits we are effectively "tossing a horseshoe" at a population value. For 95 percent confidence, we expect that 19 out of 20 times we should capture the population value within the limits given. To get greater confidence with the same sample size, we would need a larger horseshoe, that is, wider limits.

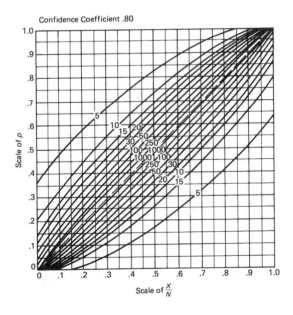

Figure 9.2   80% Confidence Belts for Proportions   Source: Clopper and Pearson charts reproduced by permission of Biometrica trustees.

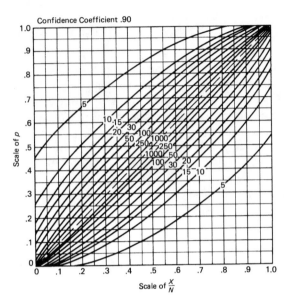

Figure 9.3   90% Confidence Belts for Proportions   Source: Clopper and Pearson charts reproduced by permission of Biometrica trustees.

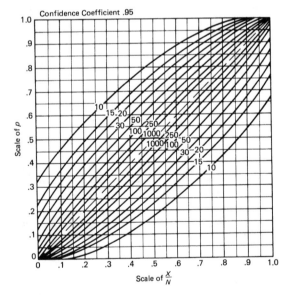

Figure 9.4   95% Confidence Belts for Proportions   Source: Clopper and Pearson charts reproduced by permission of Biometrica trustees.

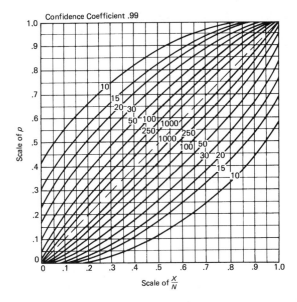

Figure 9.5   99% Confidence Belts for Proportions   Source: Clopper and Pearson charts reproduced by permission of Biometrica trustees.

For small proportions, it becomes very difficult to read the Clopper-Pearson charts accurately. An improved chart was devised by R. Ament (1977) and is shown as Figure 9.6. To use this chart, let us assume two failures out of a sample of 200 units. We read the observed sample percent defective (1.0 percent) on the abscissa as before. However, we now assign to the ordinate the actual number of failures in the sample (here, 2). Note that there are two sections to the ordinate: a numerator for the upper limit and a numerator for the lower limit. Finding the intersections of the ordinate values with the abscissa value provides, possibly with some visual interpolation, the upper and lower confidence limits for 95 percent confidence. For our case, we read 0.12 percent and 3.5 percent for the confidence limits. The very top line of the figure is used when zero failures in the sample are observed. In this case, we apply the sample size to the line marked "denominator" and read on the scale above the upper 95 percent confidence limit. For example, for zero failures out of 150 units, we have an upper 95 percent confidence limit of 2.4 percent. This limit is called *one-sided* since the lower limit cannot be less than the observed zero.

Exercise 9.7

Refer to Exercise 9.4, in which 50 devices are placed on stress for 168 hr. The probability of a device failing by 168 hr is assumed to be 0.05 or 5 percent. At the

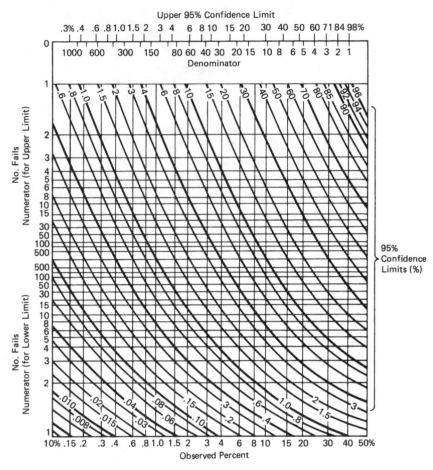

Figure 9.6   95% Confidence Limits for Proportions   Source: Reproduced by permission of the author, Richard P. Ament, Commission on Professional and Hospital Activities

end of the test there are 10 failures. Using the Clopper-Pearson chart, provide a 95 percent confidence interval for the population failure proportion. Does the interval include the assumed 5 percent failure probability? Repeat the exercise, using Figure 9.6.

NORMAL APPROXIMATION FOR THE BINOMIAL DISTRIBUTION

If np and $n(1 - p)$ are sufficiently large—say, each term greater than 5—then the fraction of defects X/n in a sample of size n will have a sampling distribution that is approximately normal, with mean p and variance $p(1-p)/n$. Alternatively, the statistic

$$\frac{\left(\dfrac{X}{n}\right)-p}{\sqrt{\dfrac{p(1-p)}{n}}}$$

has a sampling distribution that is approximately normal, with mean 0 and variance 1. This expression forms the basis for calculating confidence limits for the binomial distribution for large n based on the normal approximation. The approximation is considered acceptable (see Ross, 1994) for values of n satisfying $np(1-p) \geq 10$.

If we observe x defects in a large sample of size n, the estimate for the unknown population value p is $\hat{p} = x/n$. An approximate 95 percent confidence interval for the fraction of defects in the population will then be given by the expression

$$\hat{p} \pm z_{1-\alpha/2}\sqrt{\frac{\hat{p}(1-\hat{p})}{n}} = \frac{x}{n} \pm z_{1-\alpha/2}\sqrt{\frac{\dfrac{x}{n}\left(1-\dfrac{x}{n}\right)}{n}}$$

where the standard normal variate $z_{1-\alpha/2} = 1.645$. This formula also serves as the basis for calculating control limits for statistical process control charting of attribute data. (See section at chapter end.) However, we caution that arbitrary application of this equation can produce inaccurate results, even for very large sample sizes where low defect fraction (e.g., parts per million or PPM) situations are present.

Example 9.3   Shortcomings of the Normal Approximation

Two rejects are observed out of 80,000 units. Estimate the PPM and provide a 95 percent confidence interval (CI) based on the normal distribution.

Solution

The estimated PPM is $2/80{,}000 = 25$ PPM. A 95 percent CI based on the normal approximation is $(-9.7, 59.7)$ PPM. Negative PPMs do not have physical meaning, and so individuals often state the interval as $(0, 59.7)$. Even then, the interval is quite inaccurate when compared to the correct estimate obtained from the procedure described in the next section.

Exercise 9.8

Refer to Exercise 9.4 in which 50 devices are placed on stress for 168 hr. At the end of the test, there are 10 failures. Using the normal approximation, determine a 95 percent confidence interval for the population failure proportion. Compare to the answer in Exercise 9.7. Do you feel the normal approximation is adequate in this case?

CONFIDENCE INTERVALS FOR LOW PPM

Nearly exact confidence intervals for the parameter p of the binomial distribution for low PPM are provided by the following equations (see Trindade, 1992). If x rejects are observed in a random sample of size n, the upper 95 percent confidence limit is

$$P_U = \frac{C}{(n-x) + C}$$

where C is one-half of the 97.5th percentile of the chi-square distribution (see Appendix) with $2(x + 1)$ degrees of freedom; that is,

$$C = \frac{\chi^2_{2(x+1);97.5}}{2}$$

Similarly, the lower 95 percent confidence limit is

$$P_L = \frac{D}{(1 + n - x) + D}$$

where D is one-half of the 2.5th percentile of the chi-square distribution with $2x$ degrees of freedom; that is,

$$D = \frac{\chi^2_{2x;2.5}}{2}$$

In fact, since n is large relative to the other terms in the denominators for the upper and lower limits, further simplification of these equations is possible. Thus, the 95 percent confidence limits can be expressed as

$$P_U = \frac{C}{n} \quad \text{and} \quad P_L = \frac{D}{n}$$

where C and D are as defined previously. For other confidence levels, the appropriate percentiles of the chi-square distribution are used.

Example 9.4   Confidence Intervals for Low PPM

As in Example 9.3, two rejects are observed out of 80,000 units. Estimate the 95 percent confidence interval (CI) based on the low PPM formulas.

Solution

The degrees of freedom (df) for the upper limit are $2(x + 1) = 2(2 + 1) = 6$. The 97.5 percentile for the chi-square distribution for 6 df is 14.45. Thus, $C = 14.45/2 = 7.23$. So, $P_U = 7.23/80,000 = 90.4$ PPM. The 2.5 percentile for chi-square, with $df\, 2x = 4$, is 0.484. Thus, $D = 0.484/2 = 0.242$, and $P_L = 3.0$ PPM.

Exercise 9.9

Three rejects are observed in a random sample of 65,000 units. Estimate p in PPM and provide a 90 percent confidence interval on p.

Exercise 9.10

Refer again to Exercise 9.4 in which 50 devices are placed on stress for 168 hr. At the end of the test, there are 10 failures. Using the formulas for confidence intervals for low ppm, determine a 95 percent confidence interval for the population failure proportion. Compare to the answers in Exercise 9.7 and Exercise 9.8.

SIMULATING BINOMIAL RANDOM VARIABLES

Simulation of the binomial random variable X (where X represents the number of failures in n independent trials, each trial having failure probability p) is very simple. We generate n random numbers $U_1, U_2, \ldots, U_n$ uniformly distributed in the unit interval $(0, 1)$. Then, we set X equal to the number of the U_i that at are less than or equal to p. The event $U_i \leq p$ (that is, the ith trial generates a uniform random variate than is less than p) occurs with probability p. Hence, the number of occurrences of $U_i \leq p$ in n trials results in a binomial (n, p) random variable. Ross (1993) shows other approaches that are computationally more efficient than this simple procedure.

Example 9.5   Simulation of System Reliability

It is known that a component has probability of failure of 0.01 (1 percent) after 350,000 hr (approximately 40 years) in field usage. A system consists of six such components, each of which must function for the system to work. What percent of systems are expected to fail in 40 years?

Solution

We know the exact solution for this simple problem. The system failure probability is $1 - (1 - 0.01)^6 = 0.0585$ or 5.85 percent. However, a simulation study was conducted by generating a binomial ($n = 6, p = 0.01$) random variable 10,000 times. The average outcome of all 10,000 binomial random variables was 0.0535 or 5.35 percent, in agreement with theory. While this example is simplistic, more complicated problems (that may be difficult or impossible to solve theoretically) can often be addressed successfully via simulation.

Exercise 9.11

A system consists of two boxes. Each box consists of three components. All components must work for the box to function. However, the system will work if either box is operational. If the component reliability at mission end is 0.025, find via simulation the system reliability at mission end. Compare to exact results.

HYPERGEOMETRIC AND POISSON DISTRIBUTIONS

Hypergeometric Distribution

A few words about two other important distributions are appropriate here. The hypergeometric distribution is described by the following equation which gives the probability of getting $X = x$ rejects in a sample of size n drawn from a finite lot of size N containing a total of m rejects:

$$P(X = x) = \frac{\binom{m}{x}\binom{N - m}{n - x}}{\binom{N}{n}}, \quad x = 0, 1, 2, ..., n$$

Note that the denominator is just the number of combinations of objects, taken n at a time, from N. Similarly, the numerator is the total number of ways of getting x defectives in the sample of size n when m defectives exist in the lot of size N multiplied by the number of ways of getting $n - x$ nondefectives in the remaining group of size $N - m$. Because of the factorial terms, this formula is computationally difficult for large numbers. However, in the common situation where N is large relative to n (say, over ten times greater), the binomial distribution gives an accurate approximation to the hypergeometric distribution.

Example 9.6   Hypergeometric Distribution

The failure analysis lab has just received ten units reported defective in an accelerated stress experiment. The electrical characteristics of the rejects are very similar, and the engineer does not have time to analyze all ten units. He randomly chooses four parts for the failure autopsy. If there were actually three units with one mode of failure (say, type A) and seven with another reason for failing (say, type B), what's the probability that none of the three examined units are type A?

Solution

Use the hypergeometric distribution with $N = 10$, $n = 4$, $m = 3$ to solve for $X = 0$:

$$P(X = 0) = \frac{\binom{3}{0} \binom{10-3}{4-0}}{\binom{10}{4}} = \frac{1 \times [(7 \times 6 \times 5) / (3 \times 2 \times 1)]}{(10 \times 9 \times 8 \times 7) / (4 \times 3 \times 2 \times 1)} = 0.167$$

Hence, there's about a 17 percent chance of not detecting both modes.

Exercise 9.12

A sock drawer contains 12 socks, 4 black and 8 red, all mixed together. In the process of randomly selecting two socks in darkness, what's the probability of getting a matching pair? Hint: A match occurs when either two red or two black are drawn.

Poisson Distribution

We purposely distinguish between a defect and a defective unit. A unit is defective if it has one or more defects; a defect is defined as a nonconformance to specifications. Thus, the concept that a unit is defective is treated separately from the idea that there are a number of defects per unit or per area. The former case can be handled by the binomial (or hypergeometric) distribution. For calculations involving density (e.g., defects per wafer, failures per period, accidents per hour, etc.), the Poisson distribution is often employed. The equation for the Poisson distribution is:

$$P(X = x) = \frac{\lambda^x e^{-\lambda}}{x!}, \qquad (x = 0, 1, 2, \ldots)$$

where, λ = the average density or expected value. This simple formula can also be used numerically to approximate the binomial distribution, even though the methods of deriving the distributions are different. The approximation will be accurate when N is large compared to n, and the probability p that a unit is defective is small—say, less than 0.10. To use the Poisson as an approximation to the binomial, we use the average density, 1, set equal to the product np.

Example 9.7   Poisson Distribution

A sample of 250 units is periodically drawn and tested. If the process producing these units generates a defect level of 1 percent, what's the probability of getting less than 2 failures in the sample tested?

Solution

Note that calculations using the binomial distribution could be computationally difficult because of large factorial terms. Instead, we use the Poisson distribution with $\lambda = 250 \times 0.01 = 2.5$ The probability of less than two failures is the probability of 0 and 1 failure, so we calculate for $X = 0$:

$$P(X = 0) = \frac{\lambda^x e^{-\lambda}}{x!} = e^{-2.5} = 0.082$$

and for $X = 1$:

$$P(X = 1) = \frac{2.5^1 e^{-2.5}}{1!} = 0.205$$

The total probability is thus 0.287, or there is roughly only a 1 in 3.5 (i.e., the reciprocal of 0.287) chance of getting less than two failures.

Exercise 9.13

Compare the results using the binomial distribution to the answer given for the Poisson approximation in Example 9.6.

The Poisson distribution also has a useful relationship to the exponential life distribution as follows:

The time between emergency repair requests coming into a central office follows an exponential life distribution with parameter λ. Then, the probability of exactly k emergency repair calls in the time interval t is given by the Poisson distribution with parameter λt. In other words,

$$P(k \text{ repair calls}) = \frac{(\lambda t)^k e^{-\lambda t}}{k!}$$

As a confirmation of this relationship, consider that the probability of no calls in the time interval t is the probability for $k = 0$, or

$$P(\text{no repair calls}) = e^{-\lambda t}$$

As we showed in Chapter 1, the complementary event to zero occurrences is at least one occurrence, and this probability is given by

$$P(\text{at least one repair call}) = 1 - e^{-\lambda t}$$

This last expression is just the CDF $F(t)$ for the exponential distribution with parameter λ. So a discrete process (for example, the number of repair calls) described by a Poisson distribution with parameter λt has a probability of occurrence in the time interval t described by the continuous exponential distribution with parameter λ.

Exercise 9.14

High energy rays strike a measuring device. Occasionally, a ray of sufficient intensity causes the device to cease functioning, and a reset of the instrumentation is required. Each reset involves replacement of a seal. If the time between the arrival of the disabling rays follows an exponential distribution with $MTTF = 2190$ hr, what is the probability that five or more seals will be needed in inventory for repairs during the year?

The expected value λ for a Poisson distribution is estimated by the observed number of occurrences. Confidence limits for the expected value of a Poisson distribution can be obtained using percentiles of the chi-square distribution with appropriate degrees of freedom. For x observed occurrences, the lower and upper limits for $100(1 - \alpha)$ percent confidence are given by

$$\left(\frac{1}{2}\right)\chi^2_{2x;100\,(\alpha/2)}$$

and

$$\left(\frac{1}{2}\right)\chi^2_{2\,(x+1)\,;100\,(1-\alpha/2)}$$

respectively.

Example 9.8   Confidence Limits for Expected Value of a Poisson Distribution

The manufacturer wishes to assess the level of marking defects on integrated circuit packaging. He randomly samples 100 pieces and finds 10 units with one defect and 2 with two defects. Find the average defect density per package and provide 90 percent confidence limits.

Solution

The total number of defects observed is $10 + 2(2) = 14$. Thus, $\hat{\lambda} = 0.14$. Ninety percent confidence limits are obtained from the chi-square table in the Appendix. Thus, the 5th percentile for the chi-square distribution with $2(14) = 28$ degrees of freedom is 16.9. Similarly, the 95th percentile for $2(14 + 1) = 30$ df is 43.8. Thus, the 90 percent confidence interval is (8.45 to 21.9). The 90 percent confidence interval on average defect density is then (0.0845 to 0.219).

Exercise 9.15

Using the $\hat{\lambda}$ in Example 9.8, find the probability of 0, 1, 2, 3 defects using the Poisson distribution. How many units in a sample of 100 are expected to have 0, 1, 2, 3 or more defects. Compare to the data. Suppose a second sample of 100 is taken and 16 total defects are observed. Provide a 99 percent confidence interval for the expected number of defects per package based on the 200 combined samples.

TYPES OF SAMPLING

Let's consider a lot consisting of a very large number of packaged integrated circuits. Suppose we are concerned about a particular mode of failure. Assume that the population consisting of the entire lot would suffer a percent defective failure total of some value, say 1 percent, if the entire lot were stressed a certain length of time, say 100 hours, under normal operating conditions (voltages and temperature and humidity). However, we wish to test only a small sample for the same time under similar operating conditions to estimate the population percent defective and determine if the lot has acceptable reliability. How do we choose the appropriate sample size?

We will make a decision based on the results of looking at one sample of size n. This type of analysis is called "single-sampling." It is also possible to perform double sampling wherein the results of the first sample drawn from a lot generate one of three possible outcomes: (1) accept the lot, (2) reject the lot, or (3) take an additional sample and combine the results to reach a final decision. In fact, multiple sampling plans are possible in which more than two draws from the population are performed before a decision is reached. For further discussion of multiple sampling plans, the reader should consult Burr (1976) or Schilling (1982). Also, MIL STD 105D (1963) contains actual single, double, and multiple sampling schemes for inspection by attributes.

In this chapter, only single-sample plans will be discussed. Since one either accepts or rejects a lot or process in single-sampling plans, the probability of acceptance plus the probability of rejection is equal to one. Hence, rejection and acceptance probabilities will be used interchangeably.

RISKS

To fully understand the sampling process, one must comprehend the risks involved with the decision to either accept or reject a lot. A straightforward approach might be to establish an acceptable level or maximum level of allowed percent defective, say Y percent. Then, we could choose a convenient sample size for stressing, say 50 units, and set the criteria on the maximum number of failures in the sample on stress such that we have a specified probability of accepting or rejecting the lot. If we have more failures than the acceptance number (call it c), we reject the lot; if we have c or less fails, we accept the lot.

The acceptance number is based on the fact that a large lot with a fixed fraction defective produces varying results for each sample drawn, even if the sample size is the same. However, we can calculate an acceptance number such that if a lot has an excessive fraction defective, it is likely that the number of rejects in the sample will exceed the acceptance number. On the other hand, if the lot fraction defective is not excessive, how likely are we to exceed this number in a given sample? If the lot fraction is excessive, how likely are we not to exceed this number? Let's consider these associated risks in more detail.

First, one must realize that there exists a matrix of possible correct and incorrect decisions, shown as Table 9.1. If the lot defective is truly less than Y percent defective, and we make a decision based on the sample results (less than or equal to c failures from n units) to accept the entire lot, we have made a correct decision. If similarly, the lot is greater than Y percent defective and we reject the lot based on the sample results, we again have made the correct decision. If the lot, however, is less than Y percent defective and we end up rejecting the lot ($c + 1$ or more failures out of n units), then we have committed an error (called a Type I

TABLE 9.1   Matrix of Possible Choices

Decision on Lot	Population Value (% Defective)	
	$\leq Y\%$	$> Y\%$
Accept	Correct	Type II error
Reject	Type I error	Correct

error), and the chance of reaching this wrong decision is referred to as an "α" (alpha) risk. Similarly, a Type II error at a "β" (beta) risk level occurs when we accept the lot, based on sample results, when the lot is actually greater than Y percent defective.

Alternatively, the terms "producer's risk" and "consumer's risk" are often applied to the alpha and beta risks, respectively, since Type I error refers to the rejection of a good lot and Type II error designates the acceptance of a bad lot. Obviously, it is costly to a producer to throw away or rescreen acceptable material due to Type I error. Similarly, the cost to the user of taking product with a high reject rate, because of Type II error, may be considerable if, for example, he has to replace or repair defective systems cause by bad components. Let us now consider the calculation of the probabilities of these possible choices.

OPERATING CHARACTERISTIC (OC) CURVE

If we are sampling from a finite lot, the distribution of the possible outcomes is correctly described by the hypergeometric model. However, when the sample size selected is small relative to the lot size, say less than ten percent of the lot size, the binomial distribution is used as an excellent approximation to the hypergeometric distribution. We'll assume this is the case; i.e., the lot size is at least ten times as great as the sample size.

Our first task is to calculate the probability of getting an acceptable lot based on c or less failures out of the n, say 50, units on stress. Let's assume that the lot percent defective is actually 2 percent. Let's further state that we wish to have an acceptance number c such that at least 95 percent of the time we accept the lot if the true percent defective is 2 percent or less; that is, our risk of Type I error is 5 percent. We now calculate the cumulative binomial probabilities for 0, 1, 2, and so on up to some value of c failures, for $n = 50, p = 0.02$, until the probability of getting c or less failures exceeds 95 percent or 0.95. Then 95 percent of the time, we'll accept a lot with 2 percent defective if the decision to accept is based on c or less failures, because c or less failures are expected at least 95 percent of the time for a sample of size $n = 50$ drawn from a large lot with percent defective of 2 percent or less.

BINOMIAL CALCULATIONS

The formula for the cumulative binomial probability is

$$P(X \le c) = \sum_{x=0}^{c} \binom{n}{x} p^x (1-p)^{n-x}$$

So we require

$$\sum_{x=0}^{c} \binom{50}{x} (0.2)^x (0.98)^{50-x} \geq 0.95$$

Note: This calculation can be performed simply on a scientific calculator with memory by using the following recursive formula to calculate successive binomial terms, starting with $(1-p)^n$:

$$P(X = x+1) = \frac{(n-x)\, p\, P(X = x)}{(x+1)\,(1-p)}$$

Thus,

$$P(X = 0) = (1 - 0.02)^{50} = 0.3642$$

$$P(X = 1) = \frac{50\,(0.02)\,P(X = 0)}{1\,(0.98)} = 0.3716$$

$$P(X = 0) + P(X = 1) = 0.7358$$

$$P(X = 2) = \frac{49\,(0.02)\,P(X = 1)}{2\,(0.98)} = 0.1858$$

$$P(X = 0) + P(X = 1) + P(X = 2) = 0.9216$$

$$P(X = 3) = \frac{48\,(0.02)\,P(X = 2)}{3\,(0.98)} = 0.0607$$

$$P(X \leq 3) = P(X \leq 2) + P(X = 3) = 0.9822$$

At this point we stop, since the cumulative binomial probability, that is, $P(X \leq 3)$, is greater than 0.95. Alternatively, instead of calculating these terms individually, we could have used a table such as those in the CRC Standard Mathematical Tables (1965) which provide the cumulative binomial terms for various parameters. By either method, the desired c number is 3; that is, we accept a lot if 3 or less units fail out of the 50 sampled and reject the lot if 4 or more fail. Our risk of committing a Type I error will then be less than 5 percent.

Now what about the risk of a Type II error, that is, accepting a bad lot defined as over 2 percent defective? To calculate the risks associated with the Type II error it is not sufficient to state only one defect level; one needs to specify the various alternative percent defective values and perform each calculation separately. For example, one may be interested in the probability of getting 3 or less failures out of 50 units sampled (and thereby accepting the lot) if the true population value is 3 percent, 5 percent, 7 percent, 10 percent or 15 percent, and so on. We have done such calculations, and the numbers are shown as Table 9.2.

EXAMPLES OF OPERATING CHARACTERISTIC CURVES

This table may be better represented in the form of a graph, called the operating characteristic (OC) curve, which details the probability of accepting a lot based on c allowed rejects in a sample of size n. Such a curve is shown, for $n = 50$, $c = 3$, as Figure 9.7.

The operating characteristic curve provides the total picture of the sampling plan. We note, for example, that if a lot having 7 percent defective units is presented for sampling, there is about a fifty-fifty chance of its being accepted; we call this percent defective the "point of indifference." Also, we have about a 10 percent chance of accepting a lot offered at around 12.5 percent defective, or over six times the 2 percent value we targeted in setting up this sampling plan. Stated another way, the consumer's risk is 10 percent that a lot at 12.5 percent will have three or fewer failures in a sample of 50 and thus be accepted.

How can the Type II or consumer's risk be reduced? Obviously one can decrease the acceptance number to $c = 2$, 1, or even 0, but what would that change mean to the OC curve? Figure 9.8 shows the OC curve for various acceptance numbers with the sample size fixed at 50. We see that the probability of acceptance has been reduced at all incoming percent defective levels. At the target value of 2 percent, the probability of acceptance goes down from 98.2 percent to 92.2 percent, to 73.6 percent, and to 36.4 percent as we change c from 3, to 2, to 1, and to 0, respectively. Thus, we have greatly increased the probability of rejecting an acceptable lot (that is, the producer's risk) by lowering the acceptance number while keeping the sample size fixed.

TABLE 9.2 Probability of 3 or Less Failures in Sample of Size n = 50 for Various Lot Percent Defective Values

Lot Percent Defective (p)	Probability $[P(X \leq 3)]$
0.01	0.9984
0.02	0.9822
0.03	0.9372
0.05	0.7604
0.07	0.5327
0.10	0.2503
0.13	0.0958
0.15	0.0461

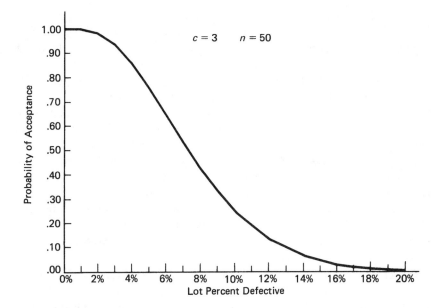

Figure 9.7   Operating Characteristic Curve

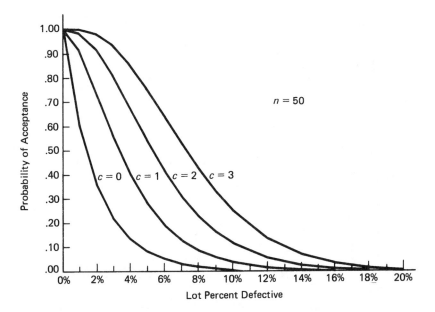

Figure 9.8   Operating Characteristic Curves for Various Accept (c) Numbers

Similarly, if only the sample size increases and the acceptance number c is held constant, then again the consumer's risk can be reduced, but only at the expense of the producer's risk. Figure 9.9 shows the OC curve for samples of size $n = 50, 100, 300,$ and 500 and acceptance number $c = 3$. The effect of increasing the sample size while the acceptance number remains fixed is to significantly reduce the probability of a lot acceptance at all percent defective values. Thus, the consumer's risk is dramatically driven lower, but the producer's risk becomes very high. For example, the probability of accepting a lot at 2 percent defective, allowing three rejects in sample of size $n = 50$, is about 98 percent; hence, the Type I risk is only 2 percent. However, a sample of size 100, 300, and 500 will have a probabilities of acceptance of about 86 percent, 15 percent, and 1 percent, respectively, for the same acceptance number $c = 3$. So the producer's risk is correspondingly 14 percent, 85 percent, and 99 percent. However, for the consumer's risk held at 10 percent, the corresponding percent defective values are approximately 13 percent, 6.5 percent, 2.2 percent, and 1.1 percent for samples of size 50, 100, 300, and 500, respectively.

In fact, the only way to decrease both Type I and II errors simultaneously is to increase the sample size drawn and adjust the acceptance number. What we would like to do, indeed, is to specify a high probability of acceptance, say 95 percent, at a desirable or acceptable quality level and simultaneously require a

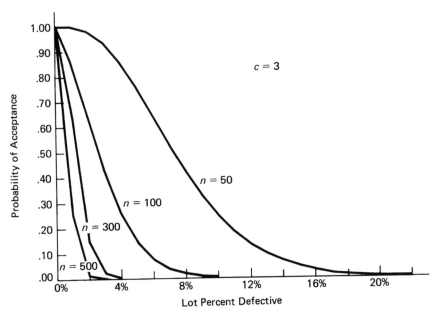

Figure 9.9   Operating Characteristic Curves (fixed acceptance no., various SS)

high probability of rejection, say 90 percent, at an undesirable or rejectable quality level. For individual lots, it is common to call the percent defective value at the 95 percent acceptance probability the "AQL," for acceptable quality level; similarly, the 10 percent acceptance probability is referred to as the "LTPD," or lot tolerance percent defective or even sometimes as the "RQL," for rejectable quality level.

Note, however, that the term AQL is often applied in the literature in the context of an expected quality level without being tied to a given probability point, such as 95 percent. Furthermore, military documents, such as MIL STD 105D (1963), refer to an AQL value assured in terms of an overall sampling scheme with different acceptance probabilities for a given AQL, depending on lot sizes and other factors. Also, there are other schemes which are based on a specific characteristic such as the LTPD for a single lot or the average outgoing quality limit (called the AOQL) for a series of lots when it is feasible to screen out defects. The book by Dodge and Romig (1959) provides extensive tables for both the LTPD and AOQL procedures. Full information on the subject of acceptance sampling can be found in the book by Schilling (1982). The next section presents several ways of generating sampling plans to provide specific risk protection at given acceptable and rejectable defect levels.

Exercise 9.16

Generate the OC curve for $n = 100$ and $c = 2$.

Exercise 9.17

Generate the family of OC curves for $n = 50, 100, 300,$ and 500 for $c = 2$.

Exercise 9.18

Generate the family of OC curves for $c = 0, 1, 2, 3$ and $n = 100$.

GENERATING A SAMPLING PLAN

A sampling plan is uniquely determined by the lot size, N, the sample size, n, and the acceptance number, c. For a large lot or for sampling from a process, the N parameter can be ignored. We shall do so here. Thus, n and c uniquely specify the sampling plan.

The simplest way to get n and c for a given set of percent defectives and matching α, β risks is to find a computer program that does the calculations. Many applications software packages for acceptance sampling are available for personal computers. The format is simple: one typically provides the acceptable and rejectable quality levels (AQL and RQL) and the associated α, β risks, and the program then generates the sampling plan, that is, the sample size and acceptance number. Also, the program may list the OC curve and provide plots if requested. An example of the output from one program (STATLIB written in APL by Jack Prins in 1984) is shown as Figure 9.10.

Another simple procedure is to employ the nomograph shown as Figure 9.11. To use the graph, two vertical scales are provided: the scale on the left refers to the fraction defective values (the x-axis on the OC curve); the right scale designates the fractional probability of acceptance (the y-axis on the OC curve). Draw two lines: for example, the first may extend from the AQL value on the left vertical scale to the matching 0.95 probability of acceptance on the right; the second line may go from the LTPD value on the left to the matching 0.10 probability on the right. The intersection of the two lines in the center grid determines the sample size n and the acceptance number c (called number of occurrences here). For the case where the AQL = 2 percent and the LTPD = 8 percent, we get $n = 98$ and $c = 4$.

Exercise 9.19

Using any available software or the nomograph in Figure 9.11, determine the sample size and acceptance number for an acceptable quality level of 0.01 with alpha risk 0.025 and a rejectable quality level of 0.05 with beta risk 0.10.

Exercise 9.20

Find a sampling plan to achieve AQL = 0.01, RQL = 0.03, $\alpha = 0.01$, $\beta = 0.10$.

Guenther (1974) presents a very simple iterative procedure, using the percentiles $\chi^2_{df;100 \times area}$ of the chi-square distribution. (See Appendix.) The equation for determining the minimum sample size n and the accompanying acceptance number c is

$$0.5 \times \left[\chi^2_{2c+2;100 \times (1-\beta)} \times \left(\frac{1}{p_2} - 0.5 \right) + c \right] \leq n$$

$$\leq 0.5 \times \left[\chi^2_{2c+2;100 \times \alpha} \times \left(\frac{1}{p_1} - 0.5 \right) + c \right]$$

For a given set of conditions $(p_1, p_2, \alpha, \beta)$, one simply tries a c value, using the chi-square percentiles for $2c + 2$ degrees of freedom and the respective probabilities, and determines if the inequalities for the chosen sample size n are mathematically satisfied. If so, then the minimum n is chosen. If not, then the c value is

```
          TRYPLAN
FIND A SAMPLING PLAN WITH NO MORE THAN 10 PERCENT CHANCE OF
ACCEPTING A LOT WHOSE TRUE PERCENT DEFECTIVES IS 12 OR WORSE,
AND 5 PERCENT CHANCE OF REJECTING WHEN THE TRUE PERCENT DEFECTIVES
IS 2.5 OR BETTER, FOR EXAMPLE:
     P₁ = .025

     P₂ = .12

     ALPHA,THE PRODUCER'S RISK = .05
     BETA, THE CONSUMER'S RISK = .10

THE PROGRAM IS ABOUT TO EXECUTE:
     .025 .12 SAMPLINGPLAN .05 .10

THE SAMPLE SIZE IS        55
THE ACCEPTANCE NUMBER IS   3
DO YOU WANT TABULAR OUTPUT?  Y/N
Y
PROBABILITY  PROPORTION
    OF           OF
ACCEPTANCE   DEFECTIVES
  .9977        .0100
  .9622        .0230
  .8641        .0360
  .7170        .0490
  .5533        .0620
  .4011        .0750
  .2756        .0880
  .1807        .1010
  .1138        .1140
  .0691        .1270
  .0406        .1400
  .0231        .1530
  .0128        .1660
  .0069        .1790
  .0036        .1920
  .0019        .2050
  .0009        .2180
  .0005        .2310
  .0002        .2440
  .0001        .2570
  .0000        .2700
DO YOU WANT AN OC CURVE?  Y/N
Y
```

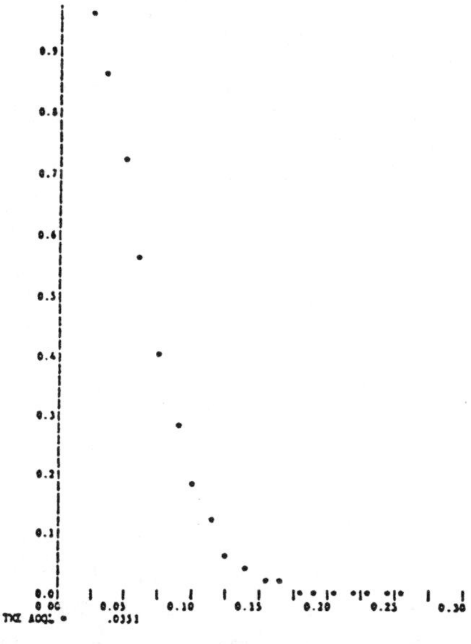

Figure 9.10 Computer-Generated Sampling Plan

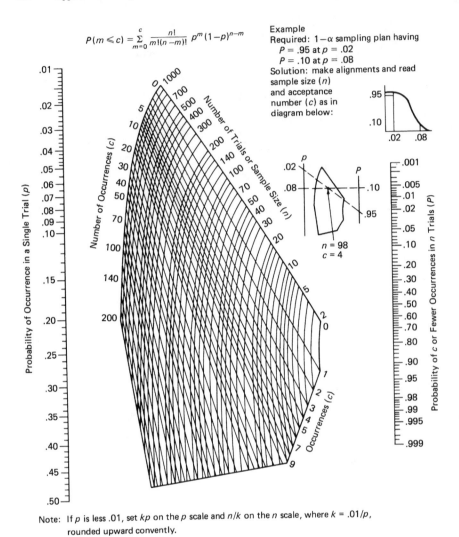

$$P(m \leqslant c) = \sum_{m=0}^{c} \frac{n!}{m!(n-m)!} \, p^m (1-p)^{n-m}$$

Example
Required: $1-\alpha$ sampling plan having
$P = .95$ at $p = .02$
$P = .10$ at $p = .08$
Solution: make alignments and read sample size (n) and acceptance number (c) as in diagram below:

$n = 98$
$c = 4$

Note: If p is less .01, set kp on the p scale and n/k on the n scale, where $k = .01/p$, rounded upward convently.

Figure 9.11 Nomograph of the Cumulative Binomial Distribution   Source: Reprinted with permission from AT&T Technologies, Inc.

incremented by one until the interval for n contains at least one integer. For example, let us find a sampling plan to accept 95 percent of the time product at 2 percent defective and reject 90 percent of the time product at 8 percent defective. If we try $c = 0, 1, 2,$ or 3 in the above formula, we find that no solutions exist. To illustrate for $c = 3$, we get

$$0.5 \times \left[\chi_{8;90}^2 \times \left(\frac{1}{0.08} - 0.5 \right) + 3 \right] \leq n$$

$$\leq 0.5 \times \left[\chi_{8;5}^2 \times \left(\frac{1}{0.02} - 0.5 \right) + 3 \right]$$

$$0.5 \times [13.36\,(12) + 3] \leq n \leq [0.5 \times [2.73\,(49.50) + 3]]$$

$$81.66 \leq n \leq 69.07$$

Therefore, no solution exists. However, for c = 4, we discover

$$0.5 \times \left[\chi_{8;90}^2 \times (12) + 4 \right] \leq n \leq 0.5 \times \left[\chi_{10;5}^2 \times (49.50) + 4 \right]$$

$$0.5 \times [15.99\,(12) + 4] \leq n \leq 0.5 \times [3.94\,(49.50) + 4]$$

$$97.94 \leq n \leq 99.52$$

Hence, the minimum sample size is 98.

Exercise 9.21

Using the same AQL, RQL, and alpha and beta risk levels as in Exercise 9.19, apply the Guenther method to determine the sampling plan; that is, n and c.

Exercise 9.22

Repeat the previous exercise using the same AQL, RQL, and alpha and beta risk levels as in Exercise 9.20.

Alternatively, one may use a specially prepared table such as the one from MIL-S-19500G (1965), shown as Table 9.3, for LTPD sampling plans. The use of this table is quite simple. The LTPD is listed across the top as a percentage. The acceptance number c is the vertical left column. The intersection of a specified LTPD and acceptance number is the minimum sample size required. Below each sample size n in parentheses is the AQL (that is, the defect level accepted 95 percent of the time) for that (n, c) pair. For example, to provide an LTPD of 2 percent with an acceptance number of 2, one would need at least 266 units to test. The AQL for this sampling plan of $n = 266$ and $c = 2$ is 0.31 percent.

Exercise 9.23

For the same AQL, RQL and alpha and beta risk levels as in Exercise 9.19, use Table 9.3 to determine a suitable sampling plan, that is, n and c.

Table 9.3. LTPD sampling plans[1,2]

Minimum size of sample to be tested to assure, with a 90 percent confidence, that a lot having percent-defective equal to the specified LTDP will not be accepted (single sample).

Minimum Sample Sizes
(For device-hours required for life test, multiply by 1000)

Max. Percent Defective (LTPD) or λ — Acceptance Number (c) (r = c + 1)	50	30	20	15	10	7	5	3	2	1.5	1	0.7	0.5	0.3	0.2	0:15	0.1
0	5 (1.03)	8 (0.64)	11 (0.46)	15 (0.34)	22 (0.23)	32 (0.16)	45 (0.11)	76 (0.07)	116 (0.04)	153 (0.03)	231 (0.02)	328 (0.01)	461 (0.01)	767 (0.007)	1152 (0.005)	1534 (0.003)	2303 (0.002)
1	8 (4.4)	13 (2.7)	18 (2.0)	25 (1.4)	38 (0.94)	55 (0.65)	77 (0.46)	129 (0.28)	195 (0.18)	258 (0.14)	390 (0.09)	555 (0.06)	778 (0.045)	1296 (0.027)	1946 (0.018)	2592 (0.013)	3891 (0.009)
2	11 (7.4)	18 (4.5)	25 (3.4)	34 (2.24)	52 (1.6)	75 (1.1)	105 (0.78)	176 (0.47)	266 (0.31)	354 (0.23)	533 (0.15)	759 (0.11)	1065 (0.080)	1773 (0.045)	2662 (0.031)	3547 (0.023)	5323 (0.015)
3	13 (10.5)	22 (6.2)	32 (4.4)	43 (3.2)	65 (2.1)	94 (1.5)	132 (1.0)	221 (0.62)	333 (0.41)	444 (0.31)	668 (0.20)	953 (0.14)	1337 (0.10)	2226 (0.062)	3341 (0.041)	4452 (0.031)	6681 (0.020)
4	16 (12.3)	27 (7.3)	38 (5.3)	52 (3.9)	78 (2.6)	113 (1.8)	158 (1.3)	265 (0.75)	398 (0.50)	531 (0.37)	798 (0.25)	1140 (0.17)	1599 (0.12)	2663 (0.074)	3997 (0.049)	5327 (0.037)	7994 (0.025)
5	19 (13.8)	31 (8.4)	45 (6.0)	60 (4.4)	91 (2.9)	131 (2.0)	184 (1.4)	308 (0.85)	462 (0.57)	617 (0.42)	927 (0.28)	1323 (0.20)	1855 (0.14)	3090 (0.085)	4638 (0.056)	6181 (0.042)	9275 (0.028)
6	21 (15.6)	35 (9.4)	51 (6.6)	68 (4.9)	104 (3.2)	149 (2.2)	209 (1.6)	349 (0.94)	528 (0.62)	700 (0.47)	1054 (0.31)	1503 (0.22)	2107 (0.155)	3509 (0.093)	5267 (0.062)	7019 (0.047)	10533 (0.031)
7	24 (16.6)	39 (10.2)	57 (7.2)	77 (5.3)	116 (3.5)	166 (2.4)	234 (1.7)	390 (1.0)	589 (0.67)	783 (0.51)	1178 (0.34)	1680 (0.24)	2355 (0.17)	3922 (0.101)	5886 (0.067)	7845 (0.051)	11771 (0.034)
8	26 (18.1)	43 (10.9)	63 (7.7)	85 (5.6)	128 (3.7)	184 (2.6)	258 (1.8)	431 (1.1)	648 (0.72)	864 (0.54)	1300 (0.36)	1854 (0.25)	2599 (0.18)	4329 (0.108)	6498 (0.072)	8660 (0.054)	12995 (0.036)
9	28 (19.4)	47 (11.5)	69 (8.1)	93 (5.9)	140 (3.9)	201 (2.7)	282 (1.9)	471 (1.1)	709 (0.76)	945 (0.57)	1421 (0.38)	2027 (0.27)	2842 (0.19)	4733 (0.114)	7103 (0.077)	9468 (0.057)	14206 (0.038)
10	31 (19.9)	51 (12.1)	75 (8.2)	100 (6.1)	152 (4.1)	218 (2.9)	306 (2.0)	511 (1.2)	769 (0.80)	1025 (0.60)	1541 (0.40)	2199 (0.28)	3082 (0.20)	5133 (0.120)	7704 (0.080)	10268 (0.060)	15407 (0.040)
11	33 (21.0)	54 (12.6)	83 (8.3)	111 (6.2)	166 (4.2)	238 (2.9)	332 (2.1)	555 (1.2)	832 (0.83)	1109 (0.62)	1664 (0.42)	2378 (0.29)	3323 (0.21)	5546 (0.125)	8313 (0.083)	11082 (0.062)	16638 (0.042)
12	36 (22.3)	59 (13.0)	89 (8.6)	119 (6.5)	178 (4.3)	254 (3.0)	356 (2.2)	593 (1.3)	889 (0.86)	1184 (0.65)	1781 (0.43)	2544 (0.30)	3562 (0.22)	5936 (0.130)	8904 (0.086)	11872 (0.065)	17808 (0.043)
13	38 (22.3)	63 (13.4)	95 (8.9)	126 (6.8)	190 (4.5)	271 (3.1)	379 (2.2)	632 (1.3)	948 (0.89)	1264 (0.67)	1896 (0.44)	2709 (0.31)	3793 (0.22)	6321 (0.134)	9482 (0.089)	12643 (0.067)	18964 (0.045)
14	40 (23.1)	67 (13.8)	101 (9.2)	134 (6.9)	201 (4.6)	288 (3.2)	403 (2.3)	671 (1.4)	1007 (0.92)	1343 (0.69)	2015 (0.46)	2878 (0.32)	4029 (0.23)	6716 (0.138)	10073 (0.092)	13431 (0.069)	21134 (0.046)
15	43 (23.7)	71 (14.1)	107 (9.4)	142 (7.1)	213 (4.7)	298 (3.3)	426 (2.4)	711 (1.4)	1066 (0.94)	1422 (0.71)	2133 (0.47)	3046 (0.33)	4265 (0.24)	7108 (0.141)	10662 (0.094)	14216 (0.071)	21324 (0.047)
16	45 (24.1)	74 (14.6)	112 (9.7)	150 (7.2)	225 (4.8)	321 (3.4)	449 (2.4)	748 (1.4)	1124 (0.96)	1499 (0.72)	2249 (0.48)	3212 (0.34)	4497 (0.24)	7495 (0.144)	11244 (0.096)	14992 (0.072)	22487 (0.048)
17	47 (24.9)	79 (14.7)	118 (9.86)	157 (7.36)	236 (4.93)	337 (3.44)	472 (2.46)	787 (1.48)	1180 (0.98)	1574 (0.74)	2361 (0.49)	3372 (0.344)	4721 (0.246)	7869 (0.148)	11803 (0.098)	15737 (0.074)	23606 (0.049)
18	50 (24.6)	83 (15.0)	124 (10.0)	165 (7.54)	248 (5.02)	353 (3.51)	495 (2.51)	825 (1.51)	1239 (1.0)	1652 (0.75)	2476 (0.50)	3536 (0.351)	4956 (0.251)	8260 (0.151)	12390 (0.100)	16520 (0.075)	24780 (0.050)
19	52 (25.5)	86 (15.5)	130 (10.2)	173 (7.76)	259 (5.12)	370 (3.58)	518 (2.56)	864 (1.53)	1296 (1.02)	1728 (0.77)	2591 (0.52)	3702 (0.358)	5183 (0.256)	8638 (0.153)	12957 (0.102)	17276 (0.077)	25914 (0.051)
20	54 (26.1)	90 (15.0)	135 (10.4)	180 (7.82)	270 (5.19)	379 (3.65)	541 (2.60)	902 (1.56)	1353 (1.04)	1803 (0.78)	2705 (0.52)	3864 (0.364)	5410 (0.260)	9017 (0.156)	13526 (0.104)	18034 (0.078)	27051 (0.052)
25	65 (27.0)	109 (16.1)	163 (10.9)	217 (8.19)	326 (5.38)	406 (3.76)	652 (2.69)	1086 (1.61)	1629 (1.07)	2173 (0.807)	3259 (0.538)	4656 (0.376)	6518 (0.269)	10863 (0.161)	16295 (0.108)	21726 (0.081)	32569 (0.054)

[1] Sample sizes are based upon the Poisson exponential binomial limit.
[2] The minimum quality (approximate AQL) required to accept on the average 19 of 20 lots is shown in parenthesis for information only.

TABLE 9.3 LTPD Sampling Plans

Exercise 9.24

Repeat previous exercise using the same AQL, RQL, and alpha and beta risk levels as in Exercise 9.20.

FURTHER GRAPHICAL TECHNIQUES FOR OBTAINING AN OC CURVE

While it is possible to search through the literature to find specific OC curves or sampling plans, it is very useful to have a simple graphical procedure that allows the engineer to quickly calculate an OC curve, assess sample requirements, or check risks for many different situations. It is for these reasons that we have developed Figures 9.12–9.15. Table 9.4 contains the actual values used to plot these curves. The description of the use of these figures will reveal their flexibility and convenience in handling many sampling problems.

There are four separate figures provided, one for each acceptance number $c =$ 0, 1, 2, and 3. It is our experience in reliability work that, because of the cost of destructive testing of units, sampling plans are generally restricted to low acceptance numbers. Hence, these figures should suffice for most situations. The abscissa (x-axis) is the sample size n required; the scale is from 10 to 100,000 units. The ordinate (y-axis) is the population percent defective p; the lowest scale

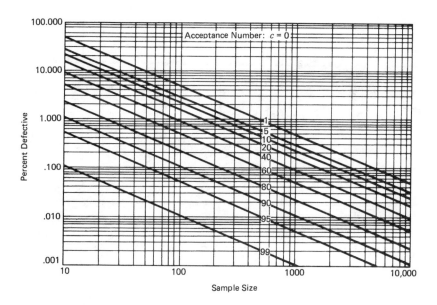

Figure 9.12 Graph of Percent Defective versus Sample Size. Diagonal lines represent various acceptance probabilities. Acceptance number c = 0.

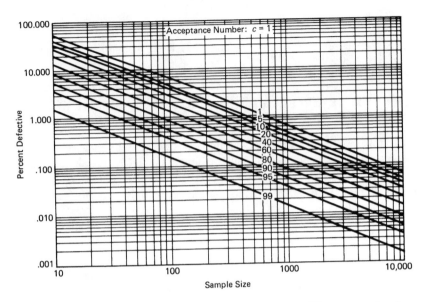

Figure 9.13 Graph of Percent Defective versus Sample Size. Diagonal lines represent various acceptance probabilities. Acceptance number c = 1.

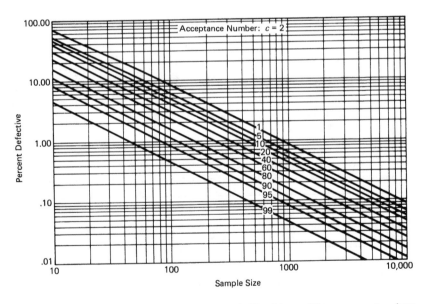

Figure 9.14 Graph of Percent Defective versus Sample Size. Diagonal lines represent various acceptance probabilities. Acceptance number c = 2.

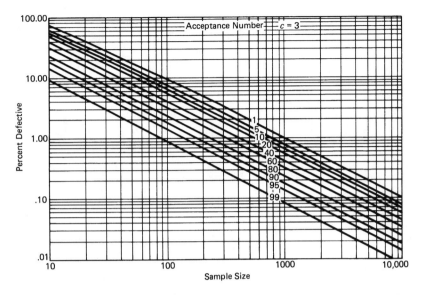

Figure 9.15 Graph of Percent Defective versus Sample Size. Diagonal lines represent various acceptance probabilities. Acceptance number c = 3.

is from 0.001 percent (10 parts per million or PPM) to 100 percent. The lines running diagonally across the graph from upper left to lower right represent the constant probability of acceptance for the intersecting (n, p) points for the specific acceptance number c; the probability range runs from 1 percent to 99 percent. The 95 percent and 10 percent lines correspond to typical AQL and LTPD specifications, respectively. These curves are calculated directly from the binomial distribution and thus may be applied in the situations where the sample size is small relative to the lot size or where we are sampling from a continuous process.

To generate an OC curve for a given sample size and acceptance number, one simply selects the figure corresponding to the acceptance number and finds the sample size on the x-axis. Then, the percent defective p values are read off the vertical axis matching the probability of acceptance values 99 percent, 95 percent (AQL), 90 percent, 80 percent, 60 percent, 40 percent, 20 percent, 10 percent (LTPD), 5 percent, and 1 percent. Alternatively, Table 9.4 may be used. For example, for $n = 100$ and $c = 1$, the respective p values (for the horizontal axis of the OC curve we'll form) are 0.149 percent, 0.357 percent, 0.533 percent, 0.825 percent, 1.37 percent, 2.01 percent, 2.96 percent, 3.83 percent, 4.66 percent, and 6.46 percent. From these values we generate the OC curve shown as Figure 9.16.

To compare risk levels of various sampling plans or to decide on alternative sampling plans while holding a characteristic such as the LTPD constant, these

TABLE 9.4 Tables of Percent Defective versus Sample Size for Given Acceptance Number

c = 0

Sample Size	Probability of Acceptance												
	1%	5%	10%	20%	30%	40%	50%	60%	70%	80%	90%	95%	99%
10	37.0000	25.8870	20.5680	14.8670	11.3440	8.7557	6.6968	4.9800	3.5039	2.2068	1.0490	0.5117	0.1005
25	16.8240	11.2929	8.7990	6.2350	4.7018	3.5989	2.7346	2.0226	1.4166	0.8887	0.4206	0.2050	0.0402
50	8.7990	5.8156	4.5008	3.1677	2.3792	1.8159	1.3768	1.0165	0.7109	0.4453	0.2105	0.1025	0.0201
100	4.6000	2.9514	2.2763	1.5966	1.1968	0.9122	0.6908	0.5095	0.3560	0.2229	0.1054	0.0513	0.0101
500	0.9169	0.5974	0.45595	0.3214	0.2406	0.1831	0.1385	0.1021	0.0713	0.0446	0.0211	0.0103	0.0020
1000	0.4595	0.2991	0.2300	0.1608	0.1203	0.0916	0.0693	0.0511	0.0357	0.0223	0.0105	0.0051	0.0010
5000	0.0921	0.0599	0.0461	0.0322	0.0241	0.0183	0.0139	0.0102	0.0071	0.0045	0.0021	0.0010	0.0002
10,000	0.0461	0.0300	0.0230	0.0161	0.0120	0.0092	0.0069	0.0051	0.0036	0.0022	0.0011	0.0005	0.0001

c = 1

Sample Size	Probability of Acceptance												
	1%	5%	10%	20%	30%	40%	50%	60%	70%	80%	90%	95%	99%
10	50.4400	39.4170	33.6850	27.0989	22.6946	19.2150	16.2270	13.5134	10.9284	8.3260	5.4529	3.6780	1.5539
25	23.7490	17.6130	14.6868	11.5089	9.4799	7.9256	6.6240	5.4648	4.3814	3.3098	2.1478	1.4404	0.6047
50	12.5530	9.1399	7.5590	5.8704	4.8808	4.0035	3.3350	2.7426	2.1927	1.6518	1.0687	0.7154	0.2997
100	6.4550	4.6560	3.8340	2.9646	2.4218	2.0130	1.6727	1.3738	1.0969	0.8252	0.5331	0.3566	0.1492
500	1.3300	0.9453	0.7757	0.5977	0.4872	0.4041	0.3355	0.2752	0.2195	0.1649	0.1064	0.0711	0.0297
1000	0.6620	0.4735	0.3885	0.2991	0.2438	0.2021	0.1678	0.1376	0.1097	0.0825	0.0532	0.0356	0.0149
5000	0.1327	0.0949	0.0778	0.0599	0.0488	0.0404	0.0336	0.0275	0.0220	0.0165	0.0106	0.0071	0.0030
10,000	0.0664	0.0474	0.0389	0.0300	0.0244	0.0202	0.0168	0.0138	0.0110	0.0082	0.0053	0.0036	0.0015

TABLE 9.4   continued

c = 2

Probability of Acceptance

Sample Size	1%	5%	10%	20%	30%	40%	50%	60%	70%	80%	90%	95%	99%
10	61.1800	50.7000	44.9700	38.0938	32.2967	29.3615	25.8575	22.5516	19.2620	15.7635	11.5826	8.7270	4.7600
25	29.6000	23.1040	19.9136	16.3490	13.9994	12.1487	10.5533	9.0899	7.6708	6.2006	4.4914	3.3520	1.8020
50	15.7800	12.0620	10.2960	8.3646	7.1149	6.1424	5.3123	4.5574	3.8312	3.0848	2.2244	1.6552	0.8870
100	8.1420	6.1620	5.2346	4.2305	3.5865	3.0883	2.6651	2.2819	1.9146	1.5387	1.1071	0.8226	0.4395
500	1.6705	1.2538	1.0609	0.8539	0.7220	0.6204	0.5345	0.4569	0.3828	0.3072	0.2206	0.1637	0.0874
1000	0.8380	0.6282	0.5314	0.4274	0.3613	0.3104	0.2673	032285	0.1914	0.1536	0.1103	0.0818	0.0436
5000	0.1690	0.1259	0.1064	0.0856	0.0723	0.0622	0.0535	0.0458	0.0383	0.0307	0.0220	0.0164	0.0087
10,000	0.8404	0.0629	0.0532	0.0428	0.0362	0.0311	0.0267	0.0229	0.0191	0.0154	0.0110	0.0082	0.0044

c = 3

Probability of Acceptance

Sample Size	1%	5%	10%	20%	30%	40%	50%	60%	70%	80%	90%	95%	99%
10	70.2890	60.6630	55.1740	48.3658	43.4480	39.2996	35.5100	31.8397	28.0900	23.9450	18.7570	15.0030	9.3300
25	34.8790	28.1730	24.8019	20.9630	18.3900	16.3100	14.4925	12.7949	11.1137	9.3259	7.1670	5.6570	3.4480
50	18.7300	14.7838	12.8757	10.7543	9.3570	8.2525	7.2950	6.4096	5.5413	4.6280	3.5348	2.7788	1.6837
100	9.6980	7.5720	6.5586	5.4459	4.7204	4.1508	3.6598	3.2080	2.7671	2.3060	1.7559	1.3777	0.8324
500	1.9950	1.5434	1.3313	1.1002	0.9508	0.8341	0.7339	0.6421	0.5529	0.4597	0.3494	0.2738	0.1660
1000	1.0010	0.7736	0.6669	0.5508	0.4759	0.4173	0.3671	0.3211	0.2765	0.2298	0.1746	0.1368	0.0825
5000	0.2008	0.1550	0.1336	0.1103	0.0952	0.0835	0.0734	0.0642	0.0553	0.0459	0.0349	0.0273	0.0165
10,000	0.1004	0.0775	0.0668	0.0551	0.0476	0.0418	0.0367	0.0321	0.0276	0.0230	0.0175	0.0137	0.0082

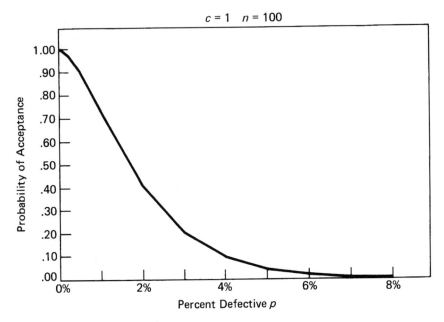

Figure 9.16 OC Curve Generated from Table 9.4

curves are especially useful. For example, let's say we are presented with a sampling plan calling for $n = 300$ units on test with $c = 3$ allowed failures. Reference to Figure 9.15 shows that this plan has an AQL of 0.45 percent and an LTPD of 2.25 percent. However, the product is very expensive, and we'd like to reduce the sample size while still retaining the LTPD (consumer's risk) at 2.25 percent. We look at Figure 9.14 for $c = 2$, find on the y-axis the LTPD value of 2.25 percent, and go across the graph to the diagonal line labeled 10 (for probability of acceptance equal to 10 percent). From the intersection of this line with $p = 2.25$ percent, we drop down to the x-axis and read $n = 223$ units. Hence, an alternative sample plan to $n = 300$, $c = 3$ that has the same LTPD is $n = 223$ and $c = 2$.

In the same manner, by referring to Figures 9.13 and 9.12, we determine that the sampling plans having $n = 165$, $c = 1$, and $n = 100$, $c = 0$ preserve the LTPD of the original plan. So we can run the experiment with a minimum sample size of 100 units, allowing no failures, and be 90 percent confident that the percent defective of the population sampled is no higher than 2.25 percent. This last statement follows since subtracting the percent value on each diagonal line from 100 percent represents the one-sided confidence level at any given percent defective p value that the number of failures in the sample size n will be less than the acceptance number.

Note, however, that by looking at the 95 percent diagonal line on each figure, the AQL has gone from 0.45 percent at $n = 300$, $c = 3$ to 0.36 percent, 0.225 percent, and 0.053 percent for the $c = 2, 1$, and 0 plans, respectively. Thus, the product sampled must have a lower percent defective to consistently pass the plans having smaller acceptance numbers. Such is the price one pays to reduce the sample size while holding the same LTPD. A nontechnical way of showing this AQL shift is to observe that the allowed fallout to pass the test goes down from 3 failures out of 300 (or 1 percent), to 2 failures in 223 (or 0.9 percent), to 1 failure in 165 (or 0.6 percent), to 0 failures in 100 (or 0 percent) as the sample size decreases. Of course, once a plan is chosen, we can immediately generate the OC curve to check the risks at other p values by the procedure described previously.

One final comment is made: if we calculate the ratio of the LTPD to the AQL for the $c = 0, 1, 2$, or 3 cases, we find the ratios equal to 43, 11, 6.5, and 4.9, respectively, independent of sample size. Thus, the larger the acceptance number, the smaller the difference between the AQL and LTPD; in other words, the OC curves become "steeper" as c increases.

Exercise 9.25

The qualification requirements allow a maximum of 2 failures on 500 devices stressed. What is the AQL and RQL at $\alpha = 0.05$ and $\beta = 0.10$ risks?

Exercise 9.26

The qualification requirements allow a maximum of 2 failures on 500 devices stressed. The engineer wishes to reduce the sample size. Determine the sample size to hold the same RQL at the same β risk for acceptance number $c = 1$. Repeat for $c = 0$. What are the AQLs for each case at $\alpha = 0.05$?

Exercise 9.27

The lot acceptance criteria allow a maximum of 3 failures on 300 devices inspected. What is the AQL and RQL at $\alpha = 0.05$ and $\beta = 0.10$ risks? The manufacturer wants to reduce the sample size for inspection. Determine the sample size to hold the same AQL at the same producer's risk for the acceptance number $c = 2$. Repeat for $c = 1$ and $c = 0$. What are the RQLs for each case at consumer's risk = 0.10?

Exercise 9.28

Use either the graphs of this section or Table 9.4 to sketch out a family of OC curves for $n = 50$ and $c = 0, 1, 2, 3$.

MINIMUM SAMPLE SIZE PLANS

Suppose we wish to protect against a Type II error using the smallest sample size possible, and Type I error is not a primary concern. This situation may occur, for example, when parts are limited in availability or are highly expensive or time consuming to test.

Minimum sampling plans are based on an acceptance number of zero, that is, $c = 0$ since any number greater than zero would permit a larger sample size to be used for the same acceptable and rejectable (AQL and RQL) percent defective levels. If we assume the sample size is small or the sample is drawn from an ongoing process, then the binomial distribution applies. Hahn (1979) treats the situation where the lot size is a factor and develops the curves according to the hypergeometric distribution; at the lower percent defective values, however, considerable "eyeball" interpolation is involved. Here, we present curves for consumer protection levels of 90 percent and 95 percent for all percent defective values over the range of 0.001 percent (10 PPM) to 100 percent (1,000,000 PPM) based on the binomial distribution.

The minimum sample size graph is shown as Figure 9.17. As is indicated by the arrows, the right-hand pair of diagonal lines refer to the right-hand vertical axis; the left hand pair refer to the left-hand vertical axis.

The derivation of the lines is quite simple. Since the acceptance number is zero, we want the probability of zero failures to be equal to the consumer risk

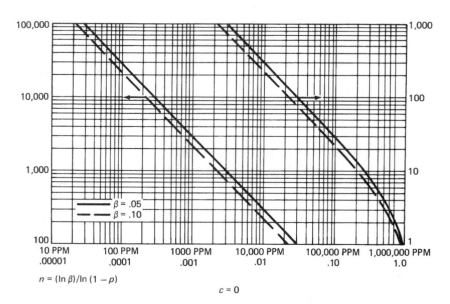

$n = (\ln \beta)/\ln (1 - p)$

$c = 0$

Figure 9.17 Graph for Minimum Sampling Plans

level, that is, the β or Type II error. Thus, we have $(1 - p)^{n} = \beta$. Solving for n gives

$$n = \frac{\ln \beta}{\ln (1 - p)}$$

This is the minimum sample size necessary to assure a maximum Type II error risk of β if that population fraction defective is no higher than p. Figure 9.17 provides two curves, one for $\beta = 0.05$ (95 percent confidence) and the other for $\beta = 0.10$ (90 percent confidence or an LTPD plan). For example, to protect against a fraction defective higher than 0.015 (i.e., $p = 1.5$ percent) with 90 percent confidence, the minimum sample size is about 150 units.

NEARLY MINIMUM SAMPLING PLANS

Some engineers (not to mention managers) might feel uncomfortable about making a decision based on no failures. They feel there is some consideration that has to be given to the producer's risk, even at the cost of additional units. Thus, we also include a graph to find the sample size for nearly minimum sampling plans based on an acceptance number of $c = 1$. Figure 9.18 shows two consumer protection levels, $\beta = 0.05$ and $\beta = 0.10$. This type of plan allows at most one failure to occur before the lot or process is rejected. Of course, the sample sizes are

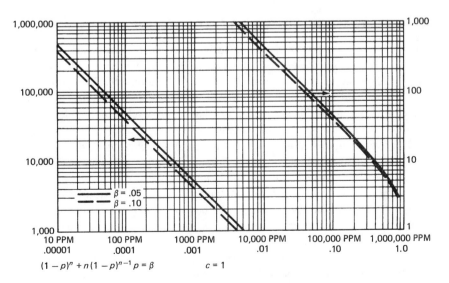

Figure 9.18 Graph for Nearly Minimum Sampling Plans

higher than the minimum sampling case ($c = 0$) for the same confidence level and fraction defective.

The derivation of the nearly minimum sampling plan is straightforward. We want the probability of zero or one failures to be set at the consumer risk level. Then, based on the binomial distribution, we require

$$(1 - p)^n + n(1 - p)^{n-1} p = \beta$$

This equation cannot be solved explicitly for n, but it is a simple procedure to calculate corresponding sample sizes n for various p values and thereby generate the lines shown in Figure 9.18. As an example, one would need approximately 315 units to assure with 95 percent confidence that the population fraction defective is no higher than 1.5 percent (0.015), assuming not more than 1 failure occurs in the sample.

Exercise 9.29

Determine the minimum sample size to provide protection at 200 PPM for a $\beta = 0.1$ risk level.

Exercise 9.30

Determine the nearly minimum sample size to provide protection at 200 PPM for a $\beta = 0.05$ risk level.

RELATING AN OC CURVE TO LOT FAILURE RATES

Lot acceptance sampling plans can be designed to evaluate and assure reliability. A sample from a lot is tested for t hours at high stress, and the proportion of reliability failures generated is the sample defect level. Intuitively, if this defect level is controlled and low enough, the product will have an acceptable failure rate. Often, however, it is difficult to quantify what defect level is acceptable and how any defect level relates to field failure rates.

To convert the proportion defective scale on an OC chart to a field failure rate scale, we need a model that relates fallout at high stress for t hours to a normal use average failure rate. In this section, we will assume that model is known to the extent that we have an acceleration factor A that converts test hours to field hours, and that we also know the life distribution. From this information we will derive the equations that can be used to transform average failure rate objectives to an OC chart scale, and vice versa.

For the exponential distribution, the calculation is simple. We convert a proportion p failures at t hours of stress to p failures at the equivalent At hours of field use and solve the following equation for λ:

$$p = 1 - e^{-\lambda At}$$

From this we obtain $\lambda = [-\ln(1-p)]/At$. This value of λ is the average failure rate (AFR)—see Chapter 2—that corresponds to p. For example, if the stress runs 48 hr, and the acceleration factor for a particular mechanism is 100, then a fraction fallout level of 0.01 corresponds to an AFR of 0.209 %/K hr. Conversely, an AFR objective of 500 FITs translates into a p value of 0.0024 for the same At.

The conversion for a Weibull or a lognormal requires an additional assumption, because these distributions have two parameters. We have to assume the shape parameter (m or σ) is known for the product and does not vary significantly from lot to lot. Lot quality causes the characteristic life c or T_{50} parameter to vary, accounting for changes in the failure rate. This is a strong assumption—but experience often has shown it to be reasonable.

To obtain the equations to go from failure proportion p under accelerated conditions to a Weibull AFR (assuming m is known), we start with the following equation:

$$p = 1 - e^{-\left(\frac{At}{c}\right)^m}$$

Solving for the characteristic life c, we get

$$c = \frac{At}{[-\ln(1-p)]^{1/m}}$$

The Weibull AFR (see Table 4.3) over the period 0 to U hr is

$$\frac{\left(\frac{U}{c}\right)^m}{U} = \frac{U^{m-1}}{c^m}$$

Substituting for c, we obtain the desired equation expressing the AFR in terms of the proportion failing p

$$AFR = \frac{U^{m-1}[-\ln(1-p)]}{(At)^m}$$

where U is the period of use lifetime of interest for the average failure rate calculation. Similarly, p can be written in terms of the AFR as

$$p = 1 - e^{-\left[\frac{(At)^m AFR}{U^{m-1}}\right]}$$

To get the equations for a lognormal failure mode (assuming σ is known), we start with:

$$p = \Phi\left[\frac{\ln(At/T_{50})}{\sigma}\right]$$

Solving for T_{50}, we obtain

$$T_{50} = Ate^{-\sigma\Phi^{-1}(p)}$$

As we saw in Chapter 2, the AFR over the period 0 to U hr is

$$AFR = \frac{-\ln[R(U)]}{U} = \frac{-\ln[1 - F(U)]}{U}$$

From Chapter 5, the lognormal CDF $F(U) = \Phi\left[\frac{\ln(U/T_{50})}{\sigma}\right]$, and so

$$AFR = \frac{-\ln\{1 - \Phi[\ln(U/T_{50})/\sigma]\}}{U}$$

where $T_{50} = Ate^{-\sigma\Phi^{-1}(p)}$. Similarly, p can be expressed in terms of the AFR as

$$p = \Phi\left[\Phi^{-1}\left(1 - e^{-U \cdot AFR}\right) - \frac{\ln(U/At)}{\sigma}\right]$$

By using the formulas in this section, lot acceptance sampling plans can be defined in terms of an AQL field average failure rate and an LTPD field average failure rate. Conversely, we can determine a sampling plan based on translating AQL and LTPD field failure rates into defect levels p occurring under accelerated stress testing.

Example 9.9   Sampling Plan for Accelerated Stress, Weibull Distribution

The acceptable (AQL at 5 percent risk) field average failure rate for 4000 hr is 330 FITs. The rejectable (LTPD) field average failure rate is 1000 FITs. If the failure distribution is Weibull with characteristic life $c = 0.8$, and the acceleration factor A is 301 for the particular failure mode of concern, what are the sample size and acceptance numbers for a stress running for 168 hr?

Solution

Using the equation above that expresses the proportion failing p in terms of the Weibull AFR, we see that AFR = 330 FITs translates to $p = 0.01$, and AFR = 1000 FITs, into $p = 0.03$. Thus, from Table 9.3, the sampling plan for these p values at risks $\alpha = 0.05$ and $\beta = 0.10$ is $n = 390$ and $c = 7$.

Exercise 9.31

The AQL field average failure rate for 40,000 hr is 180 FITs. The LTPD field average failure rate is 520 FITs. If the distribution of failures is lognormal with sigma equal to 4.0, and the acceleration factor $A = 500$, determine the sampling plan in terms of n and c for a lot acceptance test lasting $t = 48$ hours.

Exercise 9.32

The AQL field average failure rate for 24,000 hr is 50 FITs. The LTPD field average failure rate is 254 FITs. If the distribution of failures is Weibull with shape parameter equal to 0.75, and the acceleration factor $A = 205$, determine the sampling plan in terms of n and c for a lot acceptance test lasting $t = 2000$ hours.

STATISTICAL PROCESS CONTROL CHARTING FOR RELIABILITY

We have discussed procedures for lot acceptance. However, statistical process control techniques can also be used for monitoring the reliability of a process. Basically, the fraction (or percent) defective values from periodic samples are plotted on a graph. This chart, called a control chart, has a line, called an upper control limit (UCL.), that indicates when the plotted fraction defective is significantly higher than the normal process average. The difference between an UCL and an *engineering specification* is critical: the control limit is based on the recent history of the process; the engineering specification may have no relationship to the process performance. Thus, we can have a situation in which the process is "in spec" but yet out of control (and vice versa)!

The UCL is simple to calculate. If p is the historical process average, then the upper control limit is based on the normal approximation to the binomial distribution. The formula is

$$UCL = p + z_{1-\alpha}\sqrt{\frac{p(1-p)}{n}}$$

where n is the periodic sample size and $z_{1-\alpha}$ is the standard normal variate value corresponding to a desired $(1 - \alpha) \times 100$ percent confidence level. Typically $z_{1-\alpha}$ is set at 2 or 3 to roughly correspond to one sided probabilities of 97.7 percent or 99.9 percent, respectively. The interpretation is that there are only 2 chances in 100 (or 1 chance in a 1000, for the higher probability) that a sample of size n would have a percent defective above the 2 (or 3) "sigma" limit. Therefore, when the control limit is exceed, we assume the process is out of control, and corrective action takes place. Otherwise, the process is operating normally and no interference is required. The term "sigma" refers to the fact that the equation for the UCL involves the term for the standard error of the binomial distribution along with the normal distribution approximation for the probabilities.

If the periodic sample size is nearly constant (that is, any one lot is within 30 percent of the average sample size), then it is not necessary to adjust the UCL for each sample. Instead, one constant line will serve as the control limit. Figure 9.19 illustrates a typical "p-type" control chart for attribute data.

There are also control charts for variables data involving both the sample means and the sample ranges and for Poisson distributed data. Many excellent books on the subject of statistical quality control exist. In particular, see Burr (1976, 1979), Deming (1982), Duncan (1986), Grant and Leavenworth (1988), Ishikawa (1982), Juran (1988), Ott and Schilling (1990), or the *Western Electric Handbook* (1958). The interested reader involved in monitoring and improving process reliability should consult these references.

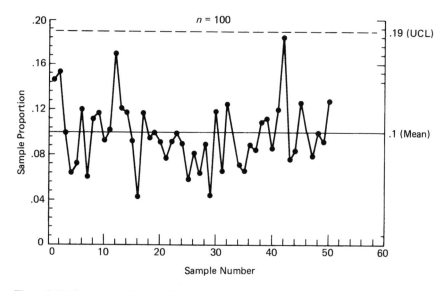

Figure 9.19 Three-Sigma Control Chart for Binomial Proportions

Example 9.10   Cumulative Count Control Charts for Low PPM

In reliability and quality work, we often encounter situations involving low defect proportions. The typical p-type control chart becomes ineffective in a low PPM environment because most samples of reasonable size yield zero defects, and a chart that shows only zeros has rather limited utility. Also, the normal approximation to the binomial distribution is not sufficiently accurate at low PPM values. For example, consider a process averaging 100 PPM. A sample of 200 pieces is routinely drawn from the process for monitoring. The three-sigma UCL based on the equation above is

$$0.0001 + 3 \sqrt{\frac{0.0001\,(0.9999)}{200}} = 0.0022$$

or 22 PPM. However, 1 reject out of 200 pieces is 0.0050 fraction defective or 50 PPM. Thus, any reject will cause the process to be considered "out-of-control." Yet, an exact calculation for the binomial distribution shows that the probability of at least one failure for $n = 200, p = 0.0001$ is about 2 percent, and not roughly 1 chance in 1000 as implied by three-sigma limits.

A superior approach concentrates on the number of good items produced instead of directing attention to the number of defective units observed. Called the *cumulative count control* (CCC) chart, the method is described by Calvin (1983) and Fasser and Brettner (1992). Basically, the accumulated number of good units from each inspection is plotted on a vertical logarithmic scale versus time. The upper and lower control limits are exact binomial limits based on the process average and are set to provide a confidence interval for zero rejects such that an accumulated point of good units will fall inside the limits when a process is under control.

For example, suppose a process is averaging 100 PPM. Then, for a sample of size 513 drawn from this process, binomial distribution calculations show there is a 5 percent chance of obtaining at least one defect and 95 percent chance of zero defects; for a sample of size 29,956, there is a 95 percent chance of getting at least one defect and a 5 percent chance of zero defects. We say the 90 percent control limit band is from 513 to 29,956. So if the number of good units accumulated is less than 513 before the first defect is observed, the data does not support a 100 PPM process capability. Similarly, if the UCL of 29,956 is exceeded by the accumulated number of good units, then we state that the process is significantly better than 100 PPM. Thus, out-of-control sequences are immediately apparent. If the accumulated number of good units is between 513 to 29,956 when the first reject occurs, we accept the results as consistent with a process at 100 PPM. We then reset the accumulated count to zero and begin accumulating good units again. An example graph for a 100 PPM process average is shown as Figure 9.20.

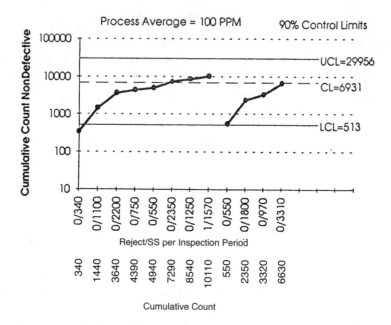

Figure 9.20 Cumulative Count Control Chart

If the process average defect level is p, the expected number of good units is $1/p$. The centerline of the control chart is set at the median (50 percent) probability level; that is, the centerline is at $-[\ln 0.5/\ln(1-p)] \approx 0.6931/p$. For example, if $p = 100$ PPM, the expected number of good units is 10,000 and the centerline is placed at 6,931. The lower control limit is set at

$$n_{LCL} = \frac{\ln(1 - \alpha/2)}{\ln(1 - p)}$$

and the upper control limit is set at

$$n_{UCL} = \frac{\ln(\alpha)}{\ln(1 - p)}$$

where $\alpha/2$ is the probability of a count less than the LCL (or above the UCL). In our example, we used $\alpha = 0.10$ for a 90 percent control limit band, $p = 100$ PPM, and found $n_{LCL} = 513$ and $n_{UCL} = 29,956$. Table 9.5 shows minimum sample sizes for various PPMs versus the associated probabilities of zero rejects. Note in the row labeled 100 PPM, we read the sample sizes 513 and 29,956 in the columns labeled 0.95 and 0.05, respectively.

TABLE 9.5   Minimum Sample Sizes for Zero Rejects at Various Probabilities

PPM	Probability of Zero Rejects																		
	0.999	0.995	0.99	0.975	0.95	0.9	0.8	0.7	0.6	0.5	0.4	0.3	0.2	0.1	0.05	0.025	0.01	0.005	0.001
1	1000	5013	10050	25318	51293	105360	223143	356675	510825	693147	916290	1203972	1609437	2302584	2995731	3688878	4605168	5298315	6907752
3.4	294	1474	2956	7446	15086	30988	65630	104904	150243	203866	269497	354109	473363	677230	881096	1084963	1354460	1558326	2031689
5	200	1003	2010	5064	10259	21072	44629	71335	102165	138629	183258	240794	321887	460516	599145	737774	921032	1059661	1381548
10	100	501	1005	2532	5129	10536	22314	35667	51082	69314	91629	120397	160943	230257	299572	368886	460515	529829	690772
20	50	251	503	1266	2565	5268	11157	17834	25541	34657	45814	60198	80471	115128	149785	184442	230256	264913	345384
50	20	100	201	506	1026	2107	4463	7133	10216	13863	18325	24079	32188	46051	59913	73776	92101	105964	138152
75	13	67	134	338	684	1405	2975	4755	6811	9242	12217	16052	21458	30700	39942	49183	61400	70642	92100
100	10	50	100	253	513	1054	2231	3567	5108	6931	9162	12039	16094	23025	29956	36887	46049	52981	69074
200	5	25	50	127	256	527	1116	1783	2554	3465	4581	6019	8046	11512	14977	18443	23024	26489	34535
500	2	10	20	51	103	211	446	713	1021	1386	1832	2407	3218	4604	5990	7376	9208	10594	13812
1000	1	5	10	25	51	105	223	356	511	693	916	1203	1609	2301	2994	3687	4603	5296	6904

Exercise 9.33

Determine the control limits for a cumulative count control chart for a process at 20 PPM. What is the expected number of good units? Determine the centerline. Assume $\alpha = 0.01$.

SUMMARY

In this chapter, we considered the relationship of quality control concepts to reliability problems. Starting with simple ideas on permutations and combinations, we developed the binomial distribution. We talked about how binomial estimates could be used to nonparametrically provide probability information in the absence of knowledge about the underlying distribution of fail times. We considered the simulation of binomial random variables. We also discussed two other discrete distributions: the hypergeometric and the Poisson. We showed the correspondence between the discrete Poisson and the continuous exponential distributions. We discussed various types of sampling, quality concepts such as AQL and LTPD, associated risks, properties of operating characteristic curves, and the selection of sample sizes. We illustrated many procedures, graphical and otherwise, to simplify the generation, implementation, and comparison of various sampling plans. The application of minimum and near-minimum sampling plans was covered. The relation of failure rates to sampling considerations was highlighted. Finally, we showed how statistical process control could be applied to the monitoring of reliability situations. In particular, we addressed the issue of monitoring low PPM situations using cumulative count control charts.

Problems

9.1 Calculate the number of permutations of four objects, A, B, C, and D. List all permutations.

9.2 Calculate the number of permutations of six objects, A, B, C, D, E, and F, taken two at a time. List all permutations.

9.3 Calculate the number of combinations of five objects, A, B, C, D, and E, taken three at a time. List all combinations.

9.4 One hundred devices are placed on stress for 10,000 hours. The probability of a device failing by 1000 hours is 0.02 or 2 percent. Assuming the devices on stress are from the same population,

a. What is the probability that all devices survive 1000 hours?

b. What is the probability that exactly two devices fail by 1000 hours?

c. What is the probability of at least two devices failing?

9.5 We run a test with 50 devices. At the end of the test, there are 15 failures. What is a 90 percent confidence interval for the population fraction defective? What is an 80 percent CI? Explain the difference.

9.6 In failure analysis, we have 20 devices to analyze. We are looking for a specific failure mechanism. If there are exactly 5 devices in the 20 with the specific mechanism, what is the probability that we get none with that mechanisms in a random sample analysis of 8 devices?

9.7 There are 5000 devices placed on stress. If the failure probability for a device at the end of test is 0.0002 (that is, 0.02 percent), what is the probability that there are no failures? At least one failure?

9.8 Fifty devices are placed on stress for 168 hr. The constant failure rate is 30 percent per K hr during the interval on stress. Using the Poisson distribution,

    a. What is the expected number of failures?

    b. What is the probability that all devices survive 168 hours?

    c. What is the probability of at least one failure?

    Compare to the results in Exercise 9.4.

9.9 One hundred fifty devices are stressed for 1000 hours. There are no failures. Using Figure 9.6, estimate an upper 95 percent confidence limit on the population percent defective.

9.10 One hundred devices are stressed for 500 hours. Two failures occur. Using Figure 9.6, estimate a 95 percent confidence interval on the population percent defective.

9.11 We wish to generate a sampling plan with a 95 percent probability of acceptance of product that is 1 percent defective and a 90 percent probability of rejecting 10 percent defective product. Using Figure 9.11, find the sample size n and the acceptance number c.

9.12 What is the minimum size of a sample to be tested to assure with 90 percent confidence that a lot having 7000 PPM defective will not be accepted? What is the sample size if the acceptance number $c = 3$?

9.13 Using Figure 9.12, sketch an OC curve for $n = 1000$, $c = 0$.

9.14 Using Table 9.4, sketch an OC curve for $n = 500$, $c = 1$.

9.15 We are given a sampling plan with $n = 500$ and $c = 2$. Using Figure 9.14, determine the AQL (95 percent) and the LTPD (10 percent). Using Figure 9.13, determine the sample size necessary for the same LTPD, but with the acceptance number $c = 1$. Check your results using Table 9.4.

9.16 What is the minimum sample size to protect against accepting a fraction defective higher than 500 PPM with 90 percent confidence?

9.17 What is the nearly minimum ($c = 1$) sample size to protect against accepting a fraction defective higher than 250 PPM with 95 percent confidence?

9.18 If the historical process average is 500 PPM, calculate the upper control limit for a periodic sample size of 1000. In an SPC mode, what happens if you get two rejects in a sample of size 1000? What's the probability of getting two or more rejects in a sample of size 1000? Explain the difference in the two approaches.

9.19 Construct a cumulative count control chart for a process average of 250 PPM. Draw the control limits and the centerline. Assume $\alpha = 0.05$.

Chapter 10

Repairable Systems I: Renewal Processes

In previous chapters, we focused on the reliability of nonrepairable components such as light bulbs and integrated circuits. The working assumption was that times to failure were a truly random sample of independent and identically distributed (i.i.d.) observations from a single population. Consequently, individual failure times could be combined for analysis, neglecting any order of occurrence of the original data. However, there are many common situations in which the sequence of failure times has significance. Consider an action that restores a failed system or process to operation. For example, replacing a circuit board fixes a computer, changing a clogged gas line filter repairs an automobile, resuming electrical power returns a factory to production following a utility failure, and calling a server reconnects an interrupted video-teleconferencing session. In all these instances, the failures occur sequentially in time, and assumptions of a single population distribution and independence for the *times between failures* are often invalid.

If the i.i.d. properties hold for the times between failures, the repair rate is stable, and the renewal process briefly mentioned in Chapter 2 is an appropriate model. This chapter focuses mainly on the characteristics and features of a renewal process. Conversely, if there is evidence of a trend such as varying (e.g., improving or worsening) repair rates, the renewal model no longer applies. Then assessing and modeling the system behavior requires consideration of the patterns of the sequential repair times. The next chapter covers the subject of nonrenewal processes in detail.

These two chapters present key concepts on analyzing data from systems undergoing repair. The topics include useful methods for describing and modeling such data. Verification of assumptions is stressed. To enhance comprehension, simulation techniques are provided. We cover many important and practical issues involving the reliability of repairable systems.

REPAIRABLE VS. NONREPAIRABLE SYSTEMS

The reliability literature covers the treatment of nonrepairable units extensively, but except for special situations such as renewal processes, these analysis procedures are not suitable for analyzing data from repairable systems. Also, the confusing use of similar terminology among authors describing very different repairable and nonrepairable situations has muddled otherwise easy-to-understand concepts. Ascher and Feingold (1984) point out that, with few exceptions, "the entire area of repairable systems has been seriously neglected in the reliability literature" even though "most real world systems are intended to be repaired rather than replaced after failure."

Consider an operating *system*—at some point in time—that fails to perform an expected function. (The concept of repairability also can be applied to a *process*, such as assembly operations in a factory. For example, if production is halted by a power outage, the repair returns the manufacturing process to full operations. However, for simplicity we shall use the term "system" for either possibility.) The system is *repairable* if it can be restored to satisfactory operation by any action, including replacement of components, changes to adjustable settings, swapping of parts, or even a sharp blow with a hammer. Obviously a television set is a repairable system which, upon failure, may be fixed possibly by replacing a failed component such as a resistor, capacitor, or transistor, or by adjustments to the sweep or synchronization settings. An automobile is another familiar type of repairable system.

Understanding repairable system behavior is important for many reasons. Maintenance schedules and spare parts provisions are affected by expected repair rates. Detection of trends is important to assure suitable performance, especially when health and safety issues are involved. In development stages, information obtained on the first prototype systems—often only a single costly, system is available for study—may be used for reliability improvement of later ones or to design better future systems. (For an example of such an application, see Usher, 1993.)

Conversely, in production mode, where many copies of a system are available, the objective may be to estimate the repair rate of the population of systems. Issues of concern may be specifying burn-in effectiveness, providing for spare parts, forecasting repair and warranty costs, or establishing a preventive maintenance schedule. (For an illustration of a comparison of two levels of in-house testing to reduce early life failures prior to shipment of systems to customer sites, see Zaino and Berke, 1992.)

For a single-component system, if repair occurs by an action that apparently restores the system to "like new" condition (such as a simple replacement of the failed component with a new component from the same population), then a renewal process seems likely. In such situations, it may be reasonable to assume that the times between successive repairs are independent and from the same component life distribution. However, replacement of a failed component, by itself, does not necessarily assure a renewal process. (For example, Usher, 1993, describes a system repaired by the replacement of a component with an identical unit from the same population. Yet, because the cooling unit of the system was degrading, the times between consecutive fails became shorter, thereby ruling out a renewal process.)

Under a renewal model, a single distribution characterizes the independent times between failures, and the repair frequency appears fairly constant. If the frequency of system repairs is increasing or decreasing with age—indicating deterioration or improvement—then we have non-renewal behavior. For a non-stable repair rate, maintenance costs and provisions for spare parts are expected to vary in time.

Consequently, it is important to distinguish renewal processes from general repair processes. Renewal processes permit simplification of analysis. Non-renewal processes are more complicated because of non-stationary characteristics. This chapter will focus on the renewal situation, and non-renewal processes will be covered in the next chapter.

The discussion of renewal processes begins with some simple graphical procedures applied to single repairable systems. Next is the analysis of many systems with different operating times (multicensored data). The graphical results motivate analytical techniques. Time will be the primary metric for repairable systems, but other units such as miles or cycles between repairs may be used. The methods shown easily extend to these alternative scales.

GRAPHICAL ANALYSIS OF A RENEWAL PROCESS

Many statistical procedures exist for analyzing data from repairable systems, but analysts will gain valuable insight into system performance by simple plots of the data. Such plots generally indicate which analytical methods are most appropriate. Graphical techniques may reveal trends and suggest models. It is strongly recommend that the first step in the analysis of any repair data be the plotting of the observations. We will present several chart types which are particularly useful in studying repairable system performance.

Consider a single system for which the times involved in making repairs are ignored. We observe ten failures immediately corrected by repairs at the following system ages in hours:

106   132   289   309   352   407   523   544   611   660

A line sketch shows the pattern of repairs:

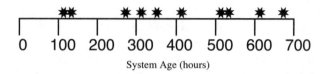

System Age (hours)

How do we treat such data to answer various questions about repair behavior? For example, "Is there any indication of the system getting better or worse or staying basically the same?"

An important measure of reliability is the cumulative number of failures $N(t)$ that occur on a system by time t. For any t, the function $N(t)$ is a random variable. A *stochastic process* is a collection of random variables that describes the evolution in time of a process. Since $N(t)$ exists for each system in a population, the number of repairs $N(t)$ occurring by time t for a system is an example of a stochastic *counting* process.

A common data graph is the cumulative plot: the cumulative number of repairs is plotted against the system age t. Each cumulative plot for a sample system is viewed as one observation from a population of possible curves. The population mean cumulative number of repairs at age t is denoted by $M(t)$, the expected number of repairs per system by time t.

Figure 10.1 is a cumulative plot of the above data. The repair history plots as a staircase function, with a step rise at each repair, using either connected or

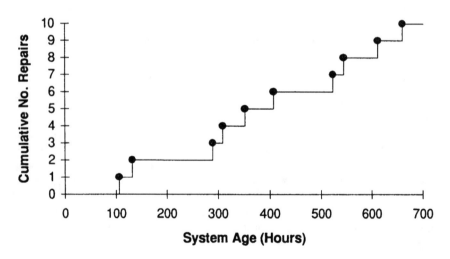

Figure 10.1 Cumulative Plot

unconnected points. The stepwise nature of the plot is made evident here, but the stairs will be omitted for simplicity in subsequent plots.

A linear appearance indicates the repair rate is staying basically constant. Alternatively, curvature in this plot can reveal whether the system reliability is improving or deteriorating.

For a single, randomly chosen system, the actual number of repairs by time t provides an unbiased estimate of the population mean number of failures per system $M(t)$, which we denote by $\hat{M}(t)$. The plot is a sample observation from a potential population of such curves. Note, however, that there are many situations in industry in which only one system is available (e.g., development activity using a prototype). Thus the entire "population" consists only of that single system, but a conceptual population may be future versions of the system.

Figure 10.1 shows that the cumulative number of repairs versus time appears to be close to linear. That is, the total number of repairs is proportional to age t, and the *rate of repairs* [that is, the derivative $dM(t)/dt$] is fairly constant in time. The repair process is called *stationary* because there is no apparent trend associated with system improvement or deterioration. This stationary characteristic suggests the possibility of a renewal process, that is, independent and identically distributed times between failures. (See Cox, 1962, for further discussion.) For a renewal process, the graph will necessarily appear linear over time. However, an observation interval that is too short may not be sufficient to reveal any pattern. We shall discuss general analytical procedures later in this chapter, but for now, let us assume a renewal process and analyze the data accordingly.

Under a renewal process, the times between failures are independent and identically distributed (i.i.d.). Let the successive ages at failure be denoted by t_i, $i = 1$, $2, \ldots$, Let X_1 be the time to the first failure (that is, $X_1 = t_1$), let X_2 be the time between the first and second failure (i.e., $X_2 = t_2 - t_1$), and so on. The X_i times, called *interarrival times*, are then:

106	26	157	20	43	55	116	21	67	49

Similar to statistical process control (SPC) charts, a second useful and revealing graph type for repairable systems, especially for detecting trends, is a plot of the interarrival times against the system age at repair. Figure 10.2 is such a graph, illustrating that there is no discernible trend in the data, as expected under a renewal process.

Because of the assumed renewal process, the X_i times can now be treated as a sample of ten i.i.d. observations from the same population, just as if they were derived from ten nonrepairable units. For example, using the probability plotting approach of Chapter 6, we sort the data from smallest to largest, assign median

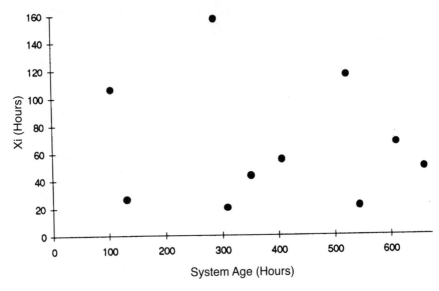

Figure 10.2 Interarrival Times versus System Age

rank plotting positions, and plot the data on an appropriate probability paper, such as Weibull or lognormal. Alternatively, we can employ maximum likelihood procedures to estimate the population parameters. In doing so, we assume the interarrival times are random variables from the same, single population. Based on physical considerations for this system, we fit a lognormal distribution to times between repairs. A lognormal probability plot of the interarrival times, shown as Figure 10.3, supports the supposition.

For estimating distribution parameters, renewal data from a single repairable system is complete, that is, noncensored data up to the time of the last failure. If we have some running time since the last repair, we should include that data as a single censored observation. For the lognormal distribution, it is now relatively simple to estimate the parameters for complete data. To estimate T_{50}, we use the mean of the logarithms of the failure times:

$$\hat{T}_{50} = e^{\hat{\mu}}$$

where the mean, μ, is estimated from the average of the ten times between failures. Similarly, the shape parameter, σ, for the lognormal distribution may be estimated from the standard deviation of the logarithms of the failure times. (See discussion in Chapter 5.) Thus,

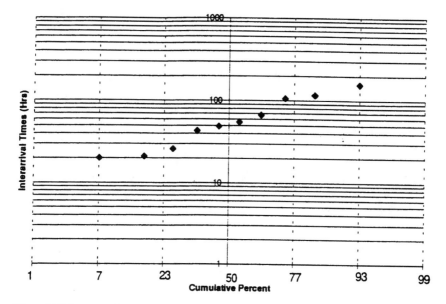

Figure 10.3 Lognormal Probability Plot

$$\hat{T}_{50} = \exp\left\{ \sum_{i=1}^{n} \ln X_i / n \right\} = (X_1 X_2 ... X_n)^{1/n}$$

$$= (106 \times 26 \times ... \times 49)^{1/10} = 52.65$$

Note that T_{50} is what is called a geometric mean of the times; that is, a geometric mean of n values is the nth root of the product of the n values. The shape parameter sigma is estimated using the equation

$$\hat{\sigma} = \sqrt{\frac{\sum_{i=1}^{n} (\ln X_i - \hat{\mu})^2}{n-1}}$$

$$= \sqrt{(\ln 106 - \ln 52.65)^2 + (\ln 26 - \ln 52.65)^2 + ... + (\ln 49 - \ln 52.65)^2}$$

$$= 0.72$$

It is always a good idea to plot the fitted model against the empirical CDF to assess the adequacy of the fit. Figure 10.4 is such a plot. The fit appears reasonable. Having a model allows us to compare actual to anticipated results, to estimate the renewal rate with a smooth function, to predict not only the future behavior of the observed system but also the expected performance of the population of similar systems, and potentially to gain insight into the repair causes. In addition, for a single distribution, we can express reliability in terms of a mean time between failures, *MTBF*, for planning purposes or comparison to objectives. Finally, a possible model also allows us to apply statistical procedures specific to the model, including tests of parameters, assumptions, and so on.

With a model, we could also predict possible future repair scenarios for similar renewal processes by simulating interarrival times by using randomly generated variables from a lognormal distribution having the estimated parameters. Each simulation of a set of interarrival times would represent one possible cumulative plot from the population of all possible such plots. The distribution of the plots at any system age *t* would suggest a degree of uncertainty in the expected number of repairs for any extrapolation.

Exercise 10.1

The following ten consecutive repair times are recorded for a system: 47, 90, 180, 208, 356, 377, 399, 461, 477, 652. Make a cumulative plot. Is there evidence

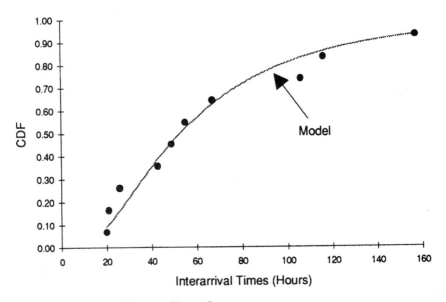

Figure 10.4 CDF Model Fit versus Observed

of a renewal process? Make a probability plot assuming a lognormal distribution. Estimate the parameters T_{50} and σ. Check the model distribution fit against the empirical CDF.

ANALYSIS OF A SAMPLE OF REPAIRABLE SYSTEMS

We have shown, for a single system, the cumulative plot that depicts the cumulative number of repairs versus the system age. However, we may be concerned with systems consisting of many identical, independent subsystems (sometimes called "sockets" in the reliability literature) for which the system behavior is the combined repairs of all subsystems.

Alternatively, we may be interested in the overall performance of many, identical systems where possibly each system may have a different operating time under use conditions, thereby producing multicensored data. (See discussion of multicensored data in Chapter 6.) For example, consider a manufacturer of televisions. The company may collect sample data on the repair frequency of a new model type installed over the last calendar year. Some TVs will have been in the field for nearly the full year and others may have only recently been purchased and placed in use. How do we estimate the mean repair rate for all TV's? What is the variation in the mean number of repairs for all TV's in the field at a given time? What is the expected time to the first repair? To the kth repair? Are the costs of repairs increasing or decreasing? What is the mean repair cost? Are spare parts adequate? Is burn-in or run-in (that is, burn-in at operating conditions) necessary? How long should burn-in be done? How cost effective is burn-in?

Both renewal and non-renewal processes may involve multiple systems. Several analyses of multi-system or multi-socket data have appeared in reliability literature. One graphical tool involves comparing distributions of interarrival times or some other numerical measure for the nth repair in a system, as measured across many systems. For example, histograms of the times to first repair among the systems can be compared to histograms of the times between first and second repairs, between second and third repairs, and so on. In this way, we are checking for trends and whether the various interarrival times are identically distributed. Obviously, this approach requires many systems for generating empirical distributions.

This method was used by Davis (1952) to analyze the number of miles between successive major failures of bus engines. Davis generated histograms showing interarrival miles to the first failure, to the second failure, and so on. In addition to depicting the average miles to the ith repair and displaying the variation in the data, the shape of the histograms provided useful information for investigating the repair frequency. For example, Davis found average inter-repair times to be decreasing, early interarrival times nearly normally distributed, but later interarrival times distributed roughly exponentially. See Ascher and Feingold (p. 86, 1984) for further discussion and display of the histograms from the Davis study. While the histogram approach has value, one requires considerable data on repairs for valid comparisons. Also, the method does not provide a model for the mean number of repairs, repair rate, or other statistical measures as a function of, for example, time or miles.

Another direction proposed by O'Connor (1991) is to display a graphical matrix structure in which each row represents the system repair history in a format similar to a dot-plot where the horizontal scale represents running time. In this manner, one can study the incidence of repairs against overhaul actions and across various systems, looking for possible patterns of common behavior. Again, there is no associated method provided for model development or testing, and multicensored data may limit the comparisons among systems.

The methodology developed by Nelson (1988a, b) and extended by Nelson and Doganaksoy (1989) also is graphical, but it has several advantages over other approaches. Nelson provides simple and informative plots of censored data on numbers and *costs* of repairs on a sample of systems. In addition, he shows how to generate a nonparametric graphical estimate of the mean cumulative cost or number of repairs per system versus age. We shall illustrate Nelson's procedure later in this section, after first developing some elementary concepts.

Consider a population of systems subject to repair actions. We can represent individual repair histories using connecting lines between repairs. A composite graph of the cumulative histories for all systems in the population could theoretically be drawn. For systems installed at different dates and consequently having different ages at the point of analysis, the referencing of all systems back to time zero would generate multicensored data. Each system repair history curve would extend only as far as the censoring time for that system. Figure 10.5 is such a plot, depicting five system histories with random censoring. The lines should be staircase functions but are shown connected for easier viewing. Table 10.1 gives the actual repair and censoring times.

TABLE 10.1   Repair Age Histories (Hours)

Repair Number	System 1	System 2	System 3	System 4	System 5
1	222	273	125	63	91
2	584	766	323	195	427
3	985	1054		325	761
4	1161				1096
5					1796
Censoring Times	1901	1316	442	636	2214

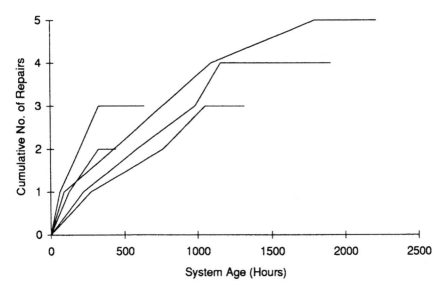

Figure 10.5 Repair History for Five Systems

As mentioned in an earlier section, we can envision a single curve, denoted by $M(t)$, that gives the average or expected cumulative number of repairs per system at time t. $M(t)$ is called the *mean cumulative repair function*. Note that $M(t)$ is derived from a vertical "slice" in Figure 10.5 of average system histories at time t. (In contrast, a horizontal "slice" would relate to a parameter such as the mean time to the ith repair.) How do we estimate $M(t)$ for multicensored data? Since the cumulative population repair function $M(t)$ is defined as the population mean cumulative number of repairs per system by time t, an unbiased estimator of $M(t)$, for a sample single system, is just the total repair count by time t. Similarly, for a collection of systems, an unbiased estimator is $\hat{M}(t)$ = average cumulative number of repairs among all systems by time t, assuming no censoring prior to time t.

How do we estimate $M(t)$ for multiple systems at times when one or more systems have been censored? (See Trindade and Haugh, 1979, for a discussion of this estimation problem applied to field renewal data.) First, consider the case of two systems. We want to estimate $M(t)$ for t greater than the censoring time L_1 of system 1. An unbiased estimator with some desirable properties, including smaller variance than other methods under censoring, is called the pooled summation estimator, developed for the general repair situation by Nelson (1988b). For our example, we use both systems to estimate $M(t)$ for t less than or equal to L_1, and then add the average number of repairs on system 2 for $t > L_1$ to the estimate $\hat{M}(t)$ at loss time L_1. Thus, for $t > L_1$, $\hat{M}(t)$ = average number of repairs by age L_1 on both systems *plus* the number of additional repairs on system 2 occurring in the interval from L_1 to t.

To illustrate, suppose system 1 has loss (censoring) time of L_1 = 100 hours and repairs at 50 and 90 hours. Assume system 2 has repairs at 60, 120, and 175 hours and a loss time L_2 = 200 hours. (See Figure 10.6.) For the repair times, t = 50, 60, 90, 120, and 175 hours, the pooled summation approach yields $\hat{M}$ (50) = 1/2 = 0.5, $\hat{M}$ (60) = 2/2 = 1, $\hat{M}$ (90) = 3/2 = 1.5, $\hat{M}$ (120) = 3/2 + 1/1 = 2.5, and $\hat{M}$ (175) = 3/2 + 1/1 + 1/1 = 3.5.

To extend to more than two systems, we add to the latest estimate $\hat{M}$ (t_i) the average cumulative number of repairs, on only the uncensored systems, occurring since the repair time t_i. Note that the pooled summation estimators are step functions at the repair times. The pooled estimator is non-decreasing in time, a desirable property not necessarily possessed by other estimators. At each time at which a system is censored, the pooled summation estimator is anchored at that point and subsequent repairs are summed to the previously anchored value.

Example 10.1   The Mean Repair Function

Using the data in Table 10.1, estimate the mean cumulative repair function. Plot the estimate with the original system repair histories.

Solution

We begin by combining and ordering the repair and censoring times from all systems. Since the earliest censoring occurs at 442 hours, the estimate of $M(t)$ at any system age prior to this time is just the total number of repairs, by age t, divided by five. Thus, $\hat{M}$ (63) = 1/5 = 0.2, $\hat{M}$ (91) = 2/5 = 0.4, ..., $\hat{M}$ (427) = 0/5 = 1.8. By the 10th repair, one system is censored, and $\hat{M}$ (584) = 1.8 + 1/4 = 2.05. At 636 hours, another system is censored, and at the 11th repair, $\hat{M}$ (761) = 2.05 + 1/3 = 2.38.

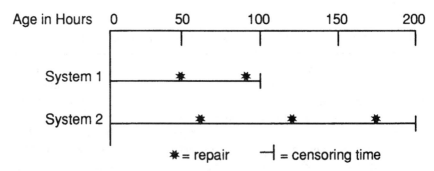

Figure 10.6 Repair Histories for Two Systems

Similarly, $\hat{M}$ (766) = 2.05 + 2/3 = 2.73,..., $\hat{M}$ (1161) = 2.05 + 6/3 = 4.05. For the final repair, $\hat{M}$ (1796) = 4.05 + 0.5 = 4.55. The estimated mean cumulative repair function is plotted as Figure 10.7.

The pooled summation approach was developed by Nelson in his 1986 and subsequent papers on the graphical analysis of system repair data. Nelson extended the concepts to estimate mean repair costs per system; in addition he reviewed the properties of the pooled sum estimator. Nelson and Doganaksoy (1989) presented a computer program for the calculation of confidence intervals for the pooled estimation method. A suitable reference on the topic of confidence limits is the expository paper by Nelson (1995).

Exercise 10.2

Field repair history has been captured on three sample systems. The first, with 500 hours total operating time, had two repairs at 145 and 368 hours. A second, with 300 hours of operation, had a repair at 247 hours. The third, with 700 hours of use, had three repairs at 33, 318, and 582 hours. Using the pooled summation approach, estimate $M(t)$ at each repair time.

In addition to graphical tools, analytical approaches are very important for developing, testing, and applying models. When the amount of data is small, or

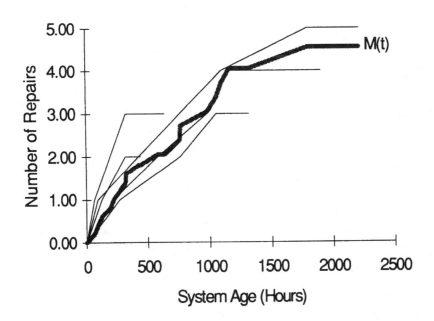

Figure 10.7 Mean Cumulative Repair Function

when we want to establish formally the significance of an effect, models are especially appropriate. We will now discuss the properties, models, and analytical methods for renewal processes.

RENEWAL PROCESSES

Let us briefly review and summarize the important considerations for a *renewal process*. Renewal processes describe certain repairable systems. In a renewal process for a single-component system, the replacements are always new items from the same population as the items being replaced. Consequently, the times between consecutive replacements are assumed independent and identically distributed. Note that we shall use the terms *components*, *units*, *items*, and *parts* interchangeably, and that a single component may be considered a *system*. Multicomponent systems are treated later in this section.

Under the renewal scenario, certain theoretical simplifications are possible, such that data analysis methods based on nonrepairable component theory are applicable. For example, times between failures may be collected for probability plotting and graphical parameter estimation. However, without the renewal process assumption, for proper analysis we need to employ methods that take the time sequence of observations into consideration. Otherwise, misleading conclusions are possible from the use of inappropriate techniques.

A system containing a single repairable component is a simple but frequently useful model. For example, a radar system may have a weak part that accounts for practically all failures during a specified period of use. The system operates until a failure occurs. The system is restored to operation (repaired) by *replacement* of the failed component. We assume the repair time to be negligible compared to the operating time. If the replacement parts are *new parts from the same population* as the original parts, and the replaced parts fail independently with a common failure distribution, we have a *renewal process*.

Generally for a repair process, two random variables are of key interest: $N(t)$, the total number of repairs by specified time t, and the total time $T(k)$ (system age) to reach the kth repair. $N(t)$ is a discrete random variable, but $T(k)$ is a continuous random variable. Both variables depend on the joint distributions of times between repairs.

$N(t)$ is a random variable which has a distribution function at any time t. The mean or expected value of $N(t)$ is called the renewal function, denoted by $M(t)$. (This value is the same as the mean cumulative number of repairs function per system discussed previously.) Thus,

$$M(t) = E[N(t)]$$

Similar to the discussion for the mean repair function, when we observe the cumulative number of replacements versus time for a renewal process for a single system, we are seeing a collection of data forming one sample (a realization) from the many possible for the random variable $N(t)$. Imagine a population of curves in which $N_i(t)$ for each system i is plotted versus time t. Then, the mean of all possible $N_i(t)$ curves is the renewal function $M(t)$.

For a renewal process, a single distribution of failure times defines the expected pattern. (Presuming the same distribution applies to *all* systems is an explicit assumption that needs to be verified.) Let X_i denote the operating time between repairs; that is, X_i is the "interarrival" time between the $(i-1)$th and the ith repairs, where $X_0 = 0$. Equivalently, $X_i = T(i) - T(i-1)$, where $T(0) = 0$. We can express $N(t)$, the number of repairs by time t, as a function of the X_i times. Thus, $N(t)$ is the maximum integer k such that the sum of the first k interarrival times are less than or equal to t; that is, $N(t) = k$ implies that

$$\sum_{i=1}^{k} X_i \le t \quad \text{and} \quad \sum_{i=1}^{k+1} X_i > t$$

For example, suppose we have a system with repairs at ages 10, 15, 27, 49, and 62 hours. The times between repairs are then 10, 5, 12, 22, and 13 hours, respectively. The observed number of repairs $N(30)$ at $t = 30$ hr is then 3, since the sum of the three interarrival times, that is $10 + 5 + 12$, is less than $t = 30$ hr; and the sum of the first four interarrival times is $10 + 5 + 12 + 22 = 49$ hours, which is greater than $t = 30$ hours.

We see that $N(t) = 0$ if the first repair time $X_1 > t$. Similarly, the time to the kth replacement can be written in terms of the X_is as

$$T(k) = \sum_{i=1}^{k} X_i$$

If we know the probability distribution of X_i, we can (at least numerically) determine the distributions for $N(t)$ and $T(k)$, along with the renewal function $M(t)$ and the corresponding population rate of renewal, $dM(t)/dt$.

Homogeneous Poisson Process

Suppose the interarrival times X_i are independent and exponentially distributed with failure rate λ; that is, the times have a PDF

$$f(x) = \lambda e^{-\lambda x} \qquad 0 \le x$$

Then the probability distributions of $N(t)$ and $T(k)$ are easy to determine. Barlow and Proschan (1975) show that $N(t)$ has a Poisson distribution with mean rate (or intensity) λ. So, the probability of observing exactly $N(t) = k$ replacements in the interval $(0, t)$ is

$$P[N(t) = k] = \frac{(\lambda t)^k e^{\lambda t}}{k!}$$

A renewal process for which the interarrival distribution is exponential is called a *homogeneous Poisson process* (or simply a Poisson process) and denoted by HPP. Based on the Poisson distribution, the expected value for $N(t)$ [that is, $M(t)$] is λt, and the variance is also λt. Similarly, based on the exponential distribution, the mean or expected value for the interarrival times is $1/\lambda$, and the variance is $1/\lambda^2$.

Note that this probability statement is consistent with interpretation of the CDF as the probability of failure in the interval $(0, t)$. The probability of at least one failure is one minus the probability of no failures. For the Poisson process, the probability of no failures in the interval $(0, t)$ is $P[N(t) = 0] = e^{-\lambda t}$. The probability of at least one failure is $1 - e^{-\lambda t}$, which is the same expression as the CDF for a exponential distribution with failure rate λ. We see that a Poisson process represents both a renewal process with independent exponentially distributed interarrival times having mean $1/\lambda$ and a counting process having the integer valued Poisson distribution with mean rate λ.

For the HPP, we can write the time $T(k)$ to the kth replacement as the sum of k independent, exponential random variables. This process is equivalent to that of a k level standby system introduced in Chapter 8. Consequently, the PDF for $T(k)$ is the gamma distribution with parameters λ and k. That is,

$$f_k(t) = \frac{\lambda^k}{(k-1)!} t^{k-1} e^{-\lambda t}$$

Since $T(k)$ has a gamma distribution, the mean time to the kth repair is k/λ, and its variance is k/λ^2.

Because of the equivalence of the two representations of a Poisson process, the CDF for $T(k)$ at time t can be viewed as the probability that k or more replacements occur by time t, which is the same as $1 -$ probability that $k - 1$ or *less* replacements occur by t. Thus, we can express the CDF using a sum of Poisson probabilities, that is,

$$F_k(t) = 1 - \sum_{i=0}^{k-1} \frac{(\lambda t)^i e^{-\lambda t}}{i!} = 1 - e^{-\lambda t} \sum_{i=0}^{k-1} \frac{(\lambda t)^i}{i!}$$

This information is important for deciding how many spare parts to stock to ensure a system operates during a mission period t. Equivalently, this CDF expression states the probability that the time $T(k)$ for the kth replacement is less than or equal to t.

Example 10.2   Spare Parts for a Poisson Process

Components are assumed to fail with an exponential distribution having failure rate $\lambda = 0.00030 = 0.030$ %/hr. The mission duration is 500 hr. To assure a 95 percent probability of successful mission completion, how many spare parts should be carried for a single component system?

Solution

The expected number of failures in 500 hr is $\lambda t = 0.003 \times 500 = 0.15$. The probability of zero failures is $e^{-0.15} = 0.861$. The probability of exactly one failure is $0.15e^{-0.15} = 0.129$. Thus, the probability of two or more failures is $1 - 0.861 - 0.129 = 0.010$ or 1.0 percent. Carrying one spare part will assure system operation during mission with nearly 99 percent probability. Note that two parts in total are required: the first is the original part, and the second is the spare.

Table 10.2 shows the probability of mission success based on the Poisson probability of observing a specified number of failures. Note the 90, 95, and 99 percent probabilities of success are delineated in this table. To use the table, determine the probability of mission success based on the intersection of the column for the assumed Poisson mean value λt with the number of spare parts. For example, for $\lambda t = 1.35$, the probability of success is 0.9518 if three spare parts are available. Again, since the table refers only to *spare* parts, the mission involves a total of four parts.

In summary, the number of repairs $N(t)$ occurring by time t is an example of a stochastic counting process. A counting process possesses independent increments if the number of events in non-overlapping intervals of time are independent. A counting process is said to have stationary increments if the distribution of the number of events which occur in an interval of time depends only on the length of the interval and not on the endpoints of the interval. (A renewal process is an example of a stationary process.)

A process with stationary and independent increments is equivalent to a process that restarts itself, in a probabilistic sense, at any point in time; that is, it has no memory. The Poisson process has stationary, independent increments since the number of repairs depends only on the elapsed time and not on the start time. For the Poisson process, the interarrival times are independent exponentially distributed random variables. (See Ross, 1993, for further discussion of counting processes.)

TABLE 10.2   Poisson Renewal Process: Mission Success Probability

Mean	0	1	2	3	4	5	6	7	8	9	10	11
					Number of Spare Parts							
0.05	0.9512	0.9988	1.0000	1.0000	1.0000	1.0000	1.0000	1.0000	1.0000	1.0000	1.0000	1.0000
0.10	0.9048	0.9953	0.9998	1.0000	1.0000	1.0000	1.0000	1.0000	1.0000	1.0000	1.0000	1.0000
0.15	0.8607	0.9898	0.9995	1.0000	1.0000	1.0000	1.0000	1.0000	1.0000	1.0000	1.0000	1.0000
0.20	0.8187	0.9825	0.9989	0.9999	1.0000	1.0000	1.0000	1.0000	1.0000	1.0000	1.0000	1.0000
0.25	0.7788	0.9735	0.9978	0.9999	1.0000	1.0000	1.0000	1.0000	1.0000	1.0000	1.0000	1.0000
0.30	0.7408	0.9631	0.9964	0.9997	1.0000	1.0000	1.0000	1.0000	1.0000	1.0000	1.0000	1.0000
0.35	0.7047	0.9513	0.9945	0.9995	1.0000	1.0000	1.0000	1.0000	1.0000	1.0000	1.0000	1.0000
0.40	0.6703	0.9384	0.9921	0.9992	0.9999	1.0000	1.0000	1.0000	1.0000	1.0000	1.0000	1.0000
0.45	0.6376	0.9246	0.9891	0.9988	0.9999	1.0000	1.0000	1.0000	1.0000	1.0000	1.0000	1.0000
0.50	0.6065	0.9098	0.9856	0.9982	0.9998	1.0000	1.0000	1.0000	1.0000	1.0000	1.0000	1.0000
0.55	0.5769	0.8943	0.9815	0.9975	0.9997	1.0000	1.0000	1.0000	1.0000	1.0000	1.0000	1.0000
0.60	0.5488	0.8781	0.9769	0.9966	0.9996	1.0000	1.0000	1.0000	1.0000	1.0000	1.0000	1.0000
0.65	0.5220	0.8614	0.9717	0.9956	0.9994	0.9999	1.0000	1.0000	1.0000	1.0000	1.0000	1.0000
0.70	0.4966	0.8442	0.9659	0.9942	0.9992	0.9999	1.0000	1.0000	1.0000	1.0000	1.0000	1.0000
0.75	0.4724	0.8266	0.9595	0.9927	0.9989	0.9999	1.0000	1.0000	1.0000	1.0000	1.0000	1.0000
0.80	0.4493	0.8088	0.9526	0.9909	0.9986	0.9998	1.0000	1.0000	1.0000	1.0000	1.0000	1.0000
0.85	0.4274	0.7907	0.9451	0.9889	0.9982	0.9997	1.0000	1.0000	1.0000	1.0000	1.0000	1.0000
0.90	0.4066	0.7725	0.9371	0.9865	0.9977	0.9997	1.0000	1.0000	1.0000	1.0000	1.0000	1.0000
0.95	0.3867	0.7541	0.9287	0.9839	0.9971	0.9995	0.9999	1.0000	1.0000	1.0000	1.0000	1.0000
1.00	0.3679	0.7358	0.9197	0.9810	0.9963	0.9994	0.9999	1.0000	1.0000	1.0000	1.0000	1.0000
1.05	0.3499	0.7174	0.9103	0.9778	0.9955	0.9992	0.9999	1.0000	1.0000	1.0000	1.0000	1.0000
1.10	0.3329	0.6990	0.9004	0.9743	0.9946	0.9990	0.9999	1.0000	1.0000	1.0000	1.0000	1.0000
1.15	0.3166	0.6808	0.8901	0.9704	0.9935	0.9988	0.9998	1.0000	1.0000	1.0000	1.0000	1.0000
1.20	0.3012	0.6626	0.8795	0.9662	0.9923	0.9985	0.9997	1.0000	1.0000	1.0000	1.0000	1.0000
1.25	0.2865	0.6446	0.8685	0.9617	0.9909	0.9982	0.9997	1.0000	1.0000	1.0000	1.0000	1.0000
1.30	0.2725	0.6268	0.8571	0.9569	0.9893	0.9978	0.9996	0.9999	1.0000	1.0000	1.0000	1.0000
1.35	0.2592	0.6092	0.8454	0.9518	0.9876	0.9973	0.9995	0.9999	1.0000	1.0000	1.0000	1.0000
1.40	0.2466	0.5918	0.8335	0.9463	0.9857	0.9968	0.9994	0.9999	1.0000	1.0000	1.0000	1.0000
1.45	0.2346	0.5747	0.8213	0.9405	0.9837	0.9962	0.9992	0.9999	1.0000	1.0000	1.0000	1.0000
1.50	0.2231	0.5578	0.8088	0.9344	0.9814	0.9955	0.9991	0.9998	1.0000	1.0000	1.0000	1.0000
1.55	0.2122	0.5412	0.7962	0.9279	0.9790	0.9948	0.9989	0.9998	1.0000	1.0000	1.0000	1.0000
1.60	0.2019	0.5249	0.7834	0.9212	0.9763	0.9940	0.9987	0.9997	1.0000	1.0000	1.0000	1.0000
1.65	0.1920	0.5089	0.7704	0.9141	0.9735	0.9930	0.9984	0.9997	0.9999	1.0000	1.0000	1.0000
1.70	0.1827	0.4932	0.7572	0.9068	0.9704	0.9920	0.9981	0.9996	0.9999	1.0000	1.0000	1.0000
1.75	0.1738	0.4779	0.7440	0.8992	0.9671	0.9909	0.9978	0.9995	0.9999	1.0000	1.0000	1.0000
1.80	0.1653	0.4628	0.7306	0.8913	0.9636	0.9896	0.9974	0.9994	0.9999	1.0000	1.0000	1.0000
1.85	0.1572	0.4481	0.7172	0.8831	0.9599	0.9883	0.9970	0.9993	0.9999	1.0000	1.0000	1.0000
1.90	0.1496	0.4337	0.7037	0.8747	0.9559	0.9868	0.9966	0.9992	0.9998	1.0000	1.0000	1.0000
1.95	0.1423	0.4197	0.6902	0.8660	0.9517	0.9852	0.9960	0.9991	0.9998	1.0000	1.0000	1.0000
2.00	0.1353	0.4060	0.6767	0.8571	0.9473	0.9834	0.9955	0.9989	0.9998	1.0000	1.0000	1.0000
2.05	0.1287	0.3926	0.6631	0.8480	0.9427	0.9816	0.9948	0.9987	0.9997	0.9999	1.0000	1.0000
2.10	0.1225	0.3796	0.6496	0.8386	0.9379	0.9796	0.9941	0.9985	0.9997	0.9999	1.0000	1.0000
2.15	0.1165	0.3669	0.6361	0.8291	0.9328	0.9774	0.9934	0.9983	0.9996	0.9999	1.0000	1.0000
2.20	0.1108	0.3546	0.6227	0.8194	0.9275	0.9751	0.9925	0.9980	0.9995	0.9999	1.0000	1.0000
2.25	0.1054	0.3425	0.6093	0.8094	0.9220	0.9726	0.9916	0.9977	0.9994	0.9999	1.0000	1.0000
2.30	0.1003	0.3309	0.5960	0.7993	0.9162	0.9673	0.9906	0.9974	0.9994	0.9999	1.0000	1.0000
2.35	0.0954	0.3195	0.5828	0.7891	0.9103	0.9673	0.9896	0.9971	0.9993	0.9998	1.0000	1.0000
2.40	0.0907	0.3084	0.5697	0.7787	0.9041	0.9643	0.9884	0.9967	0.9991	0.9998	1.0000	1.0000
2.45	0.0863	0.2977	0.5567	0.7682	0.8978	0.9612	0.9872	0.9962	0.9990	0.9998	0.9999	1.0000
2.50	0.0821	0.2873	0.5438	0.7576	0.8912	0.9580	0.9858	0.9958	0.9989	0.9997	0.9999	1.0000
2.55	0.0781	0.2772	0.5311	0.7468	0.8844	0.9546	0.9844	0.9952	0.9987	0.9997	0.9999	1.0000
2.60	0.0743	0.2674	0.5184	0.7360	0.8774	0.9510	0.9828	0.9947	0.9985	0.9996	0.9999	1.0000
2.65	0.0707	0.2579	0.5060	0.7251	0.8703	0.9472	0.9812	0.9940	0.9983	0.9996	0.9999	1.0000
2.70	0.0672	0.2487	0.4936	0.7141	0.8629	0.9433	0.9794	0.9934	0.9981	0.9995	0.9999	1.0000
2.75	0.0639	0.2397	0.4815	0.7030	0.8554	0.9392	0.9776	0.9927	0.9978	0.9994	0.9999	1.0000
2.80	0.0608	0.2311	0.4695	0.6919	0.8477	0.9349	0.9756	0.9919	0.9976	0.9993	0.9998	1.0000
2.85	0.0578	0.2227	0.4576	0.6808	0.8398	0.9304	0.9735	0.9910	0.9973	0.9992	0.9998	1.0000
2.90	0.0550	0.2146	0.4460	0.6696	0.8318	0.9258	0.9713	0.9901	0.9969	0.9991	0.9998	0.9999
2.95	0.0523	0.2067	0.4345	0.6584	0.8236	0.9210	0.9689	0.9891	0.9966	0.9990	09.997	0.9999
3.00	0.0498	0.1991	0.4232	0.6472	0.8153	0.9161	0.9665	0.9881	0.9962	0.9989	0.9997	0.9999

Example 10.3   Memoryless Property of the Poisson Process

Consider a Poisson process in which renewals occur at the rate of $\lambda = 0.5$ per month. Find the expected system age at the 5th renewal. What is the probability that the difference in system ages between the 5th and 6th renewals is greater than one month?

Solution

The expected waiting time for the 5th failure is given by the gamma distribution with mean time to the 5th repair given by

$$E[T(5)] = \frac{k}{\lambda} = \frac{5}{0.5} = 10$$

months. Since we have independent and stationary increments for the Poisson process, the probability that the elapsed time between the 5th and 6th renewal exceeds one month is the same as the probability that the time between any consecutive renewals exceeds one month. Using the time to the first renewal, we get

$$P[X_6 > 1] = P[X_1 > 1] = e^{-1\lambda} = e^{-0.5} = 0.607$$

Exercise 10.3

Suppose a component failure distribution is exponential with mean rate $\lambda = 0.3\%/K$. For a renewal process involving a single-component system, what is the probability of no replacements in 4000 hours of system operation?

Exercise 10.4

A component failure distribution is exponential with mean rate $\lambda = 6.0\%/K$ For a renewal process involving a single-component system, how many spare parts should be provided for 10,000 hours of system operation to assure that the probability of a replacement shortage is less than 5 percent?

Exercise 10.5

When the soil in an agricultural study reaches a specified level of dryness, an automatic misting follows that lasts for two minutes. Assume that the number of mistings per day is a Poisson process with rate 0.2 per hour. What is the expected number of mistings between midnight and noon? What is the expected waiting time to the 3rd misting? Given that three mistings have occurred, what is the probability that the time to the next misting is greater that 2 hours?

MTBF and MTTF for a Renewal Process

Consider a renewal process truncated at either a specified time or the last repair. If we ignore the time to make repairs, the mean time between failures (*MTBF*) can be estimated by dividing the system age by the number of failures. Also, for a renewal process, we can treat the interarrival times as if they arose from an independent sample of nonrepairable parts from a single distribution, as previously discussed in Example 10.1. To calculate the mean time to failure (*MTTF*) for the nonrepairable situation, we sum the failure times and any accumulated time since the last repair and divide by the number of failures. Since the sum of the interarrival times and accumulated time since the last repair must equal the system age, we obtain the same result for both the *MTTF* and the *MTBF* calculations. Thus, for a renewal process, the *MTTF* and the *MTBF* are equivalent. Note: it is not appropriate to talk about an *MTBF* for a non-renewal process, since the rate of repairs is changing.

If we have an HPP, in which the interarrival times are independent and exponentially distributed, then—similar to the discussion of confidence intervals for failure rates in Chapter 3—we can easily determine confidence intervals for the *MTBF*. Tables 10.3A (for failure censoring) and 10.3B (for time censoring) provide the one-sided lower bound factors for the *MTBF* at various confidence levels. Table 10.4 (for both failure and time censoring) gives the one-sided upper bound factors. To determine the specific factor, find—in the appropriate table—the intersection of the number of fails and the desired confidence level. Note that Table 10.3B also gives factors for zero failures: multiply the system age by the factor to obtain an *MTBF* lower bound.

Example 10.4   Confidence Bounds on the Population *MTBF* for an HPP

The following ages (in hours) at repair are observed for a system observed for 600 hours: 90, 133, 200, 294, 341, 360, 468, and 562. Assuming an HPP, estimate the *MTBF* and a 90 percent confidence interval on the true *MTBF*.

Solution

The *MTBF* estimate is 600/8 = 75 hr. From Table 10.3B, the single-sided, 95 percent confidence level, lower factor for 8 fails is 0.554. From Table 10.4, the upper 95 percent factor is 2.010. The 90 percent confidence interval on the true *MTBF* is thus 41.55 to 150.75 hours.

Table 10.5 is an additional table useful for determining the length of test required for a HPP in order to demonstrate a desired *MTBF* at a given confidence level if *r* failures occur. We show the use of this table in Example 10.5.

TABLE 10.3A   One-Sided Lower Confidence Bound Factors for the $MTBF$ **(failure-censored data)**

Number of Fails	Confidence Level						
	60%	70%	80%	85%	90%	95%	97.5%
1	1.091	0.831	0.621	0.527	0.434	0.334	0.271
2	0.989	0.820	0.668	0.593	0.514	0.422	0.359
3	0.966	0.830	0.701	0.635	0.564	0.477	0.415
4	0.958	0.840	0.725	0.665	0.599	0.516	0.456
5	0.955	0.849	0.744	0.688	0.626	0.546	0.488
6	0.954	0.856	0.759	0.706	0.647	0.571	0.514
7	0.953	0.863	0.771	0.721	0.665	0.591	0.536
8	0.954	0.869	0.782	0.734	0.680	0.608	0.555
9	0.954	0.874	0.791	0.745	0.693	0.623	0.571
10	0.955	0.878	0.799	0.755	0.704	0.637	0.585
12	0.956	0.886	0.812	0.771	0.723	0.659	0.610
15	0.958	0.895	0.828	0.790	0.745	0.685	0.639
20	0.961	0.906	0.846	0.812	0.772	0.717	0.674
30	0.966	0.920	0.870	0.841	0.806	0.759	0.720
50	0.971	0.935	0.896	0.872	0.844	0.804	0.772
100	0.978	0.952	0.923	0.906	0.885	0.855	0.830
500	0.989	0.978	0.964	0.956	0.945	0.930	0.918

Use for failure censored data to multiply the $MTTF$ or $MTBF$ estimate to obtain a lower bound at the given confidence level.

TABLE 10.3B   One-Sided Lower Confidence Bound Factors for the $MTBF$ **(time-censored data)**

Number of Fails	Confidence Level						
	60%	70%	80%	85%	90%	95%	97.5%
0	1.091	0.831	0.621	0.527	0.434	0.334	0.271
1	0.494	0.410	0.334	0.297	0.257	0.211	0.179
2	0.644	0.553	0.467	0.423	0.376	0.318	0.277
3	0.718	0.630	0.544	0.499	0.449	0.387	0.342
4	0.763	0.679	0.595	0.550	0.500	0.437	0.391
5	0.795	0.714	0.632	0.589	0.539	0.476	0.429
6	0.817	0.740	0.661	0.618	0.570	0.507	0.459
7	0.834	0.760	0.684	0.642	0.595	0.532	0.485
8	0.848	0.777	0.703	0.662	0.616	0.554	0.508
9	0.859	0.790	0.719	0.679	0.634	0.573	0.527
10	0.868	0.802	0.733	0.694	0.649	0.590	0.544
12	0.883	0.821	0.755	0.718	0.675	0.617	0.572
15	0.899	0.841	0.780	0.745	0.704	0.649	0.606
20	0.916	0.864	0.809	0.777	0.739	0.688	0.647
30	0.935	0.892	0.844	0.816	0.783	0.737	0.700
50	0.953	0.918	0.879	0.856	0.829	0.790	0.759
100	0.969	0.943	0.915	0.897	0.877	0.847	0.822
500	0.987	0.976	0.962	0.954	0.944	0.929	0.916

Use for time-censored data to multiply the $MTTF$ or $MTBF$ estimate to obtain a lower bound at the given confidence level. Note: for 0 failures, multiply the system operating hours by the factor corresponding to the desired confidence level.

TABLE 10.4   One-Sided Upper Confidence Bound Factors for the *MTBF*

Number of Fails	Confidence Level						
	60%	*70%*	*80%*	*85%*	*90%*	*95%*	*97.5%*
1	1.958	2.804	4.481	6.153	9.491	19.496	39.498
2	1.453	1.823	2.426	2.927	3.761	5.628	8.257
3	1.313	1.568	1.954	2.255	2.722	3.669	4.849
4	1.246	1.447	1.742	1.962	2.293	2.928	3.670
5	1.205	1.376	1.618	1.795	2.055	2.538	3.080
6	1.179	1.328	1.537	1.687	1.904	2.296	2.725
7	1.159	1.294	1.479	1.610	1.797	2.131	2.487
8	1.144	1.267	1.435	1.552	1.718	2.010	2.316
9	1.133	1.247	1.400	1.507	1.657	1.917	2.187
10	1.123	1.230	1.372	1.470	1.607	1.843	2.085
12	1.108	1.203	1.329	1.414	1.533	1.733	1.935
15	1.093	1.176	1.284	1.357	1.456	1.622	1.787
20	1.077	1.147	1.237	1.296	1.377	1.509	1.637
30	1.060	1.115	1.185	1.231	1.291	1.389	1.482
50	1.044	1.085	1.137	1.170	1.214	1.283	1.347
100	1.029	1.058	1.093	1.115	1.144	1.189	1.229
500	1.012	1.025	1.039	1.049	1.060	1.078	1.094

Use to multiply the *MTTF* or *MTBF* estimate to obtain an upper bound at the given confidence level (failure- or time-censored data).

TABLE 10.5   Test Length Guide

Number of Fails (r)	Confidence Level					
	50%	*60%*	*75%*	*90%*	*90%*	*95%*
0	0.693	0.916	1.39	1.61	2.30	3.00
1	1.68	2.02	2.69	2.99	3.89	4.74
2	2.67	3.11	3.92	4.28	5.32	6.30
3	3.67	4.18	5.11	5.52	6.68	7.75
4	4.67	5.24	6.27	6.72	7.99	9.15
5	5.67	6.29	7.42	7.91	9.27	10.51
6	6.67	7.34	8.56	9.08	10.53	11.84
7	7.67	8.39	9.68	10.23	11.77	13.15
8	8.67	9.43	10.80	11.38	12.99	14.43
9	9.67	10.48	11.91	12.52	14.21	15.71
10	10.67	11.52	13.02	13.65	15.41	16.96
15	15.67	16.69	18.49	19.23	21.29	23.10
20	20.68	21.84	23.88	24.73	29.06	30.89

Multiply desired *MTBF* by factor k to determine test time needed to demonstrated desired *MTBF* at a given confidence level, if r fails occur.

Example 10.5   Test Length Guide for an HPP

We wish to demonstrate an *MTBF* of at least 168 hours at a 90 percent confidence level. How many hours should we run the test if we allow up to three failures.

Solution

From Table 10.5, we see that 6.68×168 hours = 1,122.24 hours are required with, at most, 3 failures to demonstrate the desired *MTBF*. Obviously, the test is time censored.

Exercise 10.6

The following ages (in hours) at repair are recorded for a system truncated at the last failure: 3, 23, 32, 59, 143, and 162. Estimate the *MTBF* assuming an HPP. Provide a 95 percent confidence interval for the true *MTBF*.

Exercise 10.7

A customer requires that the system our company manufactures demonstrate an *MTBF* of at least 500 hours. How many hours do we have to run the test for confirmation at the 95 percent confidence level, assuming an HPP for repairs? Allow at most one failure.

Exercise 10.8

Consider the following cumulative replacement times for a single component system: 40, 150, 400, 430, 500, 670, 750, 800. The interarrival times X_1, (i = 1, ..., 8), are 40, 110, 250, 30, 70, 170, 80, 50. Assume a Poisson process, and estimate the failure rate λ. Give 90 percent upper and lower limits on the failure rate. (Hint: Use tables for *MTBF* and note failure rate = $1/MTBF$.)

Renewal Rates

The *renewal intensity* (*renewal rate*) at time t is the mean or expected rate (that is, number of repairs per unit time) at time t and equals the derivative of the renewal function; that is,

$$m(t) = \frac{dM(t)}{dt}$$

For a stationary process, Cox and Lewis (1966) show that $m(t)$ approaches the reciprocal of the mean of the distribution of the interarrival times as t gets large. For a Poisson process, the renewal rate is constant and equal to λ, and the independent, exponentially distributed interarrival times have mean equal to $1/\lambda$. Also, as shown by Barlow and Proschan (1975), the cumulative renewal function is linear in time t; that is, $M(t) = \lambda t$. For example, for a Poisson renewal process with intensity $\lambda = 0.025\%$/hr, the expected number of repairs in 5,000 hours is simply $0.00025 \times 5{,}000 = 1.25$. The mean of the interarrival times is $1/\lambda = 4{,}000$ hours. For a renewal process in general, one can show (see discussion in Cox, 1962) that in the limit as t goes to infinity, $M(t)$ approaches a linear function of t, and hence the renewal rate $m(t)$ becomes constant.

Exercise 10.9

A Poisson renewal process has intensity $\lambda = 0.05\%$/hr. Find the expected number of repairs in one year. What is the mean time between repairs? What is the expected number of repairs between the first and second years?

SUPERPOSITION OF RENEWAL PROCESSES

Consider a multicomponent system in which several "identical" parts can fail, causing the system to cease operation. Again, the system is restored to operation by replacement of the failed part with a new part from the same parent population. At the system level, we observe the *superposition* of the individual renewal processes. In general, a pooled process is not a renewal process; that is, the times between system repairs are usually not i.i.d. However, if each renewal process is a Poisson process (that is, the interarrival times are independent and exponentially distributed), then the superposition is also a Poisson renewal process with renewal rate equal to the sum of the component renewal rates.

Figure 10.8 is an example of the repair patterns for a system with three components. We assumed identical components, with exponential renewal rate λ, but in fact this result holds in general for the sum of any number of independent Pois-

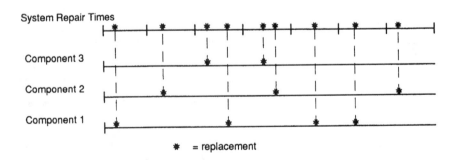

Figure 10.8 Superposition of Individual Poisson Processes

son processes with differing λ_i. This subject is treated further in Cox (1962). The system renewal rate is

$$\lambda_s = \sum_{i=0}^{c} \lambda_i$$

where c is the number of socket positions.

When the interarrival times for the individual renewal processes are not exponentially distributed, the process is not a Poisson process, and consequently the superposition will not be a Poisson process (or even a renewal process). However, for sufficiently long times, the superposition process converges to the stationary Poisson process under certain conditions described by Cox (1962).

Exercise 10.10

Use the simulation procedures for the exponential distribution (described in Chapter 3) to generate ten consecutive exponential random observations with $\lambda = 1\%/\text{Khr}$ for each socket position of a three-socket system. Illustrate the superposition pattern with a sketch. Plot the cumulative number of system repairs versus age. Does the plot appear linear? Based on the methods of Chapter 6, check the fit of the system interarrival times to the exponential model. Estimate the system mean time to repair.

CDF ESTIMATION FROM RENEWAL DATA

Consider a system that experiences a failure and is then restored to operation by some repair process. In particular, we consider a special type of restoration involving a repairable process.

Suppose we have a system (e.g., a computer) consisting of c components from the same parent population. The components fail independently with a lifetime distribution (CDF) $F(t)$. We wish to estimate $F(t)$, or equivalently the reliability function $R(t) = 1 - F(t)$, from system lifetime data. We will not assume a particular parametric form for the lifetime distribution, but will seek a nonparametric estimator of $F(t)$.

Upon failure of a component, a system is assumed to cease operation. However, the system is immediately restored to operation by replacement of the failed component with another component from the same population.

Moreover, we consider the case in which only the failure ages of the system, and not the failure times of the individual components, are recorded, a condition we refer to as "unidentified replacement." This situation may arise because records are not kept of the site of each failed component and its replacements.

For example, a drilling machine tool may have many bits that are individually replaced upon failure, but only the times of machine stoppage are recorded and not the particular bit position where replacement occurred. Hence, except for a one-component system, after the first failure, it becomes impossible to determine whether any subsequent failures occurred on original components or on components that were replacements for original units.

Since we assume the original and replacement component lifetimes are a random sample from the component lifetime distribution $F(t)$, the sequence of inter-replacement times for a single component position can be viewed as an ordinary renewal process. For a system, we then have a superposition of such renewal processes as shown in Figure 10.9. Given information on only the system failure times, how do we estimate the cumulative distribution function of the component lifetimes?

Our approach to estimating $F(t)$ relies on the fundamental renewal equation which relates $F(t)$ and the system mean cumulative renewal function, $M(t)$, via a convolution integral. $M(t)$ is the expected number of component renewals or replacements made through time t. First, $M(t)$ is estimated from system failure ages, and then $F(t)$ is estimated by numerical deconvolution of the renewal equation.

This topic has been extensively treated by Trindade and Haugh (1979, 1980). They cover the multicensored situation of N systems having different operating hours. In addition, they have investigated the statistical properties of various numerical deconvolution methods. We shall employ one such method here.

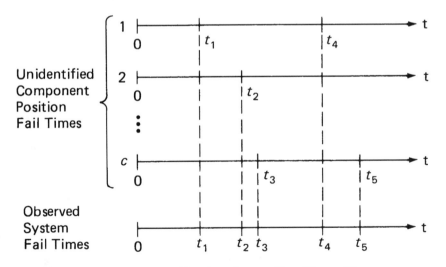

Figure 10.9 System of c Components Viewed as a Superposition of Renewal Processes

A well known relationship exists between the CDF $F(t)$ for component failure times and the mean cumulative renewal function $M(t)$. Called the *fundamental renewal equation* (Barlow and Proschan, 1975), the relation is

$$M(t) = F(t) + \int_0^t M(t-x)\, dF(x)$$

For our purposes, we write this equation in the equivalent form

$$F(t) = M(t) + \int_0^t M(x)\, dF(t-x)$$

While this equation looks difficult to solve, we will not use it directly. Instead, we use it in some statistical applications.

An unbiased estimator of $M(t)$, for a single system of c components, is just

$$\hat{M}(t) = \frac{n(t)}{c}$$

where $n(t)$ is the number of renewals for all component positions by time t. Note the estimator $\hat{M}(t)$ is a step function with jumps at the failure times. By numerically solving the fundamental renewal equation, Trindade and Haugh show that the following equations can be used to estimate the CDF at a given time t_i:

$$\hat{F}(t_1) = \hat{M}(t_1), \qquad \hat{F}(t_2) = \hat{M}(t_2 - t_1)\,\hat{M}(t_1)$$

We note at the second failure time t_2 that $\hat{F}(t_2 - t_1)$ is required, and $t_2 - t_1$ may be greater than, equal to, or less than t_1. By using the information that $\hat{F}(t_1)$ is specified at t_i, we can develop a recursive approach such that

$$\hat{F}(t_2 - t_1) = 0 \qquad \text{if } (t_2 - t_1) < t_1$$

$$= \hat{F}(t_1) \qquad \text{if } (t_2 - t_1) \geq t_1$$

In words, if the interarrival time $t_2 - t_1$ between the first and second renewal is less than the time t_1 to the first renewal, we set the CDF estimate for the time $t_2 - t_1$

equal to zero; if the interarrival time is greater than t_1, the CDF estimate is set equal to $\hat{F}(t_1)$. See Figure 10.10 for a graphical representation of this procedure. Similarly, the component CDF based on the third renewal is given by

$$\hat{F}(t_3) = \hat{M}(t_3) - \hat{F}(t_3 - t_1)\hat{M}(t_1) - \hat{F}(t_3 - t_2)\left[\hat{M}(t_2) - \hat{M}(t_1)\right]$$

where the CDF for the times between renewals is estimated by

$$\hat{F}(t_3 - t_j) = 0 \qquad \text{if } (t_3 - t_j) < t_1$$

$$= \hat{F}(t_1) \qquad \text{if } t_1 \le t_3 - t_j < t_2$$

$$= \hat{F}(t_2) \qquad \text{if } t_2 \le t_3 - t_j$$

for $j = 1, 2$. Thus, we see that differences, $t_3 - t_j$, between the given renewal time t_3 and each earlier time t_j, for $j = 1, 2$, are compared to the individual renewal times to assign a proper $\hat{F}(t_3 - t_j)$ value from the previously calculated $\hat{F}(t)$ estimates.

We will give the general equation for the estimate of the component CDF $\hat{F}(t_k)$ at each system renewal time t_k but, as we showed above, the formulas can be written without using such detailed notation shown below. So for $k \ge 2$, we have

$$\hat{F}(t_k) = \hat{M}(t_k) - \sum_{j=1}^{k-1} \hat{F}(t_k - t_j)\left[\hat{M}(t_j) - \hat{M}(t_{j-1})\right]$$

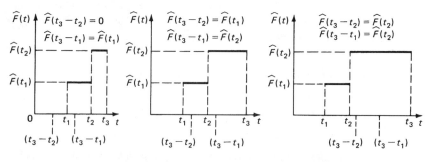

Figure 10.10   Possible Outcomes for Time Differences in Renewal Estimation

for $1 \le j \le k - 1$, with $t_0 = 0$, and $\hat{F}(0) = \hat{M}(0) = 0$. We develop recursively the CDF for times between renewals as,

$$\hat{F}(t_k - t_j) = 0 \qquad \text{if } (t_k - t_j) < t_1$$

$$= \hat{F}(t_i) \qquad \text{if } t_i \le t_k - t_j < t_{i+1}$$

$$= \hat{F}(t_{k-1}) \qquad \text{if } t_{k-1} \le t_k - t_j$$

Note the between times $t_k - t_j$ are not interarrival times except where arrivals are adjacent; that is, $k - j = 1$. The comparisons of time differences to observed failure times to locate the proper recursive values for evaluating the CDF at time t can involve considerable computational effort as the number of failures increases. Trindade and Haugh present an equal interval method of deconvolution which avoids these comparisons and speeds up computations.

Example 10.6   Renewal Data Calculation of CDF

Consider a system of 25 components. System failures were reported at 35, 79, 142, and 206 hours. Estimate the component CDF $F(t)$.

The renewal function estimates are

$$\hat{M}(35) = \frac{1}{25} = 0.04 \qquad\qquad \hat{M}(142) = \frac{3}{25} = 0.12$$

$$\hat{M}(79) = \frac{2}{25} = 0.08 \qquad\qquad \hat{M}(206) = \frac{4}{25} = 0.16$$

The CDF estimates are

$$\hat{F}(35) = \hat{M}(35) = 0.04$$

$$\hat{F}(79) = \hat{M}(79) - \hat{F}(44)\hat{M}(35) = 0.08 - 0.04(0.04) = 0.0784$$

$$\hat{F}(142) = \hat{M}(142) - \hat{F}(107)\hat{M}(35) - \hat{F}(63)[\hat{M}(79) - \hat{M}(35)]$$

$$= 0.12 - 0.784(0.04) - 0.04(0.08 - 0.04) = 0.1153$$

$$\hat{F}(206) = \hat{M}(206) - \hat{F}(171)\,\hat{M}(35) - \hat{F}(127)\,[\hat{M}(79) - \hat{M}(35)]$$

$$- \hat{F}(64)\,[\hat{M}(142) - \hat{M}(79)]$$

$$= 0.16 - 0.1153\,(0.04) - 0.0784\,(0.08 - 0.04) - 0.04\,(0.12 - 0.08)$$

$$= 0.1507$$

Exercise 10.11

Consider, using the data in Example 10.6, a fifth renewal at 306 hours. Write the expression for $\hat{F}(t_5)$. Solve for $\hat{F}(306)$.

SUMMARY

We have introduced some important concepts on repairable systems. We have emphasized the distinction between analysis of repairable and non-repairable systems. We concentrated on renewal processes. These ideas illustrate how different approaches are useful for providing for spare parts, estimating repair schedules, determining costs, and so on. We have shown graphical and modeling methods useful for analyzing data from repairable systems. We have treated multicensored and multicomponent systems. We also showed how to estimate reliability for unidentified replacement of components in a renewable system. In the next chapter, we extend repairable system concepts to non-renewal processes.

PROBLEMS

9.1 Simulate ten pseudo-random times from a lognormal distribution having mean 100 and sigma 1.5, using the methods in Chapter 5 and keeping the order of generation intact. Treat the sequential times as consecutive interarrival times of a system undergoing repair. Determine the system ages at repair. Make a line sketch of the repair pattern. Make a cumulative plot. Make a plot of the interarrival times versus system age. Make a lognormal probability plot of the interarrival times. Estimate the mean and sigma of the data. Plot the fitted model against the empirical CDF.

9.2 Repeat problem 1 using ten simulated values from a Weibull distribution, using the procedures in Chapter 4, with characteristic life of 50 hours and shape parameter 1.5.

9.3 The following repair times (in hours) are recorded on four similar systems:

System 1: 197, 241, 368, 874, 927

System 2: 56, 119, 173, 506, 771

System 3: 99, 316, 414, 663

System 4: 156, 222, 375

The system censoring times are 1000, 800, 700, and 400 hours, respectively. Estimate the mean cumulative repair function and include in a cumulative plot with the data from the four systems.

9.4 Components are assumed to fail according to an exponential distribution with failure rate 0.01%/hr. The mission duration is 7,500 hours. How many spare parts must be carried to assure at least 99 percent probability of mission success? State any assumptions you make. What is the expected waiting time to the first, second, and third repairs?

9.5 Treat the four system repair times in problem 3 above as the individual renewal times of four components in a four component system during the first 400 hours of operation. Exclude times greater than 400 hours. Estimate the renewal function at each repair point. Estimate the system renewal rate. Make a sketch of the superposition process.

9.6 Using the system renewal data from problem 5, estimate the CDF of the components from the cumulative renewal function.

Chapter 11

Repairable Systems II: Non-renewal Processes

In Chapter 10, we presented analysis techniques for repairable systems, focusing primarily on renewal processes. The times between failures were independent and identically distributed (i.i.d.) observations from a single population. If the i.i.d. assumptions do not hold, the renewal process is not a suitable model. Consequently, we need other analysis methods that deal with more general patterns of sequential repair times. For example, if a trend is present, such information would be useful for both the user and the manufacturer in determining if reliability objectives are being met. One may also ask what design or operation factors influence repair frequency, whether maintenance schedules are appropriate, what warranty costs are anticipated, whether the provisions for spare parts are adequate, and so on.

In many real-life instances, the time to a subsequent repair is generally a function of many variables, including the basic system design, the operating conditions, the environment, and the quality of the repairs (the materials used, the competency of the technician, and so on). Thus, there is a genuine possibility of non-renewal processes in which inter-repair times are neither independent nor identically distributed. For multicomponent systems and different types of repair actions, the renewal model becomes even less plausible.

This chapter continues the development of key concepts for analyzing data from systems subject to repair. We present both graphical and analytical procedures. We consider specific models and goodness of fit tests. Methods for detecting trend are presented. Simulation methods are covered. This chapter deals with the topic of non-renewal, repairable reliability having many important and practical applications.

GRAPHICAL ANALYSIS OF NON-RENEWAL PROCESSES

As in the previous chapter, we consider a single system for which the times to make repairs are ignored. Suppose that consecutive times to repair are observed at the following system ages:

20	41	67	110	159	214	281	387	503	660

A line sketch of the pattern of repairs shows:

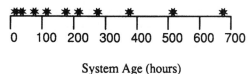

System Age (hours)

For this set of data, the cumulative plot is Figure 11.1.

We see in Figure 11.1 a flattening derivative of the curve as time increases, indicating a decreasing frequency of repairs, that is, an improving repair rate. It seems obvious that there is a trend of longer times between repairs. Yet, when we look at the times between consecutive failures, we get:

20	21	26	43	49	55	67	106	116	157

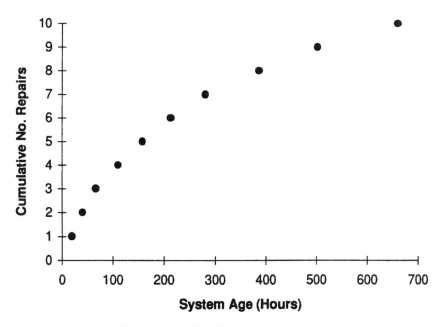

Figure 11.1 Cumulative Plot, Improving Trend

These are exactly the same interarrival times—*in a different order*—that we observed in the renewal case of Chapter 10. (See Figure 10.1.) Nevertheless, it would be inappropriate to analyze the data in the same way as we did previously for the renewal case because now the interarrival times are not independent random observations from a single distribution. There is a trend. We cannot neglect the chronological order of the data.

The plot of the interarrival times versus the system age is Figure 11.2. The graph makes the trend obvious.

On the other hand, suppose the repairs occur at the following times:

| 157 | 273 | 379 | 446 | 501 | 550 | 593 | 619 | 640 | 660 |

with associated line sketch:

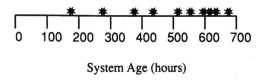

System Age (hours)

Then, the consecutive times between repairs are

| 157 | 116 | 106 | 67 | 55 | 49 | 43 | 26 | 21 | 20 |

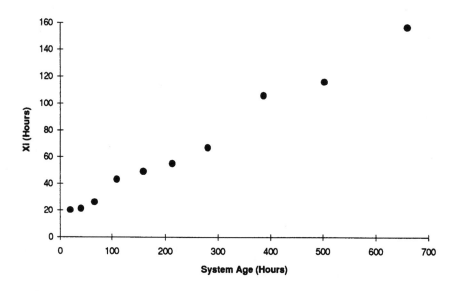

Figure 11.2 Interarrival Times versus System Age, Improving

Again, these are exactly the same interrepair times—*in a different order*—as in the two previous cases. However, the order now shows an increasing frequency, indicating a degradation in system reliability. The cumulative plot is Figure 11.3.

The plot is curving monotonically upward in time, toward a higher frequency of repairs. A trend is present. Again, it is not correct to analyze the time between repairs as if they were independent observations from a single distribution as we did for the renewal process. Chronological order of the data is important. Analyses specific to repairable systems must be used.

The plot of interarrival times versus system age is also very revealing of the degradation trend. We show the chart as Figure 11.4.

To illustrate one possible approach to analyzing data from a non-renewal process, let us attempt to model the cumulative repair function for the degradation situation using a simple equation with curvature. Two possibilities are the power relation

$$M(t) = at^b$$

and the exponential relation

$$M(t) = ae^{bt}$$

where a and b are empirically estimated constants. These equations might adequately model monotonically increasing or decreasing functions. Obviously, many other models are possible. Also, be aware that we are fitting a step function estimate, $\hat{M}(t)$, to a smooth curve, $M(t)$, similar to the relationship between an empirical distribution function and its theoretical CDF.

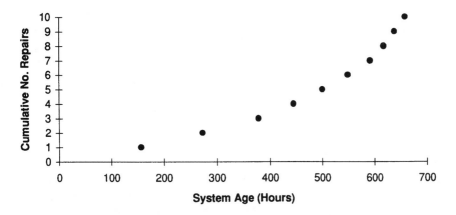

Figure 11.3 Cumulative Plot, Degrading Trend

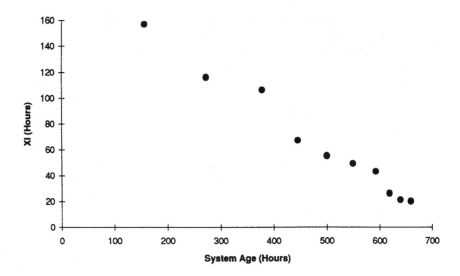

Figure 11.4 Interarrival Times versus System Age, Degrading

The first step is to rectify the equations. (Linear rectification is described in Chapter 6.) For the power relation, we take logarithms of each side to get

$$\ln M(t) = \ln a + b \ln t$$

from which we see that, if the model is adequate, the plot of $\ln \hat{M}(t)$ versus $\ln t$ on linear by linear paper (or $\hat{M}(t)$ versus t on log-log paper) should approximate a straight line with intercept $\ln a$ and slope b. The data are plotted in Figure 11.5. The plot appears to show significant curvature, indicating a potentially poor fit of the power relation model to the data.

Let's consider an alternative formula: the exponential relationship. Taking logarithms of both sides yields

$$\ln M(t) = \ln a + bt$$

If the model is adequate, the plot of $\ln \hat{M}(t)$ versus t on linear paper (or $\hat{M}(t)$ versus t on semi-log paper) should approximate a straight line with intercept $\ln a$ and slope b. The data is plotted in Figure 11.6.

The plot is a nearly straight line, indicating a reasonable fit. A least squares regression (a practical but only approximate method since least squares assumptions are not fulfilled) on the data in Figure 11.6 produced the following esti-

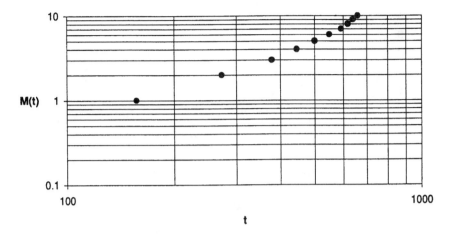

Figure 11.5 Power Relation Model Rectification

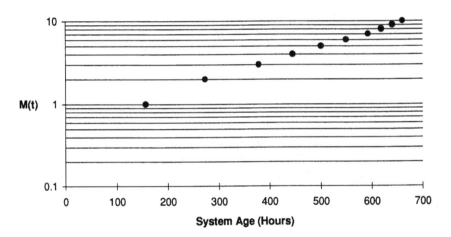

Figure 11.6 Exponential Model Rectification

mates for the empirical parameters: $\hat{a} = 0.556$, $\hat{b} = 0.00434$. The actual cumulative plot with the fitted model equation $\hat{M}(t) = 0.556e^{0.00434t}$ is shown in Figure 11.7. The model appears to provide an acceptable fit.

What does the fitted model represent? The model is an estimate of the (population) mean number of repairs per system at a specified time. Obviously the system is getting worse. Although future values of the repair function may be extrapolated forward, we may be more interested in the question, "What is the cause of the deterioration?"

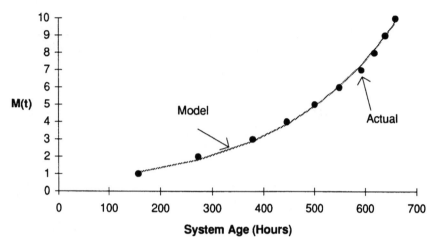

Figure 11.7 Exponential Model Fit

The model for a single system does not tell us enough about the population distribution of the cumulative number of repairs at that time. How accurate is this estimate $\hat{M}(t)$ of the expected (population) number of repairs by time t? What is the discrete distribution of the number of repairs at time t? Additional analytical tools are need to address these questions. We shall discuss these matters further in the following sections.

In summary, these examples illustrate that, when presented with system repair data, one must be careful about making an assumption of a renewal process without justification. Ascher (1981) provides a similar approach to the Weibull analysis of interarrival data. Checking for trends, independence, and distribution properties in the data is important for proper analysis.

Exercise 11.1

Perform an analysis of the improving system data in Figure 11.1. Attempt to fit power relation and exponential models to the data. Make the rectification plots. Decide on the better model and estimate the empirical parameters. Plot the model fit with the actual data in a cumulative plot. Comment on the results. Can you suggest an alternative model with a better fit?

TESTING FOR TRENDS AND RANDOMNESS

If we are developing a model, we should verify the assumptions of the model. Then, we are justified in using the model. In the previous chapter, we discussed renewal processes that involve a sequence of independent and identically distributed interarrival times. A renewal process is also a stationary process; that is, the number of events in an interval depends only on the length of the interval and not the starting point. If any trend is present, the process is nonstationary and not a renewal process. Conversely, we will assume that a process that has no trend is stationary. (Exceptions are certainly possible but this assumption is a useful simplification for many reliability applications.) Thus, *the first step in the analysis of repairable system data is to check for trend,* since a trend rules out a common distribution for the interarrival times.

How might trends or nonstationarity develop? For repairable systems, the environment, usage, part supplies, maintenance, and so on may be changing. For example, if a failed part is replaced with a new part from a different time or vintage of manufacture from the original, then the population of replacements may have a different failure curve. Consequently, we could have a nonstationary process with independent but not identically distributed arrival times. In other situations, old systems may not receive as much attention as new systems, thereby affecting the repair frequency. A system may have a subsystem, such as a cooling fan, that degrades and causes deterioration in the interarrival times. Individuals performing repairs or maintenance may learn how to do the job better and thus improve the subsequent repair frequency. Thus, there are many possible causes of trends.

Another example of nonstationarity mentioned in Chapter 10 comes from Davis (1952). He analyzed the number of miles between bus engine repairs by comparing histograms of the kth interarrival times among systems; for example, a histogram of the time to first repair, a histogram of the times between first and second repairs, and so on. He saw that the times to first failure were fairly normally distributed, but later interarrival times tended to be more skewed, appearing nearly exponential.

For single or multiple copies of a system, how do we statistically assess if interarrival times have any trend, that is, degradation or improvement? We have already mentioned that the first action should be to plot the data. A cumulative plot of number of repairs or the estimated mean cumulative repair function $M(t)$ versus system age is often revealing of trends. Plots of interarrival times versus age are also very useful. Graphs are informative, especially for checking model adequacy, but development of analytical models allows for formalized decisions, data summary, predictions, and so on.

What are the steps one should follow in the statistical analysis of repairable system data? For chronologically ordered times between repairs, one first investigates if a trend is present. Several alternative considerations are possible in checking for a trend: Are we doing a general test for any trend or a test against a specific type of trend? We will present several options in this section. If a trend is indicated, a nonstationary model is employed. We will present several nonstationary models in the next section, the most important being the nonhomogeneous Poisson process (NHPP).

If there is no trend, we will assume (as previously stated) that the process is stationarity and that the interarrival times are identically distributed but not necessarily independent. To check the interarrival times for independence, concepts from time series analysis have been suggested in the literature. For example, scatter plots of interarrival times versus the immediately proceeding interarrival time (lag 1) or earlier lags are used to check for serial correlation in the data. However, because of the typically skewed nature of reliability data, such plots are often not very revealing. In addition, large amounts of data are generally required for time series approaches. For further information on this topic, see the discussion in Cox and Lewis (1966) or the example in Crowder et al. (1991).

Ascher and Feingold (1984) point out that, even if there is dependence, it is still meaningful to reorder the interarrival times by magnitude and estimate the properties of their common distribution. If we accept that the interarrival times are independent and identically distributed, then we have a renewal process. For a renewal process, one next considers the form of the distribution of the interarrival times. In particular, is the Poisson process (HPP) appropriate? If an HPP is rejected, other distributions are possible model candidates or one may resort to distribution-free techniques.

Ascher and Feingold (1984) provide a more complete overview of statistical analysis procedures for repairable systems failure data. We mention that simple transformations may change the nature of the data. Also, using a different scale other than clock or calendar time (for example, number of units produced, length of material consumed, etc.) may affect the characteristics of the process.

Other Graphical Tools

We have already discussed the cumulative plot and the plot of interarrival times against system age as valuable graphical tools for the analysis of failure data from repairable systems. Another approach, applicable to a single system [also extendable to a pooled estimate $M(t)$], is to analyze the reciprocals of successive interarrival times to determine if a trend is present. This charting procedure is based on the *rate* of repairs $m(t)$, that is the derivative of the expected cumulative number of repairs in time. Recall that

$$m(t) = \frac{dM(t)}{dt}$$

An interarrival time measures a period between repairs, and taking the reciprocal is equivalent to estimating an average rate between consecutive repairs. Plotting these average repair rates versus total elapsed times can quickly reveal periods of increasing, decreasing, highly varying, or nearly constant repair frequencies. We refer to such charts as the average repair rate (ARR) plots. Figures 11.8 to 11.10 shows the ARR for the three sets of repair data (constant, improving, degrading) previously presented.

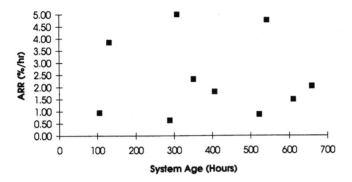

Figure 11.8 ARR versus Time, Renewable Data

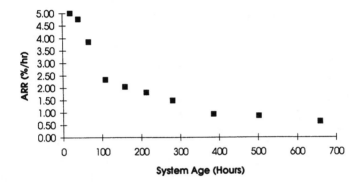

Figure 11.9 ARR versus Time, Improving

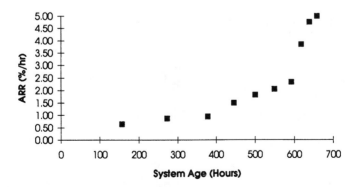

Figure 11.10 ARR versus Time, Degrading

The ARR charts visually confirm steady, decreasing, and increasing repair rates. When sufficient data is available, convenient, equal time intervals may be selected, and the number of failures during each interval can be divided by the interval length to obtain an average ROCOF plot. For smoothed repair rate plots, numerical differentiation techniques may be applied directly to cumulative plot curves as described by Trindade (1979). An alternative smoothing procedure is described by Barlow, Proschan, and Scheuer (1971).

Exercise 11.2

Using the data from Exercise 10.1, make a ARR plot. Is there evidence of any trend?

Analytical Tools

We have previously mentioned that there are two different approaches in the analysis of trends for a series of events. One may be interested only in whether a trend of any type is present. Nonparametric procedures, as discussed later in this chapter, may be useful here. Alternatively, one may check trend against a specific model for the data. (For a full discussion see Cox and Lewis, 1966.) In the latter case, the chosen model will have an important effect in establishing the significance of any trend. For example, a simple (parametric) way to check for trend involves testing whether an observed series of events is of Poisson type. The procedure tests whether observations come from a sequence of independent and identically distributed exponential random variables. Rejection implies that the HPP is not an appropriate model for the data; that is, we accept the alternative that some type of trend is present.

Let us see how such a test may be constructed. Suppose we observe n repairs for a single system during the interval 0 to T_1 (that is, we have time truncation). Ross (1993) and Parzen (1962) show that under the HPP, the first n ages for system 1, $t_{11}, t_{12}, \ldots, t_{1n}$, at which repairs occur have the same distribution as if they were the *order statistics* corresponding to n independent random variables uniformly distributed on the same interval 0 to T_1. We use the term order statistics in the sense that t_{11} is the smallest value among $U_1, U_2, \ldots, U_n$, and t_{12} is the second smallest value, and so on. Hence, one method of testing for an HPP is to test whether the observations $U_1, U_2, \ldots, U_n$ are independent and uniformly distributed over the interval 0 to T_1. According to the central limit theorem (see Chapter 5), for moderately large values of n, the sum

$$S_n = \sum_{i=1}^{n} U_i$$

of n independent random variables, all uniformly distributed on the interval 0 to T_1, may be considered normally distributed with mean

$$\frac{nT_1}{2}$$

and variance

$$\frac{nT_1^2}{12}$$

Thus, if we see $n = 10$ repairs in $T_1 = 576$ hours of observation, then the sum S_{10} of the times at which the events occur is (approximately) normally distributed with mean $10(576)/2 = 2880$ and standard deviation 525.8. Consequently, at a 95 percent level of significance, if the sum S_{10} satisfies the inequality

$$2880 - (1.96)\,525.8 \le S_{10} \le 2880 + (1.96)\,525.8$$

or

$$1849.4 \le S_{10} \le 3910.6$$

the HPP is not rejected.

This same procedure can be expressed in terms of the test statistic L_i for the ith system

$$L_i = \frac{\displaystyle\sum_{j=1}^{n} t_{ij} - \frac{nT_i}{2}}{T_i \sqrt{\dfrac{n}{12}}}$$

which approaches a standard normal variate under the HPP for moderately large n. In this form, the test has been called the Laplace or centroid test. Bates (1955) has shown the approximation adequate for the 5 percent level of significance for n at least equal to 4.

If we have repair series data from k independent systems, each system observed for time T_i, $i = 1, 2, ..., k$, (that is, time truncation) with repair counts of $n_1, n_2, ..., n_k$ and repair times for the ith system given by t_{ij} respectively, then we can test each series individually or we can pool the data into a single test statistic:

$$L = \frac{\displaystyle\sum_{j=1}^{n_1} t_{1j} + \sum_{j=1}^{n_2} t_{2j} + \dots + \sum_{j=1}^{n_k} t_{1j} - \frac{1}{2}(n_1 T_1 + n_2 T_2 + \dots + n_k T_k)}{\sqrt{\frac{1}{12}\left(n_1 T_1^2 + n_2 T_2^2 + \dots + n_k T_k^2\right)}}$$

where L has very nearly a standard normal distribution. For failure truncated data, the above test statistics are modified by replacing each n_i by $n_i - 1$ (including summations) and setting each system period of observation T_i equal to the failure truncation time. Note that all systems could have different repair rates; we are testing the null hypothesis of an HPP and not the equality of rates.

Example 11.1   Laplace Test for Trend versus a Poisson Process

Three repairable systems have experienced the following consecutive interarrival times:

System 1	98	150	37	62	15
System 2	15	77	96	14	12
System 3	55	80	48	37	

The periods of observation were 375, 220, and 225 hours, respectively. Is there any evidence against the null hypothesis of an HPP for each system? Is there any evidence based on the overall test?

Solution

For system 1, the repair ages are 98, 248, 285, 347, and 362 hr. The sum of the system ages is 1340 hr. Hence, the test statistic is

$$L_1 = \frac{1340 - \dfrac{5\,(375)}{2}}{375\sqrt{\dfrac{5}{12}}} = 1.66$$

Reference to a standard normal table shows a level of significance equal to 0.048. Similarly, the test statistics for the remaining two systems are $L_2 = 2.13$ and $L_3 = 1.10$, with respective significance levels of 0.017 and 0.14. Based on these results, there is evidence against the HPP for systems 1 and 2, but not for system 3. The overall test statistic is

$$L = \frac{1340 + 711 + 593 + \frac{1}{2}[5(375) + 4(220) + 4(225)]}{\sqrt{\frac{1}{12}[5(375)^2 + 4(220)^2 + 4(225)^2]}} = 2.698$$

with a significance level of 0.0035. Collectively, there is strong evidence against the HPP as an appropriate model for the system repair data.

Exercise 11.3

Perform the Laplace test individually on the three sets of data of Figures 10.1, 11.1, and 11.3, respectively.

Exercise 11.4

Perform the pooled Laplace test on the data from Example 11.1, assuming failure truncation at the fourth, fifth, and third repairs, respectively, for the three systems.

Reverse Arrangement Test

In testing against the HPP, which is a special type of renewal process, we did not first verify the assumption that the interarrival times were i.i.d. The test statistic is based on the i.i.d. premise. An easy-to-understand, more general test will now be described which distinguishes between i.d.d. interarrival times (that is, a renewal process) and a monotonic trend. The test has been applied by R. F. De Le Mare (1991). The procedure is a nonparametric *reverse arrangement test* (RAT) originally devised by Kendall (1938) and further developed into a table by Mann (1945).

Consider a set of n interarrival times occurring in the sequence

$$X_1, X_2, ..., X_n$$

Starting from left to right, a reversal is defined as each instance in which a lesser value occurs before any subsequent greater values following in the sequence; that is, a reversal occurs each time the following inequality holds:

$$X_i < X_j \quad \text{for } i < j$$

where

$$i = 1, \ldots, n - 1$$
$$j = i + 1, \ldots, n$$

For example, say a system has repairs at ages 25, 175, 250, and 350 hr. In the sequence of interarrival times 25,150,75,100, there are a total of $3 + 0 + 1 = 4$ reversals, since 25 is less than the three following values 150, 75, 100; and 150 is not less than any other number; and 75 is less than 100. For n items, what is the probability of a given number of reversals occurring by chance alone? A larger than expected number of reversals, associated with growing X_i numbers, would indicate an increasing trend for interarrival times. Conversely, a smaller than expected number is consistent with diminishing X_is and a decreasing trend. A tie does not count as a reversal.

Exercise 11.5

What is the number of reversals in the sequence 100, 200, 150, 50, 100, 125?

The calculation of reversal probabilities is fairly simple. The probability of a reversal is found by first determining the permutations of n quantitative objects and then counting the number of reversals for each permutation. For n variables, there are $n!$ permutations possible. (See Chapter 9.) The minimum number of reversals is 0, and the maximum number is the series

$$(n-1) + (n-2) + \ldots + 1 = \sum_{i=1}^{n-1} (n-i) = \frac{n(n-1)}{2}$$

Consider the case of $n = 4$. Designate the sequence of times between failures as the four observations as X_1, X_2, X_3, X_4. There are $4! = 24$ possible permutations. The maximum number of reversals is $4(4 - 1)/2 = 6$. We can easily show (see Exercise 11.7) that the respective probabilities of 0, 1, 2, 3, 4, 5, 6 reversals occurring by chance are (1/24), (3/24), (5/24), (6/24), (5/24), (3/24), and (1/24). The cumulative probabilities are thus 0.042, 0.167, 0.375, 0.625, 0.833, 0.958, and 1.000, respectively. The derivation of these probabilities is fairly simple. The sequence $X_1 < X_2 < X_3 < X_4$ is the only permutation of (X_1, X_2, X_3, X_4) that has six reversals. So the probability of no reversals occurring by chance is 1/24 = 0.042. There are 3 possible permutations that give 1 reversal (i.e., $X_4X_3X_1X_2$, $X_3X_4X_2X_1$, and $X_4X_2X_3X_1$), and so the probability of one reversal is 3/24 = 0.139. Consequently, there are 4 permutations that give either zero or 1 reversal for a cumulative probability of 4/24 = 0.167, and so on.

Exercise 11.6

Consider the situation of three interarrival times. How many permutations of three objects exist? List each permutation and count the number of possible reversals for each. Determine the individual and cumulative probabilities for all possible reversals.

Exercise 11.7

Consider the case of four interarrival times. How many permutations of four objects exist? List each permutation and count the number of possible reversals for each. Determine the individual and cumulative probabilities for all possible reversals.

Exercise 11.8

Consider the case of five interarrival times. How many permutations of five objects exist? List each permutation and count the number of possible reversals for each. Determine the individual and cumulative probabilities for all possible reversals.

The reverse arrangement test uses the total number of reversals in a sequence of interarrival times, comparing that number to what would be likely to occur based on pure chance. If the number of reversals is too small against some probability criteria, randomness is discredited, and a downward trend is suggested. A test for an upward trend can similarly be made by testing whether the number of reversals is too large against the upper tail probability. Tables are available which show the probability of a given number of reversals in n items occurring by chance. (See R.F. De Le Mare's 1991 paper, which provides tables that correct some errors in Mann's tables.)

Kendall (1938) showed that the expected number of reversals is $n(n-1)/4$, and the variance of the distribution of reversals is equal to $(2n+5)(n-1)n/72$. He also demonstrated that the distribution of the number of reversals R rapidly approaches the normal distribution with increasing n. Hence, the expression

$$Z = \frac{R + \dfrac{1}{2} - \dfrac{n(n-1)}{4}}{\sqrt{\dfrac{(2n+5)(n-1)n}{72}}}$$

rapidly approaches the standard normal distribution. (The $1/2$ term is added so that the normal distribution will better approximate the probabilities for the discrete distribution of R.) Thus, in addition to using tables to determine the probabilities, we can use the normal approximation for even fairly moderate values of n. For example, Table 11.1 compares the exact cumulative probabilities versus the normal approximation for $n = 4$.

Instead of calculating probabilities for all possible reversals, one is often interested in determining if an observed number of reversals is beyond random chance occurrence, indicating a possible trend toward decreasing or increasing

TABLE 11.1   Probability of R Reversals by Chance for n = 4

R	0	1	2	3	4	5	6
Exact	0.042	0.167	0.375	0.625	0.833	0.958	1.000
Normal approx.	0.045	0.154	0.367	0.633	0.846	0.955	0.991

interarrival times. Is the number of reversals statistically significant at some critical level? Table 11.2 provides the critical values for one-sided, upper and lower, statistical significance as a function of the observed sample size n. For example, if we collect a sequence of 10 interarrival times and observe 12 or fewer reversals, we would conclude that there is evidence of significant decreasing (degradation) of times at a one-sided significance level of 5 percent. Similarly, if 33 or more reversals are observed, we would suspect increasing times (improvement) at the one-sided 5 percent significance level.

Note that the results above are symmetric for any row: subtracting the lower critical value from the maximum number of reversals gives the upper critical value for any matching level of significance α; that is,

$$R_{n,\, \text{upper}\,\alpha\%} = \frac{n(n-1)}{2} - R_{n,\, \text{lower}\,\alpha\%}$$

TABLE 11.2   Critical Values $R_{n,\%}$ of the Number of Reversals for the Reverse Arrangement Test

Sample Size (n)	Single-Sided Lower Significance Level (too few reversals provide evidence of degradation)			Single-Sided Upper Significance level (too many reversals provide evidence of improvement)		
	1%	5%	10%	10%	5%	1%
4		0	0	6	6	
5	0	1	1	9	9	10
6	1	2	3	12	13	14
7	2	4	5	16	17	19
8	4	6	8	20	22	24
9	6	9	11	25	27	30
10	9	12	14	31	33	36
11	12	16	18	37	39	43
12	16	20	23	43	46	50

For n greater than 12, critical values for the number of reversals can be found by solving inversely for the upper R value in the Kendall normal approximation equation; that is,

$$R = z_{\text{critical}} \sqrt{\frac{(2n + 5)(n - 1)n}{72}} + \frac{n(n - 1)}{4} - \frac{1}{2}$$

For example, for $n = 20$, at an upper 5 percent significance level ($z_{\text{critical}} = 1.645$), the critical value is roughly 120. The maximum number of reversals is $20(19)/2 = 190$. The lower 5 percent critical value is $190 - 120 = 70$.

Example 11.2   Reverse Arrangement Test

A system experiences repairs at the following ages in hours: 155, 335, 443, 583, 718, 815, 925, 1030, 1113, 1213, 1341, 1471, 1551, 1633, 1748. Is there any statistically significant indication of a trend to shorter times between repairs?

Solution

It is always a good idea to plot the data. Figure 11.11 is a cumulative plot of the data. Although there does appear to be upward curvature in the plot (a trend to shorter times), it is difficult to assess the statistical significance of the trend, if any, by viewing only the chart. We construct a second plot (Figure 11.12) of the

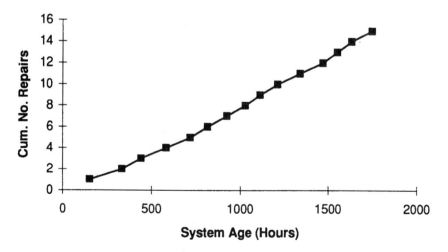

Figure 11.11   Cumulative Plot of Repair Data

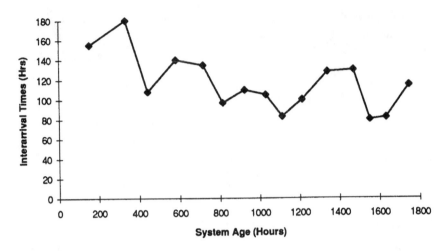

Figure 11.12   Interarrival Times versus System Age

15 observed interarrival repair times where $i = 1, \ldots, 15$ versus the system age at repair. These times are: 155, 180, 108, 140, 135, 97, 110, 105, 83, 100, 128, 130, 80, 82, 115. This plot is more revealing of the apparent decreasing trend, but again, what is the likelihood of this data sequence occurring by chance?

Exercise 11.9

Make an ARR plot (described in an earlier section) using the reciprocal interarrival times of Example 10.5 versus the system age. Comment on the results.

Applying the RAT test for randomness, we find 30 reversals in the sequence of X_i values. Is this value a statistically significant indication of a downward trend (that is, decreasing times)? For $n = 15$, the expected number of reversals is $15(14)/4 = 52.5$ and the variance is 102.08. The probability of 30 or fewer reversals is obtained by calculating $z = (30 + 0.5 - 52.5)/10.1 = 2.178$. The corresponding probability obtained from a standard normal table is 0.0147, or less than 1.5 percent. Thus, the low reversal count is a rare occurrence under chance alone. Since the number of reversals is fewer than expected, a trend towards shorter times between failures is suspected. In other words, the occurrence rate of failures appears to be increasing.

Exercise 11.10

Apply the reverse arrangement test individually to the three sets of repair data of Figures 10.1, 10.4, and 10.5, respectively ($n = 10$ for each set). What is the probability of a more extreme number of reversals occurring by chance alone for each set?

Exercise 11.11

There are 25 interarrival times observed. The number of reversals is 98. Calculate the expected number of reversals. Is there statistical evidence of a trend? Is the trend to shorter or longer interarrival times? Find the level of significance of any trend.

Exercise 11.12

Determine the upper and lower critical values of reversals for the RAT at the 5 percent significance level for 15 observed interarrival times.

Combining Data from Several Tests

If we have data on more than one system, applying the RAT to each system may be insufficient to detect a trend, but taken together, the systems might provide overall evidence of a trend. In Chapter 10, we described how to combine data from several systems into one set of data using the mean cumulative repair function estimate given by Nelson (1988b). This approach, however, does not provide an analytic test for trend.

Let us consider a procedure for combining data from a number of systems in order to get a more sensitive test for detecting trend. The procedure is from Fisher (1954) and is described by De Le Mare (1991).

Suppose that, for three independent tests, none is individually significant at some prescribed level, say 5 percent. Fisher describes the following procedure: Take the natural logarithm of each significance level, change its sign, and double it. Each value so obtained is equivalent to a chi-square (χ^2) with two degrees of freedom. Any number of these quantities may be added together for a composite test. We illustrate the method with an example.

Example 11.3   Fisher's Composite Test

Three repairable systems have experienced the following consecutive interarrival times:

System 1	98	150	37	62	15
System 2	15	77	96	14	12
System 3	55	80	48	37	

We use the RAT method and calculate that the number of reversals for the three systems, respectively, are 2, 3, and 1, with respective tail probabilities of chance occurrence equal to 14/120 = 0.117, 29/120 = 0.242, and 4/24 = 0.167. (See

Exercises 11.7 and 11.8 for probability calculations.) To test whether the aggregate should be regarded as significant, we perform the numerical steps below.

P	$-2lnP$	Degrees of Freedom
0.117	4.29	2
0.242	2.84	2
0.167	3.58	2
Sum	10.71	6

For 6 degrees of freedom, the "Percentiles of the χ^2 Distribution" table in the Appendix shows $\chi^2_{6;10}$, indicating a likelihood of less than 10 percent for the observed aggregate of data under the assumption of no trend. Thus, individually there was insufficient data to conclude significance, but the composite test provides significance at the 10 percent level for the aggregate.

Note that the composite test procedure does not assume equivalent repair rates (or even—for two-sided p-values—the same direction for trend) within the different systems; the procedure looks only at whether the individual sequences taken collectively exhibit any trend. To compare rates, a nonparametric procedure such as the Wilcoxon rank sum test for two populations or the Kruskal-Wallis test for three or more populations (see Ott, 1993) may be applied to the cumulative repair times.

Exercise 11.13

The repair histories (ages in hours) on four failure-truncated systems are shown below: System 1: 142, 309, 460; System 2: 99, 145, 300, 347; System 3: 212, 225, 273, 398, 467; and System 4: 21, 58, 150, 176. Make a graph containing the cumulative plot for each system along with the mean cumulative repair function estimate (that is, Nelson's pooled summation approach described in Chapter 10). Using the RAT procedure, calculate the probabilities of the observed reversals for each system. Perform Fisher's composite test on the aggregate. Do you believe all four systems have basically the same repair rate?

NON-HOMOGENEOUS POISSON PROCESSES

The non-homogeneous Poisson process (NHPP) is a useful generalization of the Poisson process that has wide applicability in the modeling of repairable systems. See Cox and Lewis (1966), Crow (1974, 1990, 1993), Lawless (1982), Bain and Engelhardt (1991), or Rigdon and Basu (1989) for examples. The NHPP is also employed in the modeling of software reliability. See Xie (1991) or Musa, Iannino, and Okumoto (1987).

Recall that the homogeneous Poisson process has *stationary increments,* that is, the number of events that occur in any interval of time depends only on the length of the interval and not the starting point of the interval. The interarrival times for an HPP are independent with the same exponential distribution, and the process has no memory.

The nonhomogeneous Poisson process, abbreviated by NHPP, permits nonstationary events by allowing the repair or recurrence rate at time t to be a function of age t. When the recurrence rate is a function of time, it is called the *intensity function,* λt. Define the mean cumulative function

$$M(t) = \int_0^t \lambda(\tau) d\tau$$

Then it can be shown (see Ross, 1993) that

$$P[N(t+s) - N(t) = n] = e^{-[M(t+s) - M(t)]} \frac{[M(t+s) - M(t)]^n}{n!}, \quad n \geq 0$$

This equation states that the incremental number of occurrences, $N(t+s) - N(t)$, in the interval of time from t to $t + s$, has a Poisson distribution with a mean (or expected) number of occurrences equal to $M(t+s) - M(t)$. Consequently, $M(t)$ is called the *mean value* or *mean repair function.* If the intensity function is constant, that is, $\lambda(t) = \lambda$, then we have the HPP with $M(t) = \lambda t$ and, consequently, $N(t+s) - N(t)$ is Poisson distributed with mean λs. From the above expression, we see also that the reliability $R(s)$, defined as the probability of zero occurrences in the time interval t to $t + s$, is

$$R(s) = e^{-[M(t+s) - M(t)]}$$

Example 11.4   Nonhomogeneous Poisson Process

Consider an NHPP with mean repair function modeled by the *power relation*

$$M(t) = 0.01t^{1.5}$$

What is the expected number of repairs between 80 and 100 hours? Between 180 and 200 hours? What is the reliability between 80 and 100 hours? Between 180 and 200 hours? What is the probability of at least two repairs between 80 and 100 hours? Between 180 and 200 hours?

Solution

Note that $M(80) = 0.01(80)^{1.5}$, $M(100) = 0.01(100)^{1.5} = 10.00$, $M(180) = 0.01(180)^{1.5} = 24.15$, and $M(200) = 0.01(200)^{1.5} = 28.28$. Hence, the expected number of repairs between 80 and 100 hours is $M(100) - M(80) = 10.0 - 7.16 = 2.84$, and between 180 and 200 hours is $M(200) - M(180) = 28.28 - 24.15 = 4.13$. Since the number of repairs between 80 and 100 hours is Poisson with mean 2.84, the reliability is the probability that the number of repairs is zero or

$$P[N(100) - N(80) = 0] = e^{-[M(100) - M(80)]} = e^{-2.84} = 0.0582$$

Similarly, the reliability between 180 and 200 hours is

$$P[N(200) - N(180) = 0] = e^{-[M(200) - M(180)]} = e^{-4.13} = 0.0161$$

The probability of at least two repairs between 80 and 100 hours is

$$P[N(100) - N(80) \geq 2] = 1 - P[N(100) - N(80) \leq 1]$$

$$= 1 - \left[e^{-2.84} + 2.84e^{-2.84} \right] = 0.776$$

Between 180 and 200 hours, the probability is

$$P[N(200) - N(180) \geq 2] = 1 - P[N(200) - N(180) \leq 1]$$

$$= 1 - \left[e^{-4.13} + 4.13e^{-4.13} \right] = 0.917$$

Exercise 11.14

For the model in Example 11.4, find the expected number of repairs and the reliability between 280 and 300 hours. Find the probability of at least two repairs between 280 and 300 hours.

MODELS FOR THE INTENSITY FUNCTION OF AN NHPP

What are some possible models for the intensity function of an NHPP? We consider two commonly applied, time-dependent models for the mean repair function.

Power Relation Model

The power relation model may be written as

$$M(t) = at^b$$

Its intensity function is

$$\lambda(t) = \frac{dM(t)}{dt} = abt^{b-1}$$

Note that if $0 < b < 1$, the intensity function is decreasing, that is, the rate of occurrence of failures is improving. If $b > 1$, as in Example 11.4, the rate is increasing.

We see the similarity of the intensity function to the formula for the hazard rate of the Weibull distribution. For this reason, such processes have been inappropriately and confusingly called "Weibull processes" in the reliability literature. However, we caution against using this terminology, because the power relation process is not based on the Weibull distribution, and procedures applicable to the analysis of Weibull data are not correct here.

For an NHPP with intensity rate modeled by the power relation, how do we estimate the model parameters? Earlier in this chapter we suggested a graphical technique using rectification and an eyeball fit or possibly least squares regression (only as an approximation!). Crow (1974) has developed MLEs (maximum likelihood estimates) for the power model. MLEs exist for two different forms of data truncation: by failure count or by fixed time.

Suppose a single system experiences n repairs at system ages $t_i, i = 1, 2, \ldots, n$. If the data is truncated at the nth failure, the number of repairs is fixed at n, but the time t_n to the nth failure is random. For this *failure truncated* situation, conditioned on the system age t_n, the modified MLEs for the power model (modified to provide unbiased $\hat{b}$ —see the discussion in Chapter 12) are

$$\hat{b} = \frac{n-2}{\displaystyle\sum_{i-1}^{n-1} \ln\left(\frac{t_n}{t_i}\right)}, \quad \hat{a} = \frac{N}{T^{\hat{b}}}$$

If, instead of failure truncated data, we have data censored at a fixed time T (consequently, the number of failures N by time T is random), then conditioned on the number of repairs N, the *time truncated* modified MLEs (unbiased $\hat{b}$) for the power model are

$$\hat{b} = \frac{N-1}{\displaystyle\sum_{i-1}^{N} \ln\left(\frac{T}{t_i}\right)}, \quad \hat{a} = \frac{N}{T^{\hat{b}}}$$

If we have time or failure censored data on k copies of the system operating under similar conditions, the data may be combined to improve the estimates of the model parameters. Let T_q denote the truncation time for the qth system, $q = 1, 2, \ldots, k$. Let n_q denote the total number of failures on the qth system by time T_q. Let t_{iq} denote the system age for the ith repair on the qth system. To combine both time and failure truncated data, we introduce a new variable, N_q, which equals n_q if the data on the qth system are time truncated, or equals $n_q - 1$ if the data on the qth system are failure truncated. Crow shows that, conditioned on either the number of repairs for each system or on the truncation time for each system, the unbiased modified MLE estimate for b can be expressed in closed form as

$$\hat{b} = \frac{N_s - 1}{\displaystyle\sum_{q=1}^{k} \sum_{i-1}^{N_q} \ln\left(\frac{T_q}{t_{iq}}\right)}$$

where

$$N_s = \sum_{q=1}^{k} N_q$$

The modified MLE (pooled) estimate for a (not necessarily unbiased) is

$$\hat{a} = \frac{\displaystyle\sum_{q=1}^{k} n_q}{\displaystyle\sum_{q=1}^{k} T_q^{\hat{b}}}$$

We have estimated the parameters a and b as common to all systems; that is, the systems are samples from a population having a single mean cumulative repair function $M(t)$. However, several other distinct possibilities may exist. For example, the parameter b may be common to all systems, but the parameter a may be different for each system. In that case, a would be estimated separately for each system using

$$\hat{a}_q = \frac{n_q}{T_q^{\hat{b}}}$$

where $\hat{b}$ is the common estimate. Alternatively, we may wish to test the hypothesis that the parameter b is the same for each system. We may desire confidence intervals on the parameters a or b. (Confidence intervals involving the power law relation are discussed further in Chapter 12.) There may be other censoring situations (for example, systems with left truncated data and different censoring times). Thus, there are many possible tests depending on the available data or desired objectives. For further considerations and examples, consult Crow (1974, 1990). For confidence intervals on the intensity function for the power model, see Crow (1993). For confidence intervals on the mean cumulative repair function, see Crow (1982).

To assess the suitability of the power model for a set of data (that is, to test the hypothesis that the failure times of any number of systems follow a nonhomogeneous Poisson process with a power law intensity rate) Crow (1974) derived a test statistic $C_{N_s}^2$ and tabulated the critical values. The test is based on transforming the repair times for each system. Under time truncation, divide each repair time by the corresponding system truncation time; under failure truncation, divide each repair time except the last one by the corresponding time of the last repair. Consequently, the total number of transformed values will equal N_s, previously defined above. These values are combined and ordered from smallest to largest. Call these ordered numbers Z_i, $i = 1, \ldots, N_s$. The goodness of fit statistic is

$$C_{N_s}^2 = \frac{1}{12N_s} + \sum_{i=1}^{N_s} \left[Z_i^{\hat{b}} - \frac{2i-1}{2N_s} \right]^2$$

If the test statistic is greater than the critical value of Table 11.5 (from Crow, 1990), then we reject the NHPP with power relation intensity function as an appropriate model for the data. Note if we use $b = 1$ in the goodness of fit statistic, we are testing whether the failure times follow an HPP model.

In working with the NHPP model, we have assumed that all of the system failure times are available from time zero to a time or failure truncation point. Situations may arise, however, in which only the number of failures occurring in disjoint subintervals of time and not exact failure times are available. Procedures for handling binned or grouped data are described by Crow (1988) and Engelhardt (1994).

Example 11.5   NHPP with Power Relation Intensity

Table 11.3 contains the repair history for three simulated systems. Estimate the parameters for the power law model, assuming the systems can be treated as equivalent copies. Perform a goodness of fit test using the test statistic.

Solution

The calculations are as follows:

System 1: $N_q = N_1 = 8$, and $t_q = t_1 = 1000$. Thus,

$$\sum_{i=1}^{N} \ln\left(\frac{t_1}{t_{i1}}\right) = \ln\left(\frac{1000}{15.1}\right) + \ldots + \ln\left(\frac{1000}{955.1}\right) = 14.58$$

System 2: $N_q = N_2 = 9 - 1 = 8$, and $t_q = t_2 = 981.8$. Thus,

$$\sum_{i=1}^{N_2} \ln\left(\frac{t_2}{t_{i2}}\right) = \ln\left(\frac{981.8}{15.2}\right) + \ldots + \ln\left(\frac{981.8}{797.7}\right) = 11.90$$

TABLE 11.3   Repair History in Hours (simulated data: $a = 0.25$ and $b = 0.50$)

System 1	System 2	System 3
15.1	15.2	11.3
47.1	122.7	43.3
51.1	172.1	122.8
158.7	288.1	203.3
221.2	371.9	294.0
495.5	376.2	468.0
769.2	567.2	800.7
965.1	797.7	987.4
	981.8	
Time truncated at 1,000	Failure truncated at 981.8	Time truncated at 1,000

System 3: $N_q = N_3 = 8$, and $t_q = t_3 = 1000$. Thus,

$$\sum_{i=1}^{N_3} \ln\left(\frac{t_3}{t_{i3}}\right) = \ln\left(\frac{1000}{11.3}\right) + \ldots + \ln\left(\frac{1000}{987.4}\right) = 13.53$$

Since $N_s = 8 + 8 + 8 = 24$, the unbiased modified MLE estimate for b is

$$\frac{24}{14.58 + 11.90 + 13.53} = 0.575$$

The parameter a is estimated by

$$\hat{a} = \frac{25}{1000^{0.575} + 981.8^{0.575} + 1000^{0.575}} = 0.158$$

The model for the mean cumulative repair function per system is

$$M(t) = 0.158t^{0.575}$$

To check the goodness of fit, we transform the repair times by dividing the repair ages for systems 1 and 3 by 1000, the truncation time, and for system 2 by the last repair time of 981.8. The results are shown in Table 11.4.

TABLE 11.4   Transformed Repair Times (simulated data: $a = 0.25$ and $b = 0.50$)

System 1	System 2	System 3
0.0151	0.0155	0.0113
0.0471	0.1250	0.0433
0.0511	0.1753	0.1228
0.1587	0.2934	0.2033
0.2212	0.3788	0.2940
0.4955	0.3832	0.4680
0.7692	0.5777	0.7999
0.9551	0.8125	0.9874

We now order the transformed times to get the Z values: $Z_1 = 0.0113$, $Z_2 = 0.0151, \ldots, Z_{24} = 0.9874$. Substituting these values into the goodness of fit statistic with $\hat{b} = 0.575$ and $N_s = 24$, we get

$$C_{N_s}^2 = \frac{1}{12\,(24)} + \left(0.0113^{0.5753} - \frac{1}{48}\right)^2 + \ldots + \left(0.9874^{0.575} - \frac{47}{48}\right)^2 = 0.040$$

For a test at the 0.05 significance level, Table 11.5 shows a critical value of 0.217. Since 0.040 is less than 0.217, we accept the NHPP power model.

TABLE 11.5   Critical Values for Goodness of Fit Test (from Crow, 1990) (used by permission of author)

N_s	Significance Level				
	0.20	0.15	0.10	0.05	0.01
2	0.138	0.149	0.162	0.175	0.186
3	0.121	0.135	0.154	0.184	0.23
4	0.121	0.134	0.155	0.191	0.28
5	0.121	0.137	0.160	0.199	0.30
6	0.123	0.139	0.162	0.204	0.31
7	0.124	0.140	0.165	0.208	0.32
8	0.124	0.141	0.165	0.208	0.32
9	0.124	0.142	0.167	0.212	0.32
10	0.125	0.142	0.167	0.212	0.32
11	0.126	0.143	01.69	0.214	0.32
12	0.126	0.144	0.169	0.214	0.32
13	0.126	0.144	0.169	0.214	0.33
14	0.126	0.144	0.169	0.214	0.33
15	0.126	0.144	0.169	0.215	0.33
16	0.127	0.145	0.171	0.216	0.33
17	0.127	0.145	0.171	0.217	0.33
18	0.127	0.146	0.171	0.217	0.33
19	0.127	0.146	0.171	0.217	0.33
20	0.128	0.146	0.172	0.217	0.33
30	0.128	0.146	0.172	0.218	0.33
60	0.128	0.146	0.173	0.220	0.33
100	0.129	0.147	0.173	0.220	0.34

For $N_s > 100$, use values for $N_s = 100$

Exercise 11.15

Construct a cumulative plot based on data for the three systems in Table 11.4. For the observed (combined) repair times, estimate the mean cumulative repair function and add it to the previous plot. Comment on the results.

Exercise 11.16

Apply the Crow modified MLE formulas to estimate the parameters a and b for the data of Figures 11.1 and 11.3. Plot the fitted power relationship on the cumulative plots.

Exponential Model

Consider an NHPP for which the intensity function is

$$\lambda(t) = e^{c+bt}$$

where b and c are empirically determined parameters. The mean repair function is found by simple integration

$$M(t) = \int_0^t \lambda(\tau)\, d\tau = \frac{e^c}{b}\left(e^{bt} - 1\right) = \frac{a}{b}\left(e^{bt} - 1\right)$$

where $a = e^c$. For a single system having data censored at the nth failure (*failure truncated*), for which the time t_n to the nth repair is random, Lawless (1982) provides the MLEs $\hat{b}$ and $\hat{c}$. To find $\hat{b}$, we need to substitute trial values for $\hat{b}$ into the following equation until a solution occurs

$$\sum_{i=1}^n t_i + \frac{n}{\hat{b}} - \frac{n t_n}{1 - e^{-\hat{b}t_n}} = 0$$

where t_i, $i = 1, \ldots, n$ are the chronological system ages at repair. Then, $\hat{c}$ is estimated from the expression

$$\hat{c} = \ln\left(\frac{n\hat{b}}{e^{\hat{b}t_n} - 1}\right)$$

If the data are censored at fixed time T (time truncated), the number of repairs N is random, and the MLE for $\hat{b}$ is found by trial substitution in the equation

$$\sum_{i=1}^{N} t_i + \frac{N}{\hat{b}} - \frac{NT}{1 - e^{-\hat{b}T}} = 0$$

Then, we evaluate $\hat{c}$ using the formula

$$\hat{c} = \ln\left(\frac{N\hat{b}}{e^{\hat{b}T} - 1}\right)$$

If $b = 0$ in the exponential model, then the intensity function $\lambda(t) = e^c$ is constant, indicating a Poisson process (HPP). To test whether $b = 0$ and, consequently, the data is consistent with a Poisson process, we use a test statistic from Lawless (1982). Under the null hypothesis of an HPP, the test statistic for failure truncation is

$$U = \frac{\displaystyle\sum_{i=1}^{n-1} t_i - \frac{1}{2}(n-1)t_n}{t_n\left(\dfrac{n-1}{12}\right)^{\frac{1}{2}}}$$

For time truncation at T, the test statistic from Cox and Lewis (1966) is

$$U = \frac{\displaystyle\sum_{i=1}^{N} t_i - \frac{1}{2}NT}{T\left(\dfrac{N}{12}\right)^{\frac{1}{2}}}$$

In either situation, U is approximately $N(0,1)$; that is, normally distributed with mean 0 and variance 1. We recognize this test as the Laplace's test previously described.

If we have data from several systems, the observations from all systems can be combined to provide an overall test of the null hypothesis. Under the hypothesis that each of k processes is HPP (allowing for different intensities), Cox and Lewis (1966) show that, for failure truncation on each system, the centroid test statistic

$$U = \frac{\displaystyle\sum_{l=1}^{k}\sum_{i=1}^{n_l-1} t_{li} - \frac{1}{2}\sum_{l=1}^{k}(n_l-1)\,t_{ln}}{\left(\dfrac{1}{12}\displaystyle\sum_{l=1}^{k}(n_l-1)\,t_{ln}^2\right)^{\frac{1}{2}}}$$

is approximately $N(0,1)$. Note that n_l is the number of repairs for the lth process, t_{ji} is the system age at the ith repair for the jth system, and t_{jn} is the last repair time (truncation point) for the jth system. For time truncated systems, replace each $n_l - 1$ in the above formula, including summation limits, with the number of repairs N_l, and replace each t_{ln} with the observation (truncation) age.

Example 11.6   NHPP with Exponential Intensity Model

A system experiences ten repairs at the following ages (simulated from an NHPP with $b = 0.001$ and $c = 0.005$):

59.0	207.5	284.9	484.0	552.8	636.4	826.0	988.4	1008.2	1070.1

The data stops at the 10th repair (failure truncated). Estimate the parameters b and c for the exponential model. Estimate the intensity function and the mean repair function. Check the model fit using Laplace's test.

Solution

We need to solve the equation for iteratively by substituting trial values, where $n = 10$ and

$$\sum_{i=1}^{10} t_i = 6117.4$$

That is, we seek a $\hat{b}$ such that

$$6117.4 + \frac{10}{\hat{b}} - \frac{10\,(1070.1)}{1 - e^{-\hat{b}\,(1070.1)}} = 0$$

Using the "Solver" routine in a Microsoft EXCEL spreadsheet, we get $\hat{b}$ = 0.000813. Then,

$$\hat{a} = \frac{10\,(0.000813)}{e^{0.000813\,(1070.1)} - 1} = 0.00586$$

or $\hat{c} = -5.14$. The intensity function estimate is

$$\hat{\lambda}(t) - e^{(-5.14 + 0.000813t)} = 0.00586e^{0.000813t}$$

The mean cumulative repair function estimate is

$$\hat{M}(t) = 7.21\left(e^{0.000813t} - 1\right)$$

Under failure truncation, the U statistic for Laplace's test is

$$U = \frac{\sum_{i=1}^{n-1} t_i - \frac{1}{2}(n-1)t_n}{t_n\left(\frac{n-1}{12}\right)^{\frac{1}{2}}} = \frac{5047.2 - \left(\frac{9}{2}\right)1070.1}{1070.1\sqrt{\frac{9}{12}}} = 0.250$$

The tail probability associated with $U = 0.250$ for the standard normal distribution is 0.401, which offers no evidence against the HPP.

The power and the exponential models are only two of many possible relations for the intensity function of an NHPP. For further discussion on these and other models, see Chapter 5 in Ascher and Feingold (1984).

Exercise 11.17

Using the results of Example 11.3, plot the mean cumulative repair function estimate and compare to the observed results by overlaying $\hat{M}(t)$ on the cumulative plot.

Exercise 11.18

The following repair ages were recorded for a system: 212, 459, 704, 834, 953, 1036. Observation occurred only to the sixth repair. Assuming an exponential model for the intensity $\lambda(t)$, estimate the parameters and plot the model fit against the original data. Check the model fit using Laplace's test.

Rate of Occurrence of Failures

Recall the derivative of the expected cumulative number of repairs in time,

$$m(t) = \frac{dM(t)}{dt}$$

Ascher and Feingold (1984) call $m(t)$ the *rate of occurrence of failures* (abbreviated ROCOF). Note that the instantaneous repair rate $m(t)$ is different from the observed average rate of occurrence of events defined as the total number of repairs N observed in an interval of length t divided by t, that is,

$$\bar{n} = \frac{N}{t}$$

Alternatively, when the total number of repairs is k, we may calculate an average or mean time between failures as the total time $T(k)$ divided by the cumulative number of repairs; that is,

$$MTBF_{cum} = \frac{T(k)}{k}$$

In studying reliability growth (that is, improvement in reliability as a product is improved), J.T. Duane (1964) discovered the empirical relationship that plotting $MTBF_{cum}$ versus time t on log-log paper often resulted in a straight line. The slope of this line then served as a basis for assessing reliability growth. Chapter 12 presents a discussion of the topic of reliability growth in further detail. See also O'Connor (1991) for more on this topic.

Exercise 11.19

Take the three data sets (renewal, improving, and degrading) previously presented in Chapters 10 and 11 and construct Duane plots. Do the plots look linear? Note that plotting on log-log paper tends to linearize data plots. Considering this last statement, draw some conclusions about the Duane plots.

SIMULATION OF STOCHASTIC PROCESSES

We have mentioned in previous chapters that simulation is a powerful tool for generating and understanding random events. The simulated data allows us to investigate the validity and range of applicability of analysis procedures. In situations where theory is inadequate or nonexistent, or the models are too complicated for solution, simulation studies may be the only viable approach. Additionally, simulation may be the source of data to illustrate important concepts. In this section, we discuss simulation techniques for renewable and repairable processes. In particular, we show how to simulate both the HPP and the NHPP. We provide an example of the simulation of the NHPP under the power law relation, and thereby show the application of analysis methods of previous sections.

Earlier we stated that a stochastic process is a collection of random variables describing the evolution through time of a process. To simulate a stochastic process, we thus simulate a sequence of random variables. For example, consider the times of repairs occurring by system age t for a renewal process in which the interarrival times have a specific distribution F. We simulate independent random variables $X_1, X_2, \ldots$ having this distribution F and stopping at the minimum $n + 1$ such that the sum

$$X_1 + X_2 + \ldots + X_n + X_{n+1} > t$$

Thus, the X_i, $i \geq 1$, are the simulated interarrival times, and the simulation yields n repairs occurring at system ages $X_1, X_1 + X_2, \ldots, X_1 + \ldots + X_n$.

If we want to simulate a homogeneous Poisson process having rate λ, then the distribution F for simulating the X_i, $i \geq 1$ is exponential with failure rate λ. Alternative methods for simulating a Poisson process are described in Ross (1993).

To stimulate a nonhomogeneous Poisson process having intensity $\lambda(t)$ for $0 \leq t \leq \infty$, we use an approach presented in Ross. We will simulate the successive repair times. Let $Y_1, Y_2, \ldots$ denote the system ages at repair. Since these random variables are dependent, we need to find the conditional distribution of Y_i given $Y_1, \ldots, Y_{i-1}$. Ross shows that if a repair occurs at time y, then, independent of what occurs prior to y, the time t until the next event at time $y + t$ has the distribution F_Y given by

$$F_Y(t) = 1 - \exp\left[-\int_0^t \lambda(y + \tau)\, d\tau\right]$$

Thus, we simulate the first repair time Y_1 from F_0. If the simulated value of Y_1 is y_1, we simulate Y_2 by adding y_1 to a value generated from F_{Y_1} and calling the sum y_2. Similarly, Y_3 is simulated by adding y_2 to a value generated from F_{Y_2} and calling the sum y_3, and so on. If the distribution F_Y can be inverted, the inverse transform method described in earlier chapters can be used. Let us now illustrate the approach for the case where the intensity function is described by the power relation model.

Example 11.7   Simulating an NHPP with Power Relation Intensity

The intensity is given by the model

$$\lambda(y) = aby^{b-1}$$

Then,

$$\int_0^t \lambda(y+\tau)\,d\tau = \int_0^t ab(y+\tau)^{b-1}\,d\tau = a\left[(y+t)^b - y^b\right]$$

So,

$$F_Y(t) = 1 - \exp\left\{-a\left[(y+t)^b - y^b\right]\right\}$$

The distribution F_Y can be inverted, and after some algebra to solve for t, we get

$$t = \left[y^b - \frac{1}{a}\ln(1 - F_Y)\right]^{\frac{1}{b}} - y$$

By the inverse transform method, we substitute the random unit uniform variable $1 - U$ for F_Y to get

$$F_Y^{-1}(u) = \left[y^b - \frac{1}{a}\ln(u)\right]^{\frac{1}{b}} - y$$

We can thus simulate the successive repair ages $Y_1, Y_2, \ldots$ by generating $U_1, U_2, \ldots$ and applying the formulas (where $Y_0, = 0$)

$$Y_1 = \left[-\frac{1}{a}\ln U_1\right]^{\frac{1}{b}}$$

$$Y_2 = \left[Y_1^b - \frac{1}{a}\ln U_2\right]^{\frac{1}{b}} - Y_1 + Y_1 = \left[Y_1^b - \frac{1}{a}\ln U_2\right]^{\frac{1}{b}}$$

and in general,

$$Y_1 = \left[Y_{i-1}^b - \frac{1}{a} \ln U_i \right]^{\frac{1}{b}}$$

Example 11.8   Simulating the First Six Repair Times for NHPP with Specified Power Relation Model

Let the parameters of the intensity function for the power relation model be $a = 0.25$ and $b = 0.5$; that is, the intensity function is

$$\lambda(t) = (0.25)(0.5) t^{(0.5-1)} = 0.125 t^{-0.5}$$

We will simulate the system ages for the first six repairs. Using a unit uniform pseudo-random number simulator, similar to what is available in many hand-held calculators, we generate the numbers 0.2727, 0.8305, 0.8772, 0.3344, 0.1143, and 0.7004. The first simulated repair time is (see Example 11.7)

$$y_1 = \left[y_0^{0.5} - \frac{1}{0.25} \ln(0.2727) \right]^{\frac{1}{0.5}} = 27.0$$

where $y_0 = 0$. The second simulated time is

$$y_2 = \left[(27.0)^{0.5} - \frac{1}{0.25} \ln(0.8305) \right]^{\frac{1}{0.5}} = 35.3$$

Continuing in this manner, we get the remaining repair ages 41.8, 117.6, 380.9, and 438.5.

Exercise 11.20

Simulate the first ten repair times for a system modeled by the NHPP having a power relation intensity with parameters $a = 0.10$ and $b = 0.75$. Graph the results and test the null hypothesis of an HPP.

Exercise 11.21

Simulate the first ten repair times for a system modeled by the NHPP having a exponential relation intensity with parameters $a = 0.001$ and $b = 0.005$. Graph the results and test the null hypothesis of an HPP.

SUMMARY

We have presented some keys concepts on repairable systems. While far from an exhaustive treatment, these ideas illustrate how different approaches are useful for handling data from repairable systems. Graphical tools are very important to precede analytical procedures. Methods of modeling and tests of trend for repairable system data have been discussed. We have emphasized the distinction between analysis of repairable and non-repairable systems. Different analysis situations for renewal and nonrenewal data have been illustrated. Simulation procedures for renewal processes and the NHPP have been described. We see that the subject of repairable system reliability has many important and practical applications.

Problems

11.1 The number of repairs for newly established, continuously operating measuring stations follows an NHPP with intensity linearly decreasing from 1 repair per day initially to 0.1 per day by the end of the tenth week. The rate stays at the constant level of 0.1 repairs per day thereafter. We assume the number of repairs are independent day to day. What is the probability that no repairs occur during days 15 through 21? What is the expected number of repairs after 20 weeks?

11.2 The number of accidents at an intersection follows an NHPP having an initial rate of 0.0001 per hour at 6:00 AM and increasing linearly to 0.005 per hour by 10:00 AM The rate remains constant at 0.0005 per hour until 7:00 PM From 7:00 PM, it decreases linearly from 0.0005 per hour to 0.0001 per hour by 10:00 PM, remaining constant until 6:00 AM. Assume the number of accidents occurring during disjoint time periods is independent. What is the probability of no accidents happening during the hours of 12 noon to 2:00 PM? What is the expected number of accidents during this period?

11.3 The repair histories (in hours) on four systems assumed to be independent are shown below: System 1: 142, 309, 460; System 2: 99, 145, 300, 347; System 3: 212, 225, 273, 398, 467; System 4: 21, 58, 150, 176. Is there any evidence against the null hypothesis of an HPP for each system? Is there any evidence based on the overall test?

11.4 Using the ten repair ages of Example 11.7, estimate the modified MLE parameters of the power relation model. Plot the original and the estimated intensity functions versus the system ages. Estimate the mean cumulative function for the power law model using the modified MLE parameter values. Compare to the actual data using a cumulative plot. Calculate the goodness of fit statistic $C^2_{N_s}$ and compare it to the critical values in Table 11.5.

Chapter 12

A Survey of Other Topics in Reliability

Earlier chapters have described how to analyze data and fit standard but powerful models to describe and forecast the expected reliability of components and systems. There are many other aspects of reliability analysis that are commonly found in the literature that we have not touched upon at all. For example, how can we model the improvement in reliability expected when we test a system during development and analyze all failures and take corrective actions? What method can we use to include our past experience and "engineering judgement" into a formal reliability evaluation of a product? Finally, how can we validate our reliability projections—or know when it is time to take corrective actions because the projections were much too optimistic?

Questions like these will be answered in this chapter as we briefly survey the areas of reliability growth models, Bayesian reliability analysis, and field reliability monitoring programs. We will also look briefly at predicting reliability using the U.S. government document *MIL-HDBK-217*.

RELIABILITY GROWTH MODELS

Chapter 11 discussed systems with improving repair rate within the general context of repairable systems. The models that will be described in this section were introduced in that chapter. Here, we will look at these models again from the perspective of tests or procedures aimed at improving system or equipment reliability prior to customer availability.

Consider the following situation: a complex and expensive piece of equipment has been designed, developed, prototyped, and now is undergoing reliability testing. An initial assessment of the mean time between failures (*MTBF*) is 10 hr. Since the primary customer has a minimum *MTBF* requirement of 100 hr, significant reliability improvement is needed. The equipment manufacturer decides to run a lengthy test, operating the equipment as the user would while paying careful attention to all failures. Every failure is analyzed until the root cause is discovered, and wherever possible, the equipment design or subcomponent parts selection is modified to eliminate or significantly reduce future failures of the type discovered.

This kind of equipment improvement testing is sometimes abbreviated TAAF, for *test, analyze and fix*. Another term becoming common in industry is "IRONMAN" testing. Originally, IRONMAN was an acronym for *improve reliability of new machines at night*, referring to a practice where the equipment was tested during the night and modified and improved during the day. Now, however, the term IRONMAN is used regardless of when the sequence of testing and improving is carried out.

After an extended TAAF program, it is natural to expect the equipment (or system) to operate at a higher level of reliability. In other words, the repair rate will be lower, and the *MTBF* will be higher, than it was at the start of the test. Consequently, models that estimate the rate of reliability growth are known as *reliability growth models*. An appropriate reliability growth model should be able to predict how much the *MTBF* will improve as the test time increases.

Reliability growth models assume that, once the improvement process is completed, the repair rate (or its reciprocal, the *MTBF*) is constant. Using the terminology of Chapter 10, a constant repair rate implies a homogeneous Poisson process (HPP). The goal of the reliability improvement testing program is to test long enough to increase the operational *MTBF* so that it satisfies customer requirements.

Duane (1964) studied data from several different products, all of which had been through reliability improvement (TAAF) testing. He noted consistently good linear relationships when he plotted a cumulative estimate of the *MTBF* versus system age (or "cumulative" operating time) using log-log graph paper. The cumulative estimate of the *MTBF* at the system age t_k of the *k*th failure is defined to be

$$\widehat{MTBF}_{cum}(t_k) = \frac{t_k}{k}$$

This *MTBF* is a cumulative estimate because it uses all the failures from the beginning of test just as if the failure rate had been constant throughout. Duane's empirical relationship states that

$$\ln MTBF_{cum}(T) = \alpha + \beta \ln(T)$$

Here we use natural logarithms. If base 10 logarithms are used, β is unchanged, and α is replaced by $\alpha/\ln 10$.

Using notation from Chapter 10, let $M(t)$ be the expected number of repairs up to system age t. Then we can write the definition of the cumulative *MTBF* as $MTBF_{cum}(t) = t/M(t)$, and the Duane relationship is equivalent to

$$M(t) = at^b$$

where $a = e^{-\alpha}$ and $b = 1-\beta$. This expression is just the power relationship model (sometimes misleadingly called a "Weibull process") discussed in Chapter 11. The intensity function, or instantaneous repair rate, for this process is the derivative $m(t) = M'(t) = abt^{b-1}$. The *MTBF* at the time we stop the test is just the reciprocal of the instantaneous repair rate evaluated at that time. (Remember that the *MTBF* becomes constant when the test ends.) If this instantaneous *MTBF* is denoted by $MTBF_I(t)$, it follows that

$$MTBF_I(t) = \frac{1}{m(t)} = \frac{1}{abt^{b-1}} = \frac{MTBF_{cum}(t)}{b} = \frac{MTBF_{cum}(t)}{1-\beta} = \frac{e^{\alpha}t^{\beta}}{1-\beta}$$

If we have estimates for a and b (or α and β), we can use the last equation to estimate the $MTBF = MTBF_I$ at the end of test. Since reliability has been improving throughout the test ($0 < b < 1$), the $MTBF_I$ estimate at the end of test will be greater than the $MTBF_{cum}$ estimate at the end of test.

The graphical estimation approach plots the cumulative *MTBF* estimates at each repair time against the system age at that repair time, using log-log graph paper. A "best fitting" line is drawn through these points and the slope of the line is the β estimate, and the intercept of the line is the α estimate. It is common to use a least squares regression program to eliminate the subjectivity of hand drawing the line (see Chapter 6). If the system ages at repair times are $t_1, t_2, \ldots, t_n$, then the dependent variable is $y_1 = \ln(t_1)$, $y_2 = \ln(t_2/2)$, $y_3 = \ln(t_3/3), \ldots, y_n = \ln(t_n/n)$. The independent variable is $x_1 = \ln(t_1)$, $x_2 = \ln(t_2), \ldots, x_n = \ln(t_n)$. Any standard regression program will estimate α and β using the least squares estimation formulas given in Chapter 6.

The following example illustrates the use of "Duane plots" for reliability growth estimation.

Example 12.1   Duane Reliability Growth Estimation

A multimillion-dollar, state-of-the-art semiconductor processing tool has an unacceptable reliability performance record, typically needing repairs on a daily basis. Customer requirement is for an *MTBF* of at least 100 hr. The manufacturer decides to invest in a 12 week reliability improvement testing program, during which the tool will be exercised as much as possible. All failures are carefully traced to root causes, and design and component changes are introduced during the course of the testing. Twelve failures are recorded at the following system age hours:

37	56	176	262	516	522	544	588	719	755	861	1166

Assuming the test ends after the last repair, use the Duane plot technique to estimate the *MTBF* reliability growth model and to calculate the *MTBF* of the tool at the end of the test.

Solution

The 12 $MTBF_{cum}$ (t_k) estimates corresponding to the 12 failure times are:

37	28	59	86	103	87	78	74	80	76	78	97

These are calculated by simply dividing the system age at the kth failure by k. For example, $MTBF_{cum}(t_3) = 176/3$, or approximately 59.

When these estimates are plotted against the corresponding failure times on log-log paper, we obtain the Duane plot shown in Figure 12.1. The line through the points was fit using least squares. The dependent variable is a vector of the logarithms of the $MTBF_{cum}$ estimates (or 3.61, 3.33, 4.08, 4.45, 4.64, 4.47, 4.36, 4.3, 4.38, 4.33, 4.36, 4.58). The independent variable is a vector of the logarithms of the system ages at failure (or 3.61, 4.03, 5.17, 5.57, 6.25, 6.30, 6.38, 6.58, 6.63, 6.76, 7.06).

The least squares estimate of α is 2.32 and the estimate of β is 0.32. The predicted value of the *MTBF* at the end of the test (or at 1166 hr) is ($e^{2.32}$ × $1166^{0.32}$)/(1 − 0.32) = 143 hr. By comparison, at the start of the test (for which we use 1 hr because of the log-log scale), the *MTBF* according to the model is $e^{2.32}$/ 0.68 = 15. So the Duane plot indicates a successful reliability improvement test, with the *MTBF* increasing by almost 10× and exceeding user requirements at the end of the test. Figure 12.1 shows both the least squares line through the data points and the instantaneous *MTBF* line that predicts what the improved *MTBF* estimate would be if the test had been stopped at any particular point in time. The only "real" point on this line is the one at 1166 hr. However, it is customary to plot the entire instantaneous *MTBF* line on a Duane plot so that "what if" questions can be entertained about the effects of shorter or longer test times.

As β (the slope of the reliability growth line) becomes larger, the *MTBF* improves more rapidly. O'Connor (1991) gives some guidelines for typical values of β. Generally, β will lie between 0.2 and 0.6, depending on how timely and effective the corrective actions are for eliminating failure modes found during the test.

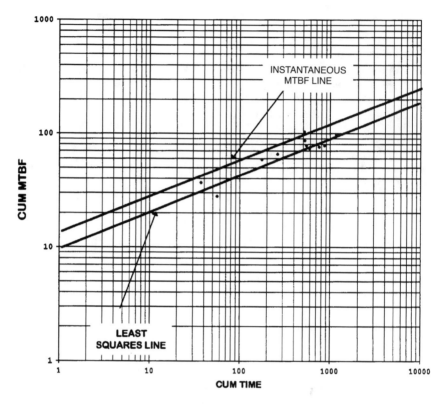

Figure 12.1 Duane Plot of Cumulative MTBF versus Cumulative Time with Least Squares Line

The Duane empirical procedure is a simple model for reliability growth that has been used successfully, as cited many times in the literature. However, it does have several disadvantages. There is no way to put valid confidence bounds around the graphical estimates or, apart from the appearance of the fit of the line, to test the whether the data is consistent with the Duane model. In addition, because the points are highly correlated and have differing statistical distributions, least squares estimates have no statistical optimality properties and are merely an objective way to fit a line through the points.

Crow (1974) developed the power relationship stochastic model for repairable systems that provides a theoretical basis for the empirical Duane model. This work was done for the U.S. Army Materials Systems Analysis Activity, and the model is often referred to as the AMSAA model. As described in chapter 10, maximum likelihood estimates for parameters $a = e^{-\alpha}$ and $b = 1 - \beta$ of the AMSAA power relationship model are given by

$$\hat{b} = \frac{n}{n-1 \over \displaystyle\sum_{i=1}^{n} \ln \frac{t_n}{t_i}}, \quad \hat{a} = \frac{n}{\frac{\hat{b}}{t_n}}$$

for the case where n repair actions take place at times $t_1, t_2, \ldots, t_n$, and the test ends right after the nth failure occurs. If the test continues beyond the time of the last failure to a total time of T, the maximum likelihood estimates are given by

$$\hat{b} = \frac{n}{\displaystyle\sum_{I=1}^{n} \ln \frac{T}{t_i}}, \quad \hat{a} = \frac{n}{T^{\hat{b}}}$$

The maximum likelihood estimate of the final $MTBF$ for a test that lasts until the nth fail is

$$\stackrel{\wedge}{MTBF}_I (t_n) = \frac{1}{\hat{a}\hat{b}t_n^{\hat{b}-1}} = \frac{t_n^{\hat{b}}}{n\hat{b}t_n^{\hat{b}-1}} = \frac{t_n}{n\hat{b}}$$

If the test is terminated at time T, where T is greater than the time of the last fail, replace t_n by T in the last equation to obtain the maximum likelihood estimate for a time censored test:

$$\stackrel{\wedge}{MTBF}_I (T) = \frac{T}{n\hat{b}}$$

The MLE for b has optimal statistical properties as the number of failures gets large. However, it has a built in bias that causes it to overestimate b (and underestimate the reliability growth slope β). This bias can be corrected by multiplying $\hat{b}$ by a factor that depends on the number of failures. An unbiased estimate of b when the test is terminated at the nth failure is given by

$$\bar{b} = \frac{(n-2)\hat{b}}{n}$$

If, instead, the test was terminated at a time T after the nth failure had occurred, an unbiased estimate of b is given by

$$\bar{b} = \frac{(n-1)\,\hat{b}}{n}$$

For n below around 50 the correction factors above will certainly have a significant effect upon the estimate of b and should be used. For large n, applying the correction factor will make little practical difference and is optional. Whenever the corrected estimate $\bar{b}$ is used, use $\bar{a} = n/T^{\bar{b}}$ for the estimate of a for consistency (even though, unfortunately, it has no guarantee of being an unbiased estimate).

We will call $\bar{b}$ and $\bar{a}$ "modified MLEs" and use them to calculate b and a estimates as well as the estimated $MTBF_I$ at the end of test. Lines plotted on a Duane plot based on these modified MLEs will be called "modified MLE lines."

Two important aspects of a reliability growth analysis remain: testing for the adequacy of the power relationship model and deriving an upper and a lower bound for the $MTBF_I$ at the end of the reliability improvement test. The C_N^2 test for the power relationship model (from Crow, 1974) was described in Chapter 10. If C_N^2 is larger than an appropriate critical value from Table 11.5, use of the power relationship model for the failure data becomes questionable. Table 11.5 gives critical values at significance levels ranging from 0.20 (an hypothesis test at the 80 percent confidence level) to 0.01 (a 99 percent confidence level test).

To calculate C_N^2 for a reliability improvement test that ends at the time of the nth failure, set $N = n - 1$ and use the equation

$$C_N^2 = \frac{1}{12N} + \sum_{i=1}^{N}\left[\left(\frac{t_i}{t_n}\right)^b - \frac{2i-1}{2N}\right]^2$$

where $\bar{b}$ is the modified MLE estimate of b.

If the test ends at a time T which is greater than the last failure time t_n, then set $N = n$ and use the equation

$$C_N^2 = \frac{1}{12N} + \sum_{i=1}^{N}\left[\left(\frac{t_i}{T}\right)^b - \frac{2i-1}{2N}\right]^2$$

A confidence interval for the $MTBF_I$ at the end of test is easily obtained using the factors in Table 12.1 and Table 12.2 (reproduced with permission from Crow, 1982). For a test that ends at the nth failure, look up the values of R_1 and R_2 from

TABLE 12.1   Values of R_1 and R_2 to Multiply the MTBF Estimate and Obtain Confidence
Bounds (test ends at nth fail)

Number of Fails (n)	Confidence Level							
	80 Percent		90 Percent		95 Percent		98 Percent	
	R_1	R_2	R_1	R_2	R_1	R_2	R_1	R_2
2	0.8065	33.76	0.5552	72.67	0.4099	151.5	0.2944	389.9
3	0.6840	8.927	0.5137	14.24	0.4054	21.96	0.3119	37.60
4	0.6601	5.328	0.5174	7.651	0.4225	10.65	0.3368	15.96
5	0.6568	4.000	0.5290	5.424	0.4415	7.147	0.3603	9.995
6	0.6600	3321	0.5421	4.339	0.4595	5.521	0.3815	7.388
7	0.6656	2.910	0.5548	3.702	0.4760	4.595	0.4003	5.963
8	0.6720	2.634	0.5668	3.284	0.4910	4.002	0.4173	5.074
9	0.6787	2.436	0.5780	2.989	0.5046	3.589	0.4327	4.469
10	0.6852	2.287	0.5883	2.770	0.5171	3.286	0.4467	4.032
11	0.6915	2.170	0.5979	2.600	0.5285	3.054	0.4595	3.702
12	0.6975	2.076	0.6067	2.464	0.5391	2.870	0.4712	3.443
13	0.7033	1.998	0.6150	2.353	0.5488	2.721	0.4821	3.235
14	0.7087	1.933	0.6227	2.260	0.5579	2.597	0.4923	3.064
15	0.7139	1.877	0.6299	2.182	0.5664	2.493	0.5017	2.921
16	0.7188	1.829	0.6367	2.144	0.5743	2.404	0.5106	2.800
17	0.7234	1.788	0.6431	2.056	0.5818	2.327	0.5189	2.695
18	0.7278	1.751	0.6491	2.004	0.5888	2.259	0.5267	2.604
19	0.7320	1.718	0.6547	1.959	0.5954	2.200	0.5341	2.524
20	0.7360	1.688	0.6601	1.918	0.6016	2.147	0.5411	2.453
21	0.7398	1.662	0.6652	1.881	0.6076	2.099	0.5478	2.390
22	0.7434	1.638	0.6701	1.848	0.6132	2.056	0.5541	2.333
23	0.7469	1.616	0.6747	1.818	0.6186	2.017	0.5601	2.281
24	0.7502	1.596	0.6791	1.790	0.6237	1.982	0.5659	2.235
25	0.7534	1.578	0.6833	1.765	0.6286	1.949	0.5714	2.192
26	0.7565	1.561	0.6873	1.742	0.6333	1.919	0.5766	2.153
27	0.7594	1.545	0.6912	1.720	0.6378	1.892	0.5817	2.116
28	0.7622	1.530	0.6949	1.700	0.6421	1.866	0.5865	2.083
29	0.7649	1.516	0.6985	1.682	0.6462	1.842	0.5912	2.052
30	0.7676	1.504	0.7019	1.664	0.6502	1.820	0.5957	2.023
35	0.7794	1.450	0.7173	1.592	0.6681	1.729	0.6158	1.905
40	0.7894	1.410	0.7303	1.538	0.6832	1.660	0.6328	1.816
45	0.7981	1.378	0.7415	1.495	0.6962	1.606	0.6476	1.747
50	0.8057	1.352	0.7513	1.460	0.7076	1.562	0.6605	1.692
60	0.8184	1.312	0.7678	1.407	0.7267	1.496	0.6823	1.607
70	0.8288	1.282	0.7811	1367	0.7423	1.447	0.7000	1.546
80	0.8375	1.259	0.7922	1.337	0.7553	1.409	0.7148	1.499
100	0.8514	1.225	0.8100	1.293	0.7759	1.355	0.7384	1.431

TABLE 12.2   Values of P_1 and P_2 to Multiply the MTBF Estimate and Obtain Confidence Bounds (test ends at time, T)

Number of Fails (n)	Confidence Level							
	80 Percent		90 Percent		95 Percent		98 Percent	
	P_1	P_2	P_1	P_2	P_1	P_2	P_1	P_2
2	0.261	18.66	0.200	38.66	0.159	78.66	0.124	198.7
3	0.333	6.326	0.263	9.736	0.217	14.55	0.174	24.10
4	0.385	4.243	0.312	5.947	0.262	8.093	0.215	11.81
5	0.426	3.386	0.352	4.517	0.300	5.862	0.250	8.043
6	0.459	2.915	0.385	3.764	0.331	4.738	0.280	6.254
7	0.487	2.616	0.412	3.298	0.358	4.061	0.305	5.216
8	0.511	2.407	0.436	2.981	0.382	3.609	0.328	4.539
9	0.531	2.254	0.457	2.750	0.403	3.285	0.349	4.064
10	0.549	2.136	0.476	2.575	0.421	3.042	0.367	3.712
11	0.565	2.041	0.492	2.436	0.438	2.852	0.384	3.441
12	0.579	1.965	0.507	2.324	0.453	2.699	0.399	3.226
13	0.592	1.901	0.521	2.232	0.467	2.574	0.413	3.050
14	0.604	1.846	0.533	2.153	0.480	2.469	0.426	2.904
15	0.614	1.800	0.545	2.087	0.492	2.379	0.438	2.781
16	0.624	1.759	0.556	2.029	0.503	2.302	0.449	2.675
17	0.633	1.723	0.565	1.978	0.513	2.235	0.460	2.584
18	0.642	1.692	0.575	1.933	0.523	2.176	0.470	2.503
19	0.650	1.663	0.583	1.893	0.532	2.123	0.479	2.432
20	0.657	1.638	0.591	1.858	0.540	2.076	0.488	2.369
21	0.664	1.615	0.599	1.825	0.548	2.034	0.496	2.313
22	0.670	1.594	0.606	1.796	0.556	1.996	0.504	2.261
23	0.676	1.574	0.613	1.769	0.563	1.961	0.511	2.215
24	0.682	1.557	0.619	1.745	0.570	1.929	0.518	2.173
25	0.687	1.540	0.625	1.722	0.576	1.900	0.525	2.134
26	0.692	1.525	0.631	1.701	0.582	1.873	0.531	2.098
27	0.697	1.511	0.636	1.682	0.588	1.848	0.537	2.068
28	0.702	1.498	0.641	1.664	0.594	1.825	0.543	2.035
29	0.706	1.486	0.646	1.647	0.599	1.803	0.549	2.006
30	0.711	1.475	0.651	1.631	0.604	1.783	0.554	1.980
35	0.729	1.427	0.672	1.565	0.627	1.699	0.579	1.870
40	0.745	1.390	0.690	1.515	0.646	1.635	0.599	1.788
45	0.758	1.361	0.705	1.476	0.662	1.585	0.617	1.723
50	0.769	1.337	0.718	1.443	0.676	1.544	0.632	1.671
60	0.787	1.300	0.739	1.393	0.700	1.481	0.657	1.591
70	0.801	1.272	0.756	1.356	0.718	1.435	0.678	1.533
80	0.813	1.251	0.769	1.328	0.734	1.399	0.695	1.488
100	0.831	1.219	0.791	1.286	0.758	1.347	0.722	1.423

Table 12.1 that correspond to n and the confidence level desired. Multiply the (unmodified) MLE of $MTBF_I(t_n)$ by R_1 and R_2 to obtain the lower and upper limits of the confidence interval.the interval is

$$\left(R_1 \times \widehat{MTBF}_I(t_n), R_2 \times \widehat{MTBF}_I(t_n) \right)$$

If a one-sided limit is desired, the lower or upper limit of a 90 percent confidence interval is a 95 percent lower or upper limit (and, in general, the lower or upper limit of a $100 \times (1 - \alpha)$ confidence interval is a $100 \times (1 - \alpha/2)$ lower or upper bound).

For a time censored test ending at $T > t_n$ (where t_n is the time of the nth failure) use the values of P_1 and P_2 from Table 12.2 that correspond to n and the confidence level desired. Again, be sure to multiply the *unmodified* MLE of $MTBF_I$ (T) by these factors to obtain the lower and upper limits of the confidence interval.

Tables 12.1 and 12.2 stop at $n = 100$. While it is unlikely that a real reliability growth test will have more than 100 failures, Crow (1982) gives formulas for approximate confidence bounds that are quite accurate for large n. For a test that ends at the nth failure, calculate an approximate $100 \times (1 - \alpha)$ percent confidence interval using

$$R_1 = \frac{1}{1 + \sqrt{\frac{2}{n}} Z_{\alpha/2}}, \quad R_2 = \frac{1}{1 - \sqrt{\frac{2}{n}} Z_{\alpha/2}}$$

where $Z_{\alpha/2}$ is the $100 \times (1 - \alpha/2)$th percentile of the standard normal distribution.

For a test ending at time $T > t_n$, calculate an approximate $100 \times (1 - \alpha)$ percent confidence interval using

$$P_1 = \left[\frac{n}{n + Z_{\alpha/2}\sqrt{\frac{n}{2}}} \right]^2, \quad P_2 = \left[\frac{n}{n - Z_{\alpha/2}\sqrt{\frac{n}{2}}} \right]^2$$

Example 12.2   Confidence Bounds and Modified MLEs

Suppose a test has 16 failures and ends at 1250 hr, which you are told is the time of the 16th failure. The MLE of b has been calculated and is $\hat{b} = 0.72$. Give an estimate and a 90 percent confidence interval for $MTBF_I(1250)$. How do the estimate and confidence interval change if you learn that the test was planned to end at 1250 hours, and the 16th failure actually occurred before that time (assume the MLE $\hat{b} = 0.72$ was calculated correctly for a time censored test)?

Solution

First we calculate the MLE for the $MTBF_f$ at the end of test. This is $1250/(16 \times 0.72) = 108.5$. To construct a confidence interval for a test that ends upon reaching the 16th failure, we look up $R_1 = 0.6367$ and $R_2 = 2.144$ from Table 12.1. The interval for $MTBF_f(1250)$ is $0.6367 \times 108.5, 2.144 \times 108.5) = (69, 232.6)$.

Instead of using $\hat{b}$ for our final estimate for b, we calculate the modified MLE estimate $\hat{b} = (14/16) \times 0.72 = 0.63$. Using this estimate of b, we recalculate the $MTBF_f(1250)$ estimate as $1250/(16 \times 0.63) = 124$. This value is our point estimate for the $MTBF$ at the end of test, while the confidence interval remains as previously calculated.

After we learn that the test was really time censored (i.e., planned to stop at 1250 hr), we recalculate the confidence interval using $P_1 = 0.556$ and $P_2 = 2.029$ from Table 12.2. The interval for $MTBF_f(1250)$ is $(0.556 \times 108.5, 2.029 \times 108.5) = (60.3, 220.1)$. Finally, we modify our MLE estimates of b and the $MTBF$ at the end of test, using $\bar{b} = (15/16) \times 0.72 = 0.675$ and $MTBF_f(1250) = 1250/(16 \times 0.675) = 115.7$.

Example 12.3   Power Relationship Model Reliability Growth

Use the maximum likelihood estimate formulas derived for the power relationship model to calculate the Duane plot slope for the data in Example 12.1. Next, estimate the achieved $MTBF$ at 1166 hr and give 90 percent confidence bounds for this $MTBF$. Is the power relationship model a reasonable assumption for this data?

Solution

The maximum likelihood estimates are $\hat{a} = 0.03$ and $\hat{b} = 0.85$. It follows that the maximum likelihood estimate for β is $1 - 0.85 = 0.15$, and the maximum likelihood estimate for α is $-\ln 0.03 = 3.5$.

The $MTBF$ at the end of the test is estimated to be $1166/(12 \times 0.85) = 114$ hr. From Table 12.1, we look up factors that multiply the $MTBF$ MLE estimate to obtain a lower 5 percent bound and an upper 95 percent bound. For 12 failures, these factors are $R_1 = 0.6067$ and $R_2 = 2.464$. Applying these factors, the estimate of 114 generates a 90 percent confidence interval for the achieved $MTBF$ of $(69.5, 282.4)$. The modified MLE estimate for b is $(10/12) \times 0.85 = 0.71$ and for β is 0.29. The modified $MTBF$ at the end of test estimate is $1166/(12 \times 0.71) = 136.9$.

Finally, we use the C_N^2 statistic defined in Chapter 11 to test the hypothesis that the failure data are consistent with the power relationship model. The test statistic is $C_{11}^2 = 0.086$, and from Table 12.1 this is not significant at even an 80 percent confidence level. Therefore, it is reasonable to use a power relationship model for the reliability growth in this example.

Figure 12.2 shows the Duane plot with the modified maximum likelihood $MTBF_{cum}(t)$ line and the corresponding modified $MTBF_I(t)$ line (here, as before, "modified" means $\bar{b}$ and $\bar{a}$ were used instead of the MLEs $\hat{b}$ and $\hat{a}$).

Examples 12.1 and 12.3 show that the Duane least squares slope and the power relationship model MLE slope may differ considerably (0.32 versus 0.15). However, the modified MLE estimates were in close agreement with the least squares estimates (0.29 versus 0.32 for β and 136.9 versus 143 for the $MTBF$ at the end of test).

A complete reliability growth data analysis begins with a Duane plot and continues with MLE and modified MLE estimates and a test for the fit of the power

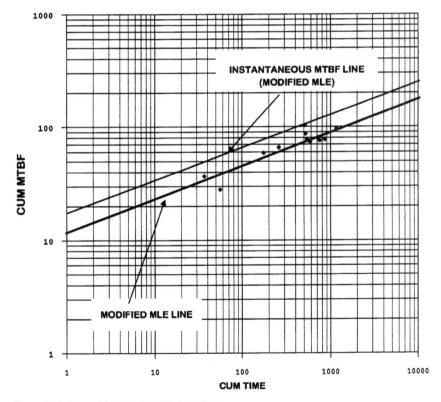

Figure 12.2 Duane Plot with Modified MLE Lines

relationship model. If the model is accepted, the final output of the analysis should be the modified MLE point estimate of the *MTBF* at the end of the test, with upper and lower confidence bounds.

Exercise 12.1

Repeat Example 12.2, this time assuming that the test ended at 1400 hr, with no more failures observed after the failure at 1166 hr.

Exercise 12.2

A prototype model of a new piece of equipment is put through a reliability improvement test. Fails are observed at the following times: 18, 20, 35, 41, 68, 211, 252, 288, 380, 382, 408, 449, and 532 hr. The test is stopped at 650 hr. Draw a Duane plot of the data and calculate MLE and modified MLE estimates for the improvement slope and $MTBF_I(650)$. Is the power relationship model a reasonable choice for the data? Give a 95 percent confidence interval for the *MTBF* at the end of the test.

Exercise 12.3

A reliability improvement test records failures at 37, 42, 44, 46, 47, and 121 hr. The test ends at 320 hr with no more failures. Draw a Duane plot of the data and calculate MLE and modified MLE estimates for the improvement slope and the *MTBF* at the end of the test. Is the power relationship model a reasonable choice for this data?

Since the Duane/AMSAA model appears in the above discussions with two different sets of parameters (a and b, or α and β) and other parameters also are used in the literature further confusing things, two key formulas are summarized below using both sets of parameters:

$$MTBF_{cum}(t) = e^{\alpha} t^{\beta} = \frac{t^{1-b}}{a}$$

$$MTBF_I(t) = \frac{e^{\alpha} t^{\beta}}{1-\beta} = \frac{t^{1-b}}{ab}$$

The equation for $MTBF_I(t)$ is useful for planning how long to run a reliability improvement test. For example, by assuming a conservative growth rate slope of $\beta = 0.3$ and a current *MTBF* value (which we set as the $MTBF_I$ at 1 hr, for convenience), it is easy to solve for the test time needed to improve to a desired final

MTBF. If previous experience indicates a more aggressive improvement slope can be achieved (i.e,. a β of 0.4 or 0.5 or even higher), then that can be used. Alternatively, previous tests on similar products may have demonstrated a given amount of improvement over the course of the test. Using the following equations, where *IF* is the improvement factor between one hr and time *T*, we can easily go from β to an improvement factor, or vice-versa.

$$IF = T^\beta \; ; \quad \beta = \frac{\log IF}{\log T}$$

Exercise 12.4

Based on past experience, a prototype of a new manufacturing tool is expected to have an *MTBF* of 50 hr at the start of a reliability improvement test and an *MTBF* of 500 at the end of test. What is the reliability growth slope? If the customer requirement is an *MTBF* of 400 hr, how long should the test be run?

Exercise 12.5

How much improvement would take place between the first and the 500th hr of testing when β is 0.5?

Much useful information concerning the practical and analytical aspects of reliability growth testing can be found in *U.S. MIL-HDBK-189 Reliability Growth Management* and *U.S. MIL-STD-1635(EC) Reliability Growth Testing*. Kececioglu (1991) has a detailed section on the Duane and AMSAA (power relationship) models that contains several worked out examples. Also note that the estimation formulas in this section only apply to one system undergoing reliability improvement testing (and the *MTBF* projection applies to all later systems that incorporate the improvements made on the one system during the test). If several systems are on test (with possibly different system ages) the estimation formulas are more complicated (see Chapter 11 and Crow, 1974, 1993).

BAYESIAN RELIABILITY ANALYSIS

Most applications of Bayesian methodology to solve applied statistical problems encounter a highly polarized reception consisting of strong supporters and equally strong detractors. To the believer, a Bayesian approach offers an intuitively pleasing way to harness past experience and "expert judgement" towards a goal of reducing costs and test time while still coming up with accurate estimates and sound decisions. This approach is especially attractive when estimating the failure rate of highly reliable components, where we have seen that sample sizes and test times can be very large if we want precise results.

On the other hand, those against this approach feel that the price you pay for what almost seems like "something for nothing" is the loss of credibility of the final results.

This section will describe the Bayesian method and show some of its advantages and disadvantages. The reader interested in learning more can consult references such as Mann, Schafer, and Singpurwala (1974) or Martz and Waller (1982). There are also many examples using Bayes techniques in the reliability literature. The annual *IEEE Reliability and Maintainability Symposium Proceedings* contain more than two dozen articles utilizing Bayes methods in the period from 1981 through 1993. For example, Kaplan, Cunha, Dykes, and Shaver (1990) use Bayes methodology to assess reliability during product development. Campodonico (1993) describes a software package to carry out Bayes analyses of survival times.

Assume we have a population of components with lifetimes that are described by an exponential distribution with parameter λ. We have shown, in Chapter 3, how to estimate λ from a sample of failure and survival data and also how to calculate an interval that we believe will contain the true (but unknown) λ with a given degree of confidence. It is worthwhile repeating the explanation given in Chapter 3 for the meaning of this confidence interval for λ to help us understand the fundamental differences between classical and Bayesian approaches to reliability.

In Chapter 3 we stated,

A 90 percent confidence interval means that if the same experiment were repeated many times, and the same method was used over and over again to construct an interval for λ, 90 percent of those intervals would contain the true λ. For the one time we actually do the experiment, our interval either does or does not contain λ.

This view is the "classical" frequency notion of probability. The probability of an event is the success ratio of that event in a repeated series of trials, as the number of repetitions approaches infinity. We can conceive of repeating a life test experiment over and over again, and we can define an event $A = $ *(the interval calculated at the end of the experiment contains the true value of the population failure rate λ)*. However, within this classical framework, we cannot even ask questions like, "What is the probability that the true λ lies within a particular calculated interval?" In the classical approach, λ is a fixed parameter that either does or does not lie within the interval. There is no repeated sequence of trials or success ratio to consider. There are only observable events that have a probability we can calculate, if we assume we know the value of λ.

The Bayesian approach uses a different concept of probability, known as subjective probability. This view is related more to our "degree of belief" in something rather than a frequency approach. An axiomatic theory can be developed for this concept. (See Savage, 1954, and Lindley, 1965). For our purposes, the Bayesian approach amounts to assuming that the unknown parameters that specify a life distribution (such as λ, T_{50}, σ, m, or c) are themselves random variables, chosen from a "prior" distribution that expresses our previous knowledge or intuition about these parameters. If we know the prior distribution, we can calculate the probability that λ lies within any particular interval. We will first give the general equations that apply for all Bayesian analyses and then focus in on the much simpler formulas that apply in the special case of an exponential life distribution.

Assume the life distribution for failure times has CDF $F(t)$ and PDF $f(t)$. Let λ denote the unknown parameter (or vector of parameters) that we have to estimate to calculate $F(t)$ and answer questions about failure rates and probabilities of survival.

Within the Bayesian framework, λ is a random variable assumed to have a CDF $G(\lambda)$ and PDF $g(\lambda)$. $G(\lambda)$ is called the *prior distribution* for λ and is a mathematical summary of all our previous knowledge or intuition about λ. After running an experiment and obtaining new data, this prior distribution is revised, using Bayes equation, to obtain a new updated distribution for λ. This new distribution is known as the posterior distribution for λ.

The general formula for the posterior distribution, $g(\lambda \mid t)$, is given by the continuous distribution version of Bayes Rule (the discrete form of Bayes Rule was discussed in Chapter 1):

$$g(\lambda|t) = \frac{g(\lambda) f(t|\lambda)}{\displaystyle\int_0^\infty f(t|\lambda) g(\lambda) \, d\lambda}$$

In this expression, t is the vector of observed failure times, and $f(t \mid \lambda)$ is the joint density, or likelihood function, of the failure times (the likelihood function was described in Chapter 4).

The mean of the posterior distribution, or expected value of λ after observing t, is known as the Bayes point estimate for λ. This expectation, $E(\lambda \mid t)$, is given by

$$E(\lambda|t) = \hat{\lambda} = \int_0^\infty \lambda g(\lambda|t) \, d\lambda$$

An upper $(1 - \alpha)$ bound for λ is obtained by solving for $\lambda_{1-\alpha}$ in the equation

$$\alpha = \int\limits_{\lambda_{1-\alpha}}^{\infty} g\,(\lambda\,|\,t)\,d\lambda$$

The above formulas show how to estimate λ for any choice of a prior distribution. In practice, it is common to choose convenient models for the prior $g(\lambda)$ so that the integral formulas above have simple solutions. In particular, if $g(\lambda)$ is a gamma distribution and $f(t\,|\,\lambda)$ is an exponential distribution, then the posterior distribution $g(\lambda\,|\,t)$ is also a gamma distribution. (When the prior and posterior distributions both belong to the same family, as in this example, then g is called a *conjugate prior* for the sampling distribution f. Conjugate priors are very popular choices for Bayesian applications.)

From now on we will restrict ourselves to the case where we have failure times modeled by the exponential distribution. The Bayesian approach has the following steps:

1. Choose a prior distribution for the life distribution parameter that matches well with the assumed sampling distribution. For exponential failure times, the typical choice is a gamma prior for λ.
2. Estimate the parameters of the gamma prior distribution $g(\lambda)$ based on either past data or a judgement call made by experts.
3. Use current data and the above formulas to calculate the posterior distribution and a point estimate with bounds for the failure rate.
4. If the objective of the testing is to confirm that a specified failure rate will be met at a given $(1 - \alpha)$ confidence level, continue testing until either the upper bound calculated in step 3 meets the requirement or a lower bound calculated from

$$1 - \beta = \int\limits_{\lambda_{\beta}}^{\infty} g\,(\lambda\,|\,t)\,d\lambda$$

is higher than the specified failure rate. Here, β is a chosen Type II error probability, or chance of rejecting good product (see Chapter 9). This kind of testing is known as sequential testing, since the stopping point is data dependent and not set in advance.

In the case of an exponential life distribution, the probability that a unit fails in the first T hours, given that the failure rate λ has the value λ_0, is expressed by

$$F\,(T\,|\,\lambda = \lambda_0) = 1 - e^{\lambda_0 T}$$

The formula for a gamma PDF (described in Chapter 8) with parameters a and b is

$$g(\lambda) = \frac{b^a \lambda^{a-1} e^{-b\lambda}}{\Gamma(a)}$$

It is well known that the mean of this distribution is given by $\mu = a/b$, and the variance is $\sigma^2 = a/b^2$. Without any testing, our estimate of λ is just μ or a/b. If our testing produced r failures in T total unit hours of test time, then the posterior distribution $g(\lambda \mid r$ fails in T hours), which is also a gamma distribution, can be shown to have new parameters

$$a' = a + r$$

$$b' = b + T$$

The new failure rate estimate using the prior knowledge and the test results is

$$\hat{\lambda} = \frac{a'}{b'} = \frac{(a+r)}{(b+T)}$$

and the *MTTF* (or *MTBF*) estimate is *MTTF = MTBF* = $(b + T)/(a + r)$.

Next we calculate an upper bound for λ by solving for $\lambda_{1-\alpha}$ in the equation

$$\alpha = \int_{\lambda_{1-\alpha}}^{\infty} g(\lambda \mid t)\, d\lambda = 1 - G(\lambda_{1-\alpha} \mid t)$$

Fortunately, the gamma CDF $G(\lambda \mid t)$ can easily be evaluated using chi-square distribution tables. If the gamma has parameters a´ and b´, then

$$\lambda_{1-\alpha} = \frac{\chi^2_{2a';100 \times (1-\alpha)}}{2b'}$$

is a $100 \times (1-\alpha)$ upper percentile point of the gamma distribution. For example, if $a' = 2$ and $b' = 500$, and we want an upper 90 percent estimate for λ (that we expect λ to be below 90 percent of the time), then we calculate

$$\frac{\chi^2_{4;90}}{2b'} = \frac{7.74}{1000} = 0.00774$$

Our point estimate for λ is 2/500 = 0.004, and we are 90 percent confident that λ is below 0.00774.

Example 12.4   Bayesian Reliability Testing

We are planning a qualification test for a complex system that is required to meet an *MTBF* specification of 1000 hr. Assuming the failure distribution is exponential, we would have to test for a minimum of 2,303 hr, with no failures, to be 90 percent confident that the system will have the specified reliability (see the section on minimum test times in Chapter 3). However, based on experience with earlier prototypes of the system and knowledge of the system subassemblies, we are willing to assume a (Bayesian) prior distribution for the failure rate that is a gamma with mean 1/1000 = 0.001. Because we feel our prior knowledge is somewhat shaky, we want the prior distribution to be "weak." We can get a weak prior by setting the standard deviation of the prior to be equal to or greater than the mean. In this case, we set $\sigma = 0.001$ ($\sigma^2 = 0.000001$).

What is the minimum Bayesian test time needed in order to be 90 percent confident the system *MTBF* is at least 1000 hr, assuming we have no failures?

Solution

We start by solving for a and b, knowing that $\mu = a/b = 0.001$, and $\sigma^2 = a/b^2 = 0.000001$. We obtain $a = 1$ and $b = 1000$. After a test of T hours with no failures, the parameters of the gamma posterior distribution will be $a' = a + 0 = 1$ and $b' = b + T = 1000 + T$. We want a value of T such that the 90th percentile of the posterior distribution exactly equals the failure rate objective. With that T and no failures, we will be 90 percent confident the system meets its reliability objective. An upper 90 percentile for the failure rate is given by $(\chi^2_{2;90})/2(1000 + T)$, and we have to set this value equal to the objective of 0.001 and solve for T. This procedure gives the equation

$$\chi^2_{2;90} = 0.002\,(1000 + T)$$

from which $4.61 = 0.002(1000 + T)$ and $T = 2305 - 1000 = 1305$.

Note that in this example the use of a Bayes prior allows us to reduce the minimum required test time from around 2300 hr to 1300 hr. The Bayes test time is more than 43 percent less than the time required by the standard methods given in Chapter 3.

Exercise 12.6

Repeat Example 12.4, this time choosing a "strong" prior with a standard deviation only one-third as large as the mean. Can you think of why a strong prior might lead to a larger sample size requirement, as happens in this example (larger, in fact, than the required non-Bayesian sample size)? Hint: a strong prior starts with the mean at close to the 50th percentile and requires a lot of "good news" to "change its mind." A weak prior may start with the mean already located at a percentile of the distribution greater than 60 (for the gamma example) and then improves rapidly with "good news."

As Example 12.4 showed, often the Bayes approach will allow failure rate confirmation of highly reliable products with smaller sample sizes and shorter test times than classical methods require. The Bayes results can be misleading, however, if either the form of the prior distribution or the choice of parameters for the prior distribution is not correct. Even when we have strong initial opinions about a product's reliability, it is difficult to express these opinions accurately using a prior distribution. In the case where we begin with only a weak guess about the reliability of a product, a Bayes approach would not be advisable.

Exercise 12.7

Assume the test in Example 12.4 is run for 1305 hr, but instead of 0 failures, 1 failure occurs. What is the new failure rate estimate and what is an upper 95 percent bound for the failure rate? How long would you have planned the test to run if you wanted to allow one failure and still be 90 percent confident of meeting the *MTBF* specification of 1000 hr?

RELIABILITY PREDICTION USING *MIL-HDBK-217*

This text has approached reliability projection from a failure mechanism "bottoms-up" approach in the earlier Chapters (the general reliability algorithm summarized in Chapter 8) and from a system-level "tops-down" approach described both in Chapter 10 and in the reliability growth section earlier in this chapter. There is a middle-ground approach that has several decades of empirical data and experience behind it. That approach is described in *U.S. MIL- HDBK-F, Reliability Prediction Of Electronic Equipment* (updated most recently in 1992). This handbook provides information for projecting electronic system failure rates starting at the component level. All components are assumed to follow an exponential distribution. The handbook formulas calculate component level failure rates, which can then be added to obtain the total system failure rate.

MIL-HDBK-217 offers two methods for predicting component failure rates: a "parts count method" and the more complicated "parts stress method." Nineteen component categories are covered, ranging from integrated circuits to discrete resistors and capacitors. The failure rate numbers come from historical data and the models are empirically derived multiplicative factors that adjust for such things as temperature, environmental conditions, degree of screening, ratio of operating power to rated power, and process maturity.

In the early design stages of a system, many of the inputs needed for the parts stress method will not yet be known. The simpler parts-count method (where default values are assumed for some of the factors) is recommended for use at that time. If system-level adjustments for redundancy or error correction or system maintenance policies are needed, the user of *MIL-217* is encouraged to follows methods described in *MIL-STD-756, Reliability Modeling And Prediction*.

Critics of *MIL-217* fall into two main camps: those who prefer an approach based on failure mechanism modeling and more accurate temperature acceleration modeling (see Pecht and Cushing, 1992, and Hakim, 1990) and those who prefer a system-level "tops-down" approach for reliability projection (see O'Connor, 1988, 990). Examples exist where *MIL-217* predictions were far off the mark, and the simplistic exponential assumption often receives a share of the blame for inaccurate or unrealistic results.

On the other hand, Morris and Reilly (1993) defend *MIL-217* based on its simplicity and ease of use and proven long-term track record. They agree that a failure mechanism approach is more accurate—however, it is also much more costly and complicated and time consuming, and not at all suited to answering early design reliability trade-off questions.

Another approach, also based on component level projections, is offered by Klinger, Nakada, and Menendez (1990). They replace *MIL-217* numbers with numbers based on AT&T field experience and use a non-constant failure rate model that better matches the typical field "bathtub" curve.

Perhaps the best answer to the inevitable question, "Which of these projection methods should I use—the general reliability algorithm method of Chapter 8, the MIL-217 approach, or the AT&T approach?" is: use them all, where appropriate, but be ready to adjust or replace any failure rate projection or model as soon as real data from field monitoring programs is available. These monitoring programs will be discussed next.

FIELD RELIABILITY MONITORING

Many key examples of reliability monitoring take place before shippable product leaves the manufacturing location. Reliability monitoring begins during the design and development stages and continues through early product testing and product qualification. Reliability knowledge gained during these critical phases of product life has enormous leverage, as the reliability growth models of the previous section illustrated. FMECA, standing for *failure mode, effects and criticality analysis,* is a widely used process for design reliability analysis described in the government standard *U.S. MIL-STD-1629.*

Reliability monitoring should continue as an integral part of the manufacturing process throughout the life of a product. The numbers of reliability defects removed from the product by burn-in operations or stress screens can be used to control the process and product's ability to meet reliability specifications (see Chapter 8 on burn-in and Chapter 9 for quality control tests for reliability). References such as O'Connor (1991) and Klinger, Nakada, and Menendez (1990), describe this kind of quality assurance monitoring of reliability in detail. A formal reliability monitoring and corrective action process is often known as a FRACAS (Failure Reporting And Corrective Action System). *US MIL-STD-781* describes what is needed in a good failure reporting system.

This section will focus on an often neglected aspect of reliability monitoring: field performance monitoring. The neglect stems from two primary considerations. First, it is usually a slow, difficult, and costly operation to get accurate data and failed parts back from a customer's office or home. Second, there is a widespread feeling that, once a product is in the hands of the ultimate customer, it is too late to take any meaningful or cost-effective actions to improve reliability. Effort should be expended early in the development stage or, as a last resort, in the manufacturing stage, where it is still possible to prevent the shipment of reliability defects.

While these arguments have a certain amount of truth and persuasiveness to them, it is still worthwhile to look at what a good field reliability monitoring program can do for you. Field reliability monitoring provides

- an opportunity to detect unexpected failure mechanisms (unexpected either by type or by magnitude of occurrence) as soon as they appear so that corrective action can be taken ("insurance policy" rationale for field monitoring)
- input to improve the design of future products and processes
- information to provide feedback to customers on what went wrong and what you have done (or are doing) to find the root cause of the failure and prevent reoccurrences
- data to validate and fine tune or correct the models and methods currently used for reliability projection

The insurance policy argument is often the most convincing reason for initiating a field reliability monitoring program. There is an ever-increasing emphasis on a manufacturer's legal and moral responsibility for the product it ships, to the point where large-scale product recalls are becoming commonplace. Good field reliability data is crucial to the processes of

- detecting customer problems
- sizing the magnitude or severity of these problems

- making well-informed decisions on proper corrective actions
- verifying corrective actions worked as anticipated

To have a good insurance policy, and also to realize the other benefits field reliability monitoring can provide, many different types of data are needed. The data elements consist of traceability information about the population of shipped components, operational information about the manufacturing process, and comprehensive information about all failed components.

Population Traceability Information

Each component should have the equivalent of a "vintage code" or "manufacturing job" code that can be used to identify when and where it was manufactured. Individualized serial numbers on each component will achieve this objective, but codes that link the component to a specific manufacturing job or batch are often good enough for less expensive, high-volume components. In addition, it is very important to be able to form a "power-on-hours" portrait of the field; that is, to know exactly how many components, by vintage groups, have been operating for any given number of hours.

Complete traceability, from a given manufacturing lot or job to a specific location in the field, along with dates shipped and estimated or exact power-on hours, will provide everything needed. Because of cost, however, this level of tracking is typically carried out for only expensive units or those that can affect human health or safety.

Manufacturing Process Information

Tracing backward from a component in the field, we need to be able to determine the process details and the specifications that were in place when the component was manufactured. This information includes key process parameter values, test results, and yields. At an even finer level of detail, it is sometimes useful to be able to trace back to the particular materials, tools, and operators involved in the manufacture of a particular component.

Failed Component Information

Apart from the backward traceability information described above, additional information for returned failed parts includes the power-on hours until failure and any helpful diagnostic information discovered during the repair operation. After the part is tested and failure analyzed down to a failure mechanism with a root cause, you have the key information needed for an effective reliability improvement program.

The information described above may be available only for a carefully selected subset of the components shipped, and the field coverage time span may only extend to the end of the customer's warrantee period. Tracking under these limitations can still provide all the benefits ascribed to field monitoring, although the risk of missing useful information is increased.

In some cases, the product may be mature and well understood, and the risk of not doing field monitoring might be judged small. Key factors involved in making this kind of risk assessment are

- the degree of innovation of the product
- the amount of prior experience with the same or very similar products
- the volume of projected shipments
- customer sensitivity to the component's failure rate performance
- human health and safety factors
- replacement costs if the failure rate is several times larger than projected
- the cost of field monitoring

When you do decide to monitor field reliability performance, many of the statistical methods described in this text can be used to analyze the field data. Failure rate estimates by month of life in the field (referred to as "time-in-field" failure rates, as opposed to "calendar time" failure rates for the entire field population) are generally calculated based on an exponential distribution assumption (or Poisson process) using

$$h(n) = \frac{\text{no. of fails during } n\text{th month of life}}{\text{no. of component power-on hours during } n\text{th month of life}}$$

for the repair or replacement rate during the nth month.

The upper and lower bound factors given in Chapter 3 (Tables 3.5 and 3.6) can be used to compute confidence intervals for each of the monthly failure rates. If the expected or targeted failure rate lies below the lower confidence interval limit (calculated from the actual data) for several months in a row, it should trigger a serious investigation.

Another sure signal to take action comes when failure analysis of a returned part indicates an unexpected failure mechanism. The insurance policy aspect of field monitoring has "paid off." All the types of data described as part of a good field reliability monitoring system now are used to make well informed decisions in problem management.

The first step is to "size" the problem:

1. How bad is it now?

2. How much product is affected?
3. Can the problem be traced to a particular vintage or collection of jobs with a definite start and stop date of manufacture?
4. Do we have any reliability data from in-house tests run during the period under investigation?
5. Is the product currently being shipped also susceptible?
6. How bad is the failure rate likely to get?
7. Do we have appropriate reliability models? Do we need to develop new models?

If the susceptible product appears to come from a restricted time period of manufacture, the questions would continue until a probable cause that is consistent with failure analysis evidence is uncovered. Next, screens or process controls would be devised to protect current product and salvage product that has not been shipped. The decision of what to do with the field population (often an economic and logistical question) would rely heavily on input about the size of the problem. This sizing would be continuously updated as new field data comes in. Confidence bounds for the field failure rate, in addition to point estimates, are a valuable input to this risk decision making process.

Finally, future field data provides the crucial final confirmation that the actions taken did, indeed, contain or eliminate the problem.

Since so much costly activity inevitably follows the signal of a potential major field problem, care must be taken to avoid overreacting to what might be a false alarm. That possibility is why using a confidence bound approach, coupled with effective failure analysis, forms the foundation of a good field monitoring program.

SUMMARY

Reliability growth modeling using the Duane-Crow-AMSAA model has been used in industry with a history of successful applications. The model applies to testing situations where failures are rapidly traced to root causes and improvements that change system design and subcomponent choices are immediately implemented. The mean time between failures for the system improves proportional to T^β, where T is the length of the reliability improvement test, and β is the "improvement slope." An alternate way of describing the model sets $M(t)$ (the mean number of failures up to time t) equal to at^b (where $b = 1 - \beta$). This is equivalent to the power relationship NHPP model discussed in Chapter 10, and MLEs and confidence bounds can be easily be calculated. An estimate of how long to test in order to reach a desired $MTBF$ level can be calculated after estimating (or assuming values for) the current $MTBF$ and the reliability growth improvement slope.

Bayesian methods allow a reliability engineer to incorporate past data and engineering judgement into the analysis of a current reliability test by assuming that the unknown failure rate is itself a random variable with a "prior" probability distribution. The past is used to "specify" the parameters of the prior distribution. The information from the current tests is used (via Bayes Rule) to update the prior distribution. If the prior distribution assumption is reasonably accurate, Bayes methods are often more efficient than the classical approach. If the prior distribution is just a guess, however, the end result may be misleading.

Bottoms-up calculations of system reliability depend on accurate subcomponent failure rate data bases that can be adjusted for operating temperature and other system application conditions. The methods described throughout this book can be used to develop these failure rates for critical parts. This approach is often time consuming and costly, however. For many sub-components, information contained in *MIL-HDBK-217* or the *AT&T Reliability Manual* (Klinger, Nakada, Menendez, 1990) will suffice. Failure rates calculated from, or validated by, extensive field data are best of all.

A good field monitoring program, while often difficult and costly, can provide the critical feedback needed to have a "closed-loop" reliability improvement process. Field monitoring provides early detection of unexpected problems as well as verification of the models and distributions used for failure rate projections. Of even more importance, by analyzing field failures down to failure modes and root causes, we can often learn how to improve the reliability of both our current and our future products.

PROBLEMS

12.1 Describe how you would plan a reliability growth test for a new tool to achieve an *MTBF* objective of 200 hr at 90 percent confidence. Assume the tool will begin the test with an *MTBF* (at 1 hr) of 80, and past experience indicates that a reliability growth slope of 0.3 can be expected. Note that the straightforward approach of using the test time where the $MTBF_I$ first reaches 200 hr is not likely to result in actual data that will confirm the objective *MTBF* as a 90 percent lower confidence limit. (Hint: make use of the simulation techniques described in Chapter 10 for the power relationship model and the confidence bound factors in Table 12.1).

12.2 The following procedure has been used to find appropriate parameters for a gamma prior distribution: pick (using expert judgment or prior test data) a likely value for λ. This will be set equal to the median (50th percentile) of the gamma prior. Pick a "worst case" or very unlikely "bad" value of λ. This will be set equal to the upper 95th percentile of the gamma prior. Now it is possible to numerically solve for the two parameters, *a* and *b*, of the gamma prior. Discuss the possible merits and disadvantages (or dangers) of using this procedure for a Bayesian analysis of reliability test data.

12.3 A team of engineers, applying the procedure described in problem 2, estimated the most likely and worst case *MTBF*s to be 300 hr and 50 hr. Verify
that the gamma prior has parameters $a = 0.6673$ and $b = 127.82$.

12.4 The team of engineers from problem 3 next ran a 168 reliability test and
observed 1 failure. Estimate the *MTBF* using both the gamma prior distribution and the test results (assuming an exponential model for failures,
given a fixed λ). Give a 90 percent confidence interval for the *MTBF*.

Answers to Selected Exercises

Chapter 1

1.5   33, 0.40, and 0.27

1.6   0.873 (assuming independence, and all components must survive for the card to survive)

1.8 $1, \mu, 0.5$, and $1/12$

1.9 $f(t) = (1 + t)^{-2}$, $T_{50} = 1$, and the mean does not exist ($\mu = \infty$)

Chapter 2

2.1   a. 0.3935
      b. 0.3127
      c. 0.2057
      d. 78.7, 62.54

2.2   0.22

2.4 $F(t) = 1 - e^{-\lambda t}$

2.5   353.553 %/K and 3,535,530 FIT, 111,803 %/K and 1,118,030 FIT, 35.355 %/K and 353,550 FIT, 11.18 %/K and 111,800 FIT, 3.356 %/K and 33,560 FIT

2.6   39.2 %/K

2.7   0.74, 0.855, 1.06, 1.70, 3.33, 6.24, 9.44, and 10 (all in %/year)

Chapter 3

3.2   2 %/K, 5, 268 hr, 3.4, 657 hr

3.3   0.24, 0.24

3.5 $\lambda = 0.007829$

3.6   0.736

3.7   b. $\lambda = 99.38/K$, $MTTF = 1006.26$ hr
        c. 697.5, 688
        d. The chi-square statistic is 3.4. When compared to a table value (90 percent level, 3 d.o.f.) of 6.25, we cannot reject an exponential model fit.
        e. 1068 hr, 1173.8 hr

3.8   (42,582) in FIT units

3.9   2,773 FIT, 1386 FIT

3.10   0.0013, 0.50

3.11   5320, 3100, 2020

3.12   22,400 hr, 6733 hr

3.13   12, 20, 28

3.14   380

3.15   11,583 hr, 4,582 hr

3.16   2940.7 hr, (1940.0 hr, 4106.7 hr)

Chapter 4

4.1   518 hr

4.2   3.1%

4.3   11,157 hr

4.4   1.10

4.5 2.52×10^5 hr

4.6   37%

4.7   2000 hr, 480 hr, 4472 hr

4.8 $c = 625$ hr, $m = 0.5$, $MTTF = 1250$ hr, median life = 300 hr, $R_s(100) = 0.4$

4.9   Assuming the Rayleigh distribution, 44 to 45 hits; for a uniform distribution, 24 to 25 hits

Chapter 5

5.1   0.00199, 0.02275, 7.36 oz

5.2   32, 6, 0.09176, 0.05705

5.3   11.25 microns, 9.75 microns

5.6   0.0215 or 2.15%

5.7   37.5%, 60.3%

5.8   59,972 hr, 335 hr

5.9   Least squares estimates are 6554 hr for median life and 0.21 for σ

5.10   1132 hr

5.11   38,290 hr

5.13 $\sigma = 1.46$

5.14 $T_{50} = 50.3 \times 10^6$ hr

Chapter 6

6.1 $m = -A/B$, $b = -C/B$, possible solutions are $A = k$, $B = -k/m$, and $C = kb/m$, where k is any constant

6.2   131° F

6.3   The $\hat{y}$ values corresponding to $x = 1, 2, 3, 4,$ and 5 are 3.06, 4.03, 5, 5.97, and 6.94.

6.4 $y = a + bS$, where $y = \ln Y$

6.5 $A = 1.02$, $B = 418°$ K (using least squares)

6.9   79.3

6.11   The ordered failure times are 105, 512, 2417, 3250, 5997, and 7012 hr. The corresponding KM CDF estimates are 0.0022, 0.0044, 0.0078, 0.0111, 0.0145, and 0.0245. Modified KM estimates are 0.0016, 0.0038, 0.0071, 0.0104, 0.0138, and 0.0238.

Chapter 7

7.2

$$t_u = AFt_S^2$$

$$F_u(t) = F_S\left(\sqrt{\frac{t}{AF}}\right)$$

$$f_u(t) = \left(\frac{1}{2\sqrt{tAF}}\right)f_S\left(\sqrt{\frac{t}{AF}}\right)$$

$$h_u(t) = \left(\frac{1}{2\sqrt{tAF}}\right)h_S\left(\sqrt{\frac{t}{AF}}\right)$$

When $t_S AF < 1$, time to fail at use is greater than time to fail at stress. The crossover time is $t_S = 1/AF$, after which times to fail at use are later than times to fail at stress.

7.3

$$t_u = AFe^{t_S}$$

$$F_u(t) = F_S\left(\ln\frac{t}{AF}\right)$$

$$f_u(t) = \left(\frac{1}{t}\right)f_S\left(\ln\frac{t}{AF}\right)$$

$$h_u(t) = \left(\frac{1}{t}\right)h_S\left(\ln\frac{t}{AF}\right)$$

7.4

$$F_u(t) = 1 - \left(\frac{t}{AF}\right)^{\lambda_S}$$

$$h_u(t) = \frac{\lambda_S}{t}$$

7.6   84.34

7.9 $\hat{m} = 1.12$, $\hat{c} = 5862$, $\hat{\lambda} = 17.67\%/K$, and the chi square accepts an exponential model (50.9% significance)

7.10   The common shape M.L.E. is 1.55, but the hypothesis of common shapes is rejected at 99.9 percent confidence ($\chi^2 = 13.35$).

7.11   Chi square = 3.619 with 5° of freedom. A value this large occurs 60.6% of the time, so a common shape is reasonable.

7.12   The three ΔH estimates are, respectively, 0.577, 0.496, and 0.476.

7.13 $\Delta H = 0.40$

7.14   The new operating temperature is 15.3° C.

7.15   The acceleration factor goes from 117 to 13,731 (a 117× increase).

7.16 ΔH for a lognormal fit is 1.25. The cum population failures at 25° C (at 100,000 hr) based on Weibull model is 0.01; based on a lognormal model, the cum failures at 100,000 hr are 0.0008. The Weibull model log likelihood is −197.3. The lognormal model likelihood is −199.1 (1.8 less than the Weibull).

7.18   Both the equal shape assumption (−2 log LIK = 3.62 with 5 d.o.f.) and the model fit (−2 log LIK = 1.62 with 3 d.o.f.) cannot be rejected.

7.19   Under a Weibull model assumption, the ΔH estimate is 0.505, and the B estimate is 1.003. The Weibull AFR(40,000) projection is 0.25%/K (versus 0.028%/K for the lognormal). The lognormal log likelihood is −428.48, however, which is 14.11 smaller than the Weibull log likelihood. In addition, Weibull plots look better.

7.20 $AF = e^{-B(V_1 - V_2)}$ and $AF = \left(\dfrac{V_1}{V_2}\right)^{-B}$

7.21   500, 2573, and 12,468 hr

7.22   The expected percent failures at the readouts are 0.1, 1, 3.1, 6.2, 15, 28, 39.7, and 49.5. This compares to the actuals of 0, 0.4, 3.2, 5.2, 6, 7.6, 22, 37, and 48.8.

7.23   A 25.5 hr burn-in is needed. AFR(720) = 28.2 FITs after a 1 hr burn-in.

7.24   Slope estimate using 200, 400, 600, and 720 hr points is 0.998. Slope estimate using 100,200, 100,400, 100,600, and 100,720 hr points has dropped to 0.88.

Chapter 8

8.1   The board failure rate is 6892 FITs, and the reliability at 40,000 hr is 0.725.

8.2   10.018 FITs and 0.697

8.3   1 FIT

8.4   0.999925

8.5   0.9850, 0.7351

8.6

$$(1 - F_1 F_2 F_3) \ (1 - F_4) \ (1 - F_5) \ (1 - F_6) \ (1 - F_7 F_8)$$

$$\times \left[6 \, (1 - F)^2 F^2 + 4 \, (1 - F)^3 F + (1 - F)^4 \right]$$

8.7

$$F_E [1 - (1 - R_A R_B) \ (1 - R_C R_D) \ (1 - R_F R_G)]$$

$$+ R_E (1 - F_A F_E F_F) \ (1 - F_B F_D F_G)$$

8.8

$$\alpha = 0.06, \, MTTF = 37.33, \, F_T(t) \ = \ 0.06 \left(1 - e^{-0.0268t} \right) + 0.94 F_N(t)$$

8.10

Time	AFR without Burn-In (FITs)	AFR with Burn-In (FITs)
1,000	3219	3349
10,000	1412	692
50,000	552	447
100,000	601	597

8.12

$$LIK \ = \ \left\{ p^r \prod_{j=1}^{k} [F(T_j - T_{j-1})]^{r_j} \right\} [1 - pF(T_k)]^{n-r}$$

8.13   The MLEs for the fraction mortal, μ and σ are 0.2482, 7.288, and 0.489, respectively. The likelihood ratio test does not, however, reject the hypothesis that all the populations is mortal.

Chapter 9

9.1   34,650

9.2 $_nC_r = \binom{n}{r}$

9.3 $_nC_r = \binom{n}{r}$

9.4   0.0769, 2.5, 0.9231

9.6   0.9995

9.7   CP: 10% to 34%, no, Fig. 9.6: 9.5% to 33%

9.8   10.7% to 29.3%

9.10   9.6% to 36.8% for simple equations, or 10.5% to 31.5% using first set of equations.

9.12   17/33

9.13   0.286

9.14   0.371

9.15   Partial answer: Expect 79, 19, 2, 0 with 0, 1, 2, 3 defects, respectively.

9.19 $n = 132, c = 3$

9.20 $n = 612, c = 12$

9.25   0.1637, 1.0609

9.29   11,513

9.30 $n = 783, c = 7$

9.31 $n = 132, c = 3$

9.33   503 to 230, 256, 50,000, 34,655

Chapter 10

10.3   0.98807

10.4   2

10.5   2.4, 15 hr, 0.670

10.8   4.4, 2000 hr, 4.4

10.9   0.875%/hr

10.12   0.1802

Chapter 11

11.3   6

11.5   Partial answers: 24 permutations, respective probabilities of 0.042, 0.167, 0.375, 0.625, 0.833, 0.958, 1.000

11.8   22, 45, 0; $Pr(R \leq 22) = 0.5, Pr(R \geq 45) = 0$

11.9   150, 0.81%

11.12   69, 36

11.16   5.1, 0.0060, 0.963

Chapter 12

12.1   The MLE estimate for $MTBF(1400)$ is 159 hr, while the estimate based on an unbiased slope is 173.4 hr.

12.2   The MLE estimate of the improvement slope is 0.35, and the modified estimate is 0.40. The fit is reasonable, and a 95 percent interval for $MTBF(650)$ is (34.8, 207.6).

12.3   The MLE estimates of the improvement slope and *MTBF*(320) are 0.455 and 97.8, respectively. The corresponding modified estimates are 0.546 and 117.4. The fit, however, is rejected by the Q test at a greater than 99 percent level.

12.4 $\beta = 0.333$, T $= 512$ hr

12.5   22.36× improvement

12.6 $T = 4000$ is the required test time.

12.7   The new failure rate estimate is 2/2305 = 0.000868 fails/hr. A 95 percent upper bound is 0.00206 fails/hr. $T = 2890$ hr if we want to succeed despite one failure.

Appendix

Percentiles of the χ^2 Distribution[*]

v	\multicolumn{9}{c}{g}								
	0.005	0.010	0.025	0.05	0.10	0.20	0.30	0.40	0.50
1	0.0^4393	0.0^3157	0.0^3982	0.0^2393	0.0158	0.0642	0.148	0.275	0.455
2	0.0100	0.0201	0.0506	0.103	0.211	0.446	0.713	1.02	1.39
3	0.0717	0.115	0.216	0.352	0.584	1.00	1.42	1.87	2.37
4	0.207	0.297	0.484	0.711	1.06	1.65	2.19	2.75	3.36
5	0.412	0.554	0.831	1.15	1.61	2.34	3.00	3.66	4.35
6	0.676	0.872	1.24	1.64	2.20	3.07	3.83	4.57	5.35
7	0.989	1.24	1.69	2.17	2.83	3.82	4.67	5.49	6.35
8	1.34	1.65	2.18	2.73	3.49	4.59	5.53	6.42	7.34
9	1.73	2.09	2.70	3.33	4.17	5.38	6.39	7.36	8.34
10	2.16	2.56	3.25	3.94	4.87	6.18	7.27	8.30	9.34
11	2.60	3.05	3.82	4.57	5.58	6.99	8.15	9.24	10.3
12	3.07	3.57	4.40	5.23	6.30	7.81	9.03	10.2	11.3
13	3.57	4.11	5.01	5.89	7.04	8.63	9.93	11.1	12.3
14	4.07	4.66	5.63	6.57	7.79	9.47	10.8	12.1	13.3
15	4.60	5.23	6.26	7.26	8.55	10.3	11.7	13.0	14.3
16	5.14	5.81	6.91	7.96	9.31	11.2	12.6	14.0	15.3
17	5.70	6.41	7.56	8.67	10.1	12.0	13.5	14.9	16.3
18	6.26	7.01	8.23	9.39	10.9	12.9	14.4	15.9	17.3
19	6.84	7.63	8.91	10.1	11.7	13.7	15.4	16.9	18.3
20	7.63	8.26	9.59	10.9	12.4	14.6	16.3	17.8	19.3
21	8.03	8.90	10.3	11.6	13.2	15.4	17.2	18.8	20.3
22	8.64	9.54	11.0	12.3	14.0	16.3	18.1	19.7	21.3
23	9.26	10.2	11.7	13.1	14.8	17.2	19.0	20.7	22.3
24	9.89	10.9	12.4	13.8	15.7	18.1	19.9	21.7	23.3
25	10.5	11.5	13.1	14.6	16.5	18.9	20.9	22.6	24.3
26	11.2	12.2	13.8	15.4	17.3	19.8	21.8	23.6	25.3
27	11.8	12.9	14.6	16.2	18.1	20.7	22.7	24.5	26.3
28	12.5	13.6	15.3	16.9	18.9	21.6	23.6	25.5	27.3
29	13.1	14.3	16.0	17.7	19.8	22.5	24.6	26.5	28.3
30	13.8	15.0	16.8	18.5	20.6	23.4	25.5	27.4	29.3
35	17.2	18.5	20.6	22.5	24.8	27.8	30.2	32.3	34.3
40	20.7	22.2	24.4	26.5	29.1	32.3	34.9	37.1	39.3
45	24.3	25.9	28.4	30.6	33.4	36.9	39.6	42.0	44.3
50	28.0	29.7	32.4	34.8	37.7	41.4	44.3	46.9	49.3
75	47.2	49.5	52.9	56.1	59.8	64.5	68.1	71.3	74.3
100	67.3	70.1	74.2	77.9	82.4	87.9	92.1	95.8	99.3

*Abridged from Table V of A. Hald, "Statistical Tables and Formulas," 1952. New York: John Wiley & Sons.

					g				
v	*0.60*	*0.70*	*0.80*	*0.90*	*0.95*	*0.975*	*0.990*	*0.995*	*0.999*
1	0.708	1.07	1.64	2.71	3.84	5.02	6.63	7.88	10.8
2	1.83	2.41	3.22	4.61	5.99	7.38	9.21	10.66	13.8
3	2.95	3.67	4.64	6.25	7.81	9.35	11.3	12.8	16.3
4	4.04	4.88	5.99	7.78	9.49	11.1	13.3	14.9	18.5
5	5.13	6.06	7.29	9.24	11.1	12.8	15.1	16.7	20.5
6	6.21	7.23	8.56	10.6	12.6	14.4	16.8	18.5	22.5
7	7.28	8.38	9.80	12.0	14.1	16.0	18.5	20.3	24.3
8	8.35	9.52	11.0	13.4	15.5	17.5	20.1	22.0	26.1
9	9.41	10.7	12.2	14.7	16.9	19.0	21.7	23.6	27.9
10	10.5	11.8	13.4	16.0	18.3	20.5	23.2	25.2	29.6
11	11.5	12.9	14.6	17.3	19.7	21.9	24.7	26.8	31.3
12	12.6	14.0	15.8	18.5	21.0	23.3	26.2	28.3	32.9
13	13.6	15.1	17.0	19.8	22.4	24.7	27.7	29.8	34.5
14	14.7	16.2	18.2	21.1	23.7	16.1	29.1	31.3	36.1
15	15.7	17.3	19.3	22.3	25.0	27.5	30.6	32.8	37.7
16	16.8	18.4	20.5	23.5	26.3	28.8	32.0	34.3	39.3
17	17.8	19.5	21.6	24.8	27.6	30.2	33.4	35.7	40.8
18	18.9	20.6	22.8	26.0	28.9	31.5	34.8	37.2	42.3
19	19.9	21.7	23.9	27.2	30.1	32.9	36.2	38.6	43.8
20	21.0	22.8	25.0	28.4	31.4	34.2	37.6	40.0	45.3
21	22.0	23.9	26.9	29.6	32.7	35.5	38.9	41.4	46.8
22	23.0	24.9	27.3	30.8	33.9	36.8	40.3	42.8	48.3
23	24.1	26.0	28.4	32.0	35.2	38.1	41.6	44.2	49.7
24	25.1	27.1	29.6	33.2	36.4	39.4	43.0	45.6	51.2
25	26.1	28.2	30.7	34.4	37.7	40.6	44.3	46.9	52.6
26	27.2	29.2	31.8	35.6	38.9	41.9	45.6	48.3	54.1
27	28.2	30.3	32.9	36.7	40.1	43.2	47.0	49.6	55.5
28	29.2	31.4	34.0	37.9	41.3	44.5	48.3	51.0	56.9
29	30.3	32.5	35.1	39.1	42.6	45.7	49.6	52.3	58.3
30	31.3	33.5	36.3	40.3	43.8	47.0	50.9	53.7	59.7
35	36.5	38.9	41.8	46.1	49.8	53.2	57.3	60.3	66.6
40	41.6	44.2	47.3	51.8	55.8	59.3	63.7	66.8	73.4
45	46.8	49.5	52.7	57.5	61.7	65.4	70.0	73.2	80.1
50	51.9	54.7	58.2	63.2	67.5	71.4	76.2	79.5	86.7
75	77.5	80.9	85.1	91.1	96.2	100.8	106.4	110.3	118.6
100	102.9	106.9	111.7	118.5	124.3	129.6	135.6	140.2	149.4

*Abridged from Table V of A. Hald, "Statistical Tables and Formulas," 1952. New York: John Wiley & Sons.

References

Abramovitz, M., and Stegun, I.A. (Ed.), (1964), *Handbook of Mathematical Functions*, National Bureau of Standards, Washington, D.C.

Aitchison, J., and Brown, J.A.C. (1957), *The Log-Normal Distribution*, Cambridge University Press, New York and London.

Ament, R. (1977), "Improved Chart of Confidence Limits for *p* in Binomial Sampling," *PAS Reporter*, 15(10).

Ascher, H. (1981), "Weibull Distribution vs. Weibull Process," *1981 Proceedings Annual Reliability and Maintainability Symposium*, pp. 426–431.

Ascher, H., and Feingold, H. (1984), *Repairable Systems Reliability*, Marcel Dekker, Inc., New York.

Bain, L.J., and Englehardt, M. (1991), *Statistical Analysis of Reliability and Life-Testing Models: Theory and Methods*, 2nd ed., Marcel Dekker, New York.

Barlow, R.E., and Proschan, F. (1975), *Statistical Theory of Reliability and Life Testing*, Holt, Rinehart and Winston, New York.

Barlow, R.E., F. Proschan, and E. Scheuer (1969), "A System Debugging Model," *Operations Research and Reliability*, ed. by Grouchko, pp. 401–420, Gordon and Breach, New York.

Bates, G.E. (1955), "Joint Distributions of Time Intervals for the Occurrence of Successive Accidents in a Generalized Polya Scheme," *Ann. Math. Stat.*, 26:705–720.

Burr, I.W. (1976), *Statistical Quality Control Methods*, Marcel Dekker, New York.

Burr, I.W. (1979), *Elementary Statistical Quality Control*, Marcel Dekker, New York.

Calvin, T.W. (1983), "Quality Control Techniques for Zero Defects," *IEEE Trans on Components, Hybrids, and Manufacturing Technology*, Vol. CHMT-6, No. 3, pp. 323–328.

Campodonico, S. (1993), "Bayes Analysis of Survival Times and Reliability Data: A Software Package," *1993 Proceedings Annual Reliability and Maintainability Symposium*.

Chace, E.F. (1976), "Right-Censored Grouped Life Test Data Analysis Assuming a Two Parameter Weibull Distribution Function," *Microelectronics Reliability,* Vol. 15 pp 497– 499.

Chambers, J.M. et al. (1983), *Grouped Methods For Data Analysis,* Wadsworth, Monterey, CA.

Chemical Rubber Company (CRC) (1965), *Standard Mathematical Tables,* 15th ed., Cleveland, OH.

Clopper, C.J., and Pearson, E.S. (1934), "The Use of Confidence or Fiducial Limits Illustrated in the Case of the Binomial," *Biometrica,* 26:404.

Cohen, A.C., Jr. (1959), "Simplified Estimators For The Normal Distribution When Samples Are Singly Censored Or Truncated," *Technometrics* 1, pp. 217 to 237.

Cohen, A.C., and Whitten, B.J. (1988), *Parameter Estimation in Reliability and Life Span Models,* Marcel Dekker, New York.

Cox, D.R. (1962), *Renewal Theory,* John Wiley & Sons, Inc., New York.

Cox, D.R., and Lewis, P.A.W. (1966), *The Statistical Analysis of Series of Events,* John Wiley & Sons, Inc., New York.

Crow, L.H. (1974), "Reliability Analysis for Complex Repairable Systems," in *Reliability and Biometry,* F. Proschan and R.J. Serfling, eds., SIAM, Philadelphia, pp 379–410.

Crow, L.H. (1975), "On Tracking Reliability Growth," *Proceedings Annual Reliability and Maintainability Symposium,* pp. 438–443.

Crow, L.H. (1982), "Confidence Interval Procedures for the Weibull Process With Applications to Reliability Growth," *Technometrics,* 24(1):67–72.

Crow, L.H. (1990), "Evaluating the Reliability of Repairable Systems," *Proceedings Annual Reliability and Maintainability Symposium,* pp. 275–279.

Crow, L.H. (1993), "Confidence Intervals on the Reliability of Repairable Systems," *Proceedings Annual Reliability and Maintainability Symposium,* pp. 126–134.

Crowder, M.J., et al. (1991), *Statistical Analysis of Reliability Data,* Chapman Hall, New York.

Cushing, M.J. (1993), "Another Perspective on the Temperature Dependence of Microelectronic-Device Reliability," 1993 *Proceedings Annual Reliability and Maintainability Symposium.*

Davis, D.J. (1952), "An Analysis of Some Failure Data," *J. Amer. Stat. Soc.,* 47:113–150.

De Le Mare, R.F. (1991), "Testing for Reliability Improvement or Deterioration in Repairable Systems," *Quality and Reliability Engineering International,* 8:123–132.

Deming, W.E. (1982), *Quality, Productivity, and Competitive Position,* Massachusetts Institute of Technology, Cambridge.

Dixon, W.J., and Massey, F.J., Jr., (1969), *Introduction to Statistical Analysis,* 3rd edition, McGraw-Hill, New York.

Dodge, H.F., and Romig, H.G. (1959), *Sampling Inspection Tables, Single and Double Sampling,* 2nd ed., Wiley, New York.

Draper, N.R., and Smoth, H. (1981), *Applied Regression Analysis,* 2nd ed., Wiley, New York.

Duane, J.T. (1964), "Learning Curve Approach to Reliability Monitoring," *IEEE Transactions on Aerospace,* 2, pp. 563–566.

Duke, S.D., and Meeker, W.Q., Jr. (1981), "CENSOR—A User Oriented Computer Program For Life Data Analysis," *The American Statistician,* 34, pp. 59 to 60.

Duncan, A.J. (1986), *Quality Control and Industrial Statistics*, 5th ed., Richard D. Irwin, Homewood, IL.

Eyring, H., Glasstones, S., and Laidler, K.J., (1941), *The Theory of Rate Processes*, McGraw-Hill, New York.

Fasser, Y., and Brettner, D., (1992), *Process Improvement in the Electronics Industry*, John Wiley & Sons, New York.

Feller, W. (1968), *An Introduction to Probability Theory and Its Applications*, Vol. 1: 3rd ed., Wiley, New York.

Fisher, R.A. (1954), *Statistical Methods for Research Workers*, 12th ed., Oliver and Boyd, London.

Gerald, C.F., and Wheatley, P.O. (1984), *Applied Numerical Analysis*, 3rd ed., Addison-Wesley, Reading, Mass.

Gnedenko, B.V., Belyayev, Y.K., and Solovyev, A.D., (1969), *Mathematical Methods of Reliability Theory*, Academic Press, New York.

Grant, E.L., and Leavenworth, R.S. (1988), *Statistical Quality Control*, 6th ed., McGraw-Hill, New York.

Gumbel, E.J., (1954), *Statistical Theory of Extreme Values and Some Practical Applications*, National Bureau of Standards, Washington, D. C.

Guenther, W.C. (1974), "Sample Size Formulas For Some Binomial Problems," *Technometrics*, 16:465–467.

Hahn, G.J. (1979), "Minimum And Near Minimum Sampling Plans," *Journal of Quality Technology*, 11(4):206–212.

Hahn, G.J., and Shapiro, S.S. (1967), *Statistical Models in Engineering*, Wiley, New York.

Hakim, E.B. (1990), "Reliability Prediction: Is Arrhenius Erroneous," *Solid State Technology*.

Hastings, N.A., and Peacock, J.B. (1974), *Statistical Distributions*, Butterworth & Co. Ltd., London.

Ishikawa, K. (1982), *Guide to Quality Control*, Asian Productivity Organization, Tokyo.

Jensen, F., and Petersen, N.E., (1982), *Burn-In: An Engineering Approach To The Design And Analysis Of Burn-In Procedures*, Wiley, New York.

Johnson, L.G. (1951), "The Median Ranks Of Sample Values In Their Population With An Application To Certain Fatigue studies," *Industrial Mathematics*, 2:1–9.

Juran, J.M., ed. (1988), Quality Control handbook, 4th ed., McGraw-Hill, New York.

Kaplan, S., Cunha, G.D.M., Dykes, A.A., and Shaver, D. (1990), "A Bayesian Methodology for Assessing Reliability During Product Development," *1990 Proceedings Annual Reliability and Maintainability Symposium*.

Kaplan, E.L., and Meier, P. (1958), "Nonparametric Estimation from Incomplete Observations," *Journal of the American Statistical Association*, 53:457–481.

Kececioglu, D. (1991), *Reliability Engineering Handbook*, Vol. 2, Prentice Hall Inc., Englewood Cliffs, New Jersey.

Kendall, M.G. (1938), "A New Measure of Rank Correlation," *Biometrika*, 30:81–93.

Kielpinski, T.J., and Nelson, W., (1975), "Optimum Censored Accelerated Life-Tests For The Normal And Lognormal Life Distributions," *IEEE Transactions on Reliability*, Vol. R-24, 5, pp. 310 to 320.

Klinger, D.J., Nakada, Y, and Menendez, M.A. (ed.) (1990), *AT&T Reliability Manual*, Van Nostrand Reinhold, New York.

Kolmogorov, A.N., (1941), "On A Logarithmic Normal Distribution Law Of The Dimensions Of Particles Under Pulverization," *Dokl. Akad Nauk*, USSR 31, 2, pp 99–101.

Landzberg, A.H., and Norris, K.C., (1969), "Reliability Of Controlled Collapse Interconnections," *IBM Journal of Research and Development* Vol. 13, 3.

Lawless, J.F. (1982), *Statistical Models and Methods for Lifetime Data*, Wiley, New York.

Lindley, D.V. (1965), *Introduction to Probability and Statistics from a Bayesian Viewpoint, Part I, Probability and Part 2, Inference*, University Press, Cambridge.

Mann, H.B. (1945), "Nonparametric Test Against Trend," *Econometrica*, 13:245–259.

Mann, N.R., Schafer, R.E., and Singpurwalla, N.D., (1974), *Methods for Statistical Analysis of Reliability and Life Data*, Wiley, New York.

Martz, H.F., and Waller, R.A. (1982), *Bayesian Reliability Analysis*, Wiley, New York.

Meeker, W.R., and Hahn, G.J. (1985), "How to Plan an Acceleration Life Test—Some Practical Guidelines," *ASQ Basic Reference in Quality Control: Statistical Techniques–Vol. 10*, ASQC, Milwaukee, Wisconsin.

Meeker, W.Q., and Nelson, W., (1975), "Optimum Accelerated Life-Tests For The Weibull And Extreme Value Distributions," *IEEE Transactions on Reliability*, Vol. R-24, 5, pp 321–332.

Meeker, W.Q., and Duke, S.D. (1981), "CENSOR–A User-Oriented Program for Life Data Analysis," *American Statistician*, Vol. 35 p 112.

Mendenhall, W., Schaeffer, R.L., and Wackerly, D.D. (1990), *Mathematical Statistics with Applications*, PWS Kent, Boston, Mass.

Michael, J.R., and Schucany, W.R. (1986), "Analysis of Data From Censored Samples," *Goodness of Fit Techniques*, ed. by D'Agostino, R.B., and Stephens, M.A., Marcel Dekker, New York.

MIL-HDBK-189 (1981), Reliability Growth Management, U.S. Government Printing Office.

MIL-HDBK-217F (1986), Reliability Prediction of Electronic Equipment, U.S. Government Printing Office.

MIL-STD-105D (1963), Sampling Procedures and Tables for Inspection by Attributes, U.S. Government Printing Office.

MIL-STD-756B (1963), Reliability Modeling & Prediction, U.S. Government Printing Office.

MIL-STD-781C, Reliability Design Qualification and Production Tests: Exponential Distribution, U.S. Government Printing Office.

MIL-STD-1629A (1980), Procedures For Performing a Failure Mode, Effects and Criticality Analysis, U.S. Government Printing Office.

MIL-STD-1635 (EC) (1978), Reliability Growth Testing, U.S. Government Printing Office.

MIL-S-19500G (1963), General Specification for Semiconductor Devices, U.S. Government Printing Office.

Montgomery, D.C. (1991), *Introduction to Statistical Quality Control*, second edition, Wiley, New York.

Morris, S.F., and Reilly, J.F. (1993), "MIL-HDBK-217—A Favorite Target," *1993 Proceedings Annual Reliability and Maintainability Symposium*.

Musa, J.D., Iannino, A., and Okumoto, K. (1987), *Software Reliability*, McGraw-Hill, New York.

Nelson, W. (1969), "Hazard Plotting For Incomplete Failure Data," *Journal of Quality Technology,* 1, pp. 27 to 52.

Nelson, W. (1972), "Theory and Application of hazard plotting for Censored Failure Data," *Technometrics,* 14: pp 945–946.

Nelson, W. (1975), "Graphical Analysis Of Accelerated Life Test Data With A Mix Of Failure Modes," *IEEE Transactions on Reliability,* Vol. R-24, 4, pp 230–237.

Nelson, W. (1982), *Applied Life Data Analysis,* Wiley, New York.

Nelson, W.B. (1986), "Graphical Analysis of Failure Data from Repairable Systems," *General Electric Co. Corp. Research & Development TIS Report 86CRD114,* Schenectady, NY 12345.

Nelson, W. (1988a), "Analysis of Repair Data," *Proceedings Annual Reliability and Maintainability Symposium,* pp. 231–233.

Nelson, W. (1988b), "Graphical Analysis of System Repair Data," *Journal of Quality Technology,* 20(1):24–35.

Nelson, W. (1995), "Confidence Limits for Recurrence Data-Applied to Cost or Number of Product Repairs" (to be published).

Nelson, W.B., and Doganaksoy, N. (1989), "A Computer Program for an Estimate and Confidence Limits for the Mean Cumulative Function for Cost or Number of Repairs of Repairable Products," *General Electric Co. Corp. Research & Development TIS Report 89CRD239,* Schenectady, NY 12345.

Nelson, W. (1990), *Accelerated Life Testing,* Wiley, New York.

Neter, J., Wasserman, W., and Kutner, M.H. (1990), *Applied Linear Statistical Models,* 3rd ed., Richard D. Irwin, Homewood, Illinois.

O'Connor, P.D.T. (1988), "Undue Faith in US Mil-HDBK-217 For Reliability Prediction," *IEEE Transactions on Reliability* 37,5 (p. 468).

O'Connor, P.D.T. (1990), "Reliability Prediction: Help or Hoax?" *Solid State Technology.*

O'Connor, P.D.T. (1991), *Practical Reliability Engineering,* John Wiley & Sons, Inc., New York.

Ostle, B., and Mensing, R.W., (1975), *Statistics in Research,* third edition, The Iowa State University Press, Ames, Iowa.

Ott, E.R., and Schilling, E.G. (1990), *Process Quality Control,* 2nd ed., McGraw-Hill, New York.

Ott, R.L. (1993), *An Introduction to Statistical Methods and Data Analysis,* Wadsworth, Belmont, CA.

Parzen, E. (1962), *Modern Probability Theory and Its Applications,* John Wiley & Sons, Inc., New York.

Pecht, M., Lall, P., and Hakim, E.B. (1992), "The Influence Of Temperature On Integrated Circuit Failure Mechanisms," *Quality and Reliability Engineering International,* Vol. 8 (pp 167–175).

Peck, D., and Trapp, O.D., (1980), *Accelerated Testing Handbook,* Technology Associates and Bell Telephone Laboratories, Portola, Calif.

Prins, J. (1984), STATLIB: Statistical Software for the IBM Personal Computer, IBM Personally Developed Software, P. O. Box 3280, Wallingford, CT.

Rigdon, S.E., and Basu, A.P. (1989), "The Power Law Process: A Model for the Reliability of Repairable Systems," *Journal of quality Technology,* 21(6): 251–60.

Ross, S, (1994), *A First Course in Probability,* 4th ed., Macmillan, New York.

Ross, S.M. (1993), *Introduction to Probability Models*, 5th ed., Academic Press, San Diego, CA.

Savage, L.J., (1954), *The Foundations of Statistics*, Wiley, New York.

Schilling, E.G. (1982), *Acceptance Sampling in Quality Control*, Marcel Dekker, New York.

Strauss, S.H. (1980), "STATPAC: A General Purpose Package For Data Analysis And Fitting Statistical Models To Data," *The American Statistician*, 34, pp. 59 to 60.

Thomas, G.B., Jr. (1960), *Calculus and Analytic Geometry*, 3rd ed., Addison-Wesley, Reading, MA.

Trindade, D.C. (1975), "An APL Program to Differentiate Data by the Ruler Method," *IBM Technical Report, TR 19.0361*, Burlington, VT.

Trindade, D.C. (1980), "Nonparametric Estimation of a Lifetime Distribution via the Renewal Function," PhD. Dissertation, University of Vermont, Burlington, VT.

Trindade, D.C. (1991), "Can Burn-in Screen Wearout Mechanisms? Reliability Models of Defective Sub-populations—A Case Study," *29th Annual Proceedings of Reliability Physics Symposium*.

Trindade, D.C. (1992), "Confidence Intervals for PPM," AMD Corporation Internal Memo, Sunnyvale, California.

Trindade, D.C., and Haugh, L. (1979), "Nonparametric Estimation of a Lifetime Distribution via the Renewal Function," *IBM Technical Report, TR 19.0463*, Burlington, VT.

Trindade, D.C., and Haugh, L. (1980), "Estimation of the Reliability of Computer Components from Field Renewal Data," *Microelectronics and Reliability*, 20:205–218.

Tummula, R.R., and Rymaszewski, E.J. (Ed), (1989), *Microelectronics Packaging Handbook*, Van Nostrand Reinhold, New York.

Turnbull, B.W. (1976), "The Empirical Distribution Function With Arbitrarily Grouped, Censored, and Truncated Data," *Journal Royal Statistical Society* B, pp. 290–295.

Usher, J.S. (1993), "Case Study: Reliability Models and Misconceptions," *Quality Engineering*, 6(2):261–271.

Weibull, W., (1951), "A Statistical Distribution Function Of Wide Applicability," *Journal of Applied Mechanics*, Vol. 18, pp 293–297.

Western Electric Company (1958), *Statistical Quality Control Handbook*, 2nd ed., Mack, Easton, PA.

Wilks, S.S.,(1962), *Mathematical Statistics*, Wiley, New York.

Xie, M. (1991), *Software Reliability Modeling*, World Scientific, New Jersey.

Zaino Jr., N.A., and Berke, T.M. (1992), "Determining the Effectiveness of Run-in: A Case Study in the Analysis of Repairable-System Data," *Proceedings Annual Reliability and Maintainability Symposium*, pp. 58–70.

Index

415